2nd audio anthology

COMPILED FROM

AUDIO ENGINEERING

From January 1950 to July 1952

by

C. G. McProud
Editor

The Marshall-Jones Co.
Publishers
Francestown, New Hampshire

Preface

C. G. McProud's second collection of articles taken from the pages of *Audio Engineering* magazine published between 1950 and 1952 are in a neat balance of nearly equal quarters. The first segment offers nine power amplifiers followed by nine on power amp characteristics including feedback, loudspeaker damping, cathode followers, mismatch problems, and the mysteries of cathode bias.

Preamps follow in the next quarter including two for the then new magnetic phonograph pickups, DC power for preamp tube heaters and a whole array of articles attempting to bring some order and understanding to the jungle of recording equalization curves being used by record makers. And tucked into the middle of these offerings is a first article on stereophonic reproduction and how it might be realized by filtering of the mono signal.

Then McProud pauses to offer a landmark article on construction practice that is still valid today even when etched circuit cards have displaced terminal boards and beeswaxed cord lacing for cable harnesses.

Next the subject switches to loudspeakers starting off with resonant enclosures and port sizes needed for them. Reflexed cabinets, how to deal with resonance peaks and practical realizations of Bass Reflex cabinets are all present, along with a survey of horn types and how to build several bass enhancing versions.

The volume closes with two tape recording accessories by McProud in the form of a portable interview amplifier including a mike, level control and meter—designed to work with the old Magnecord tape machines—and a tape playback preamp.

All this equipment was designed to be built by the amateur in a day when there were few stores you could walk into to buy these items—at least not in many cities and not at a price the ordinary mortal could afford. Many of the designs were ahead of the manufacturers who were not yet producing much in the way of serious equipment for home use. McProud had been alone as an audio publisher until 1951 when the first issue of Milton Sleeper's and Charles Fowler's *High Fidelity* magazine appeared.

But all through the decade of the fifties, McProud kept the ''engineering'' in his magazine's name and at the same time worked diligently and successfuly to help found the Audio Engineering Society. These pioneering articles had a profound effect on audio's gestation period in postwar America. They enshrine a standard of excellence which deserves study even now—perhaps most especially now.

Edward T. Dell, Editor/Publisher
Audio Amateur and *Speaker Builder* Magazines

CONTENTS

Foreword

One of the penalties of popularity is that demands are made upon time in the case of a person, or upon stocks of back numbers in the case of a magazine. AUDIO ENGINEERING has never been able to maintain a complete stock of back issues, even though what appeared to be an adequate number of copies were ordered each month.

The first **audio anthology** succeeded in fufilling the demands of many who had not become acquainted with the magazine during its early years, and who had not had the opportunity to build up their own file of back copies. In particular, those articles which appealed to the audio hobbyist and experimenter were assembled in one volume so that they could serve as a constant reference to the material which had been most popular.

This same plan has been followed in assembling **the 2nd audio anthology**, and it will be found that most of the articles about amplifiers, loudspeakers and enclosures, and phonograph equipment that were originally printed in AUDIO ENGINEERING over the period from January 1950 to July 1952 are reprinted herein. It is believed that any errors which appeared in the original printing have been corrected, and some of the articles have been brought up to date by minor changes and additions. To those who have searched for unavailable back issues and who have so regularly requested reprints, this volume is appreciatively dedicated.

C. G. McPROUD, *Editor*
AUDIO ENGINEERING

Mineola, N. Y.
December, 1952

An Ultra-Linear Amplifier

DAVID HAFLER and HERBERT I. KEROES

Presenting a new output-stage connection in an otherwise conventional amplifier which provides a degree of listenability which is well above average.

IT HAS BEEN CLAIMED that there is no more room for improvement of power output stages since other elements of a complete sound system—particularly the electro-mechanical ones—are far inferior. There is a prevalent belief that "one good amplifier is only marginally different from another." The proponents of this line of thought imply that significant improvement in power amplifiers is extremely difficult to achieve, and with this idea the authors agree, but the authors disagree as to the *need* for further improvement. Obviously, the weaker links do need improvement, but this alone is no reason for abandoning the further development of stronger links in the chain of audio reproduction—the power amplifier and primarily the power output stage which is the prime generator of distortion in the purely electronic part of the audio system.

Present thinking is very parallel to the views of the 1935 era when it was felt that the principle need was for better program sources and that the transducers and audio amplifiers had reached a stage of near perfection which could hardly be improved. Now, what audiophile would be satisfied with the reproduction standards of sixteen years ago when playing the new LP's or high grade tape? By analogy, therefore, as well as for the never-ending search for a never-attainable perfection, we must continue to seek improvement in *every* link of the audio chain.

The old standards for evaluating amplifier quality have fallen into disrepute. It can be audibly demonstrated that a wide pass band and low harmonic content do not necessarily mean that the amplifier satisfies the critical listener. Newer criteria have been developed such as intermodulation distortion analysis and square wave testing, both of which simulate dynamic conditions to some extent and take into consideration that music and speech are not of a static nature. These new tests produce higher correlation between experimental data and listener preference. Therefore, modern amplifiers sound better than the ones of a few years ago as a general rule. However, these tests do not always separate the wheat from the chaff. Amplifiers which measure well do not necessarily sound well although an amplifier which shows up as poor on measurements will not sound well. Excellent measurements are a necessary but not a sufficient condition for quality of sound. This means that the listening test is the one of most importance—it is the most stringent test of all.

On the basis of listening tests (definitely not on the basis of measurements) the audio school has been divided into two camps—triodes versus tetrodes. There has been shifting between the popularity of the two, but there has always been a distinct cleavage. When the triode-without-feedback was judged superior to the tetrode-without-feedback, the tetrode school added feedback and reaffirmed the merits of this tube type. This was again superseded by the triode-with-feedback, but the beam tetrode still has its followers, presently in the category of a defensive minority among the audio elite.

The very fact that each tube type has ardent supporters is evidence that each has definite points of merit. Possibly the devotees of each type listen for *different* qualities of reproduction, and this causes divergence of opinion. The triode fan usually emphasizes "smoothness" or "sweetness" of sound. The beam power advocates seek "crispness" or "clean sound." Each group obviously desires sound which simulates the original, but each rejects the elusive and unmeasureable distortions which characterize the tube type preferred by the opposition camp. A new type of tube, none of which has been put on the market for many years, might be the thing which could reconcile these diverse views of listeners who all look for the same thing but seek it in different ways.

The requisites for such a new tube can be listed readily:

1. Low internal impedance, such as is offered by the triode.
2. High power sensitivity of the tetrode so as to minimize drive problems.
3. Lower harmonic and intermodulation distortion than either triode or tetrode at both high and low levels of operation.
4. Sufficiently high efficiency to permit adequate output without undue bulk or cost.

Since no such tube is available, the only recourse is to seek a mode of operation of existing type tubes to approximate the desired qualities and then to see whether the theory is justified by listening tests.

Linearizing the Output Stage

The physical difference between the triode and tetrode is, of course, the screen grid. This gives the tetrode its efficiency on the one hand, but also increases the plate resistance and contributes toward the "tetrode sound" which is so violently disliked by triode favorers. Therefore, the screen grid seems to be the element which gives the tetrode its advantages and its disadvantages compared to the triode. In fact, when the screen is connected to the plate, the resultant tube is a triode which is excellent in many respects though handicapped by limited power output and low permissible dissipation. Control of the screen is a logical step toward extracting the favorable attributes of the tube and discarding the unfavorable.

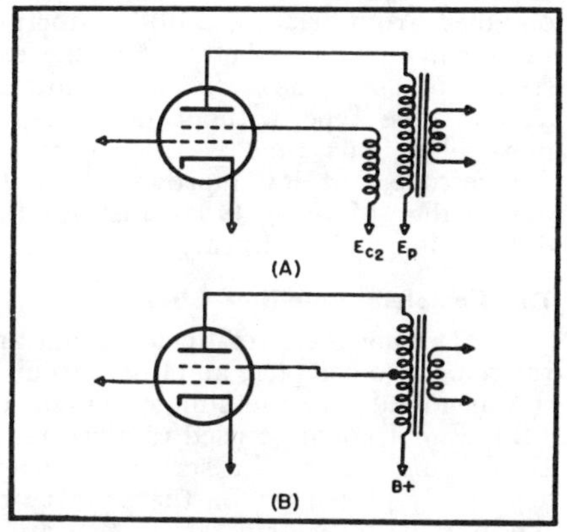

Fig. 1. Arrangements for energizing the screen grid to improve tube linearity.

Experimentally it was found that the goal of improved operation could be achieved through energizing the screen with d.c. through a special winding on the output transformer and combining the effects of both plate and screen current in the output transformer. This is illustrated at (A) in *Fig.* 1 with an alternative and simpler method shown at (B). It has been found that the screens must be fed from a low-impedance source or the benefits of this arrangement cannot be realized. This eliminates the possibility of doing the same job with resistive bridge networks or voltage dividers.

The screen load impedance is somewhat critical if optimum results are desired. As the ratio of screen load impedance varies from zero (tetrode operation) to unity (triode operation), important effects show up:

1. The internal impedance takes a sharp drop and then levels off.
2. Maximum undistorted output drops slightly at first, then decreases rapidly.
3. Intermodulation distortion at high level operation drops to a minimum and then soars upward.
4. Low level IM decreases somewhat and then holds almost level.

The situation is demonstrated graphically in *Fig.* 2 where it can be seen

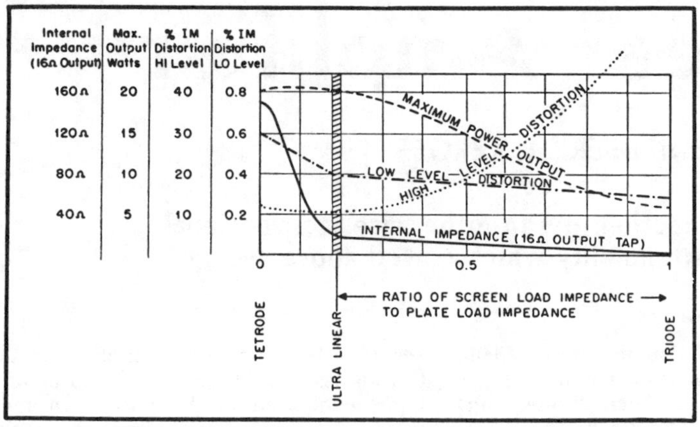

Fig. 2. Comparison of Ultra-Linear operation with triode and tetrode operation using a push-pull stage without feedback.

that over a narrow band of operation where screen load impedance is about 18.5 per cent of plate load impedance, the new arrangement provides the high power output of tetrodes with low internal impedance such as is normally obtained from triodes, while distortion figures are equal or better than the extremes of operation. We have achieved a new tube type without designing a new tube. This tube is neither triode nor tetrode, but its improved linearity over either of those types justifies the designation "ultra linear."

The Complete Amplifier Circuit

In applying the ultra-linear output arrangement to complete amplifier circuits, it was found that the simple version of (B), *Fig.* 1 could be used to advantage. By feeding d.c. to the screens through a properly placed tap on the primary of the output transformer, the operating conditions are preserved, and the close coupling between screen and plate is advantageous when feedback is carried around the stage. The disadvantage of this simpler arrangement is that screen and plate must operate at the same d.c. potential. In the particular arrangement used the screen and plate are operated at the same potential (350 volts plate to cathode) without exceeding dissipation requirements, either quiescent or at maximum output. This new output coupling arrangement reduces screen dissipation at high levels and is a safe mode of operation with respect to tube life.

A circuit arrangement has been designed to take full advantage of the ultra-linear output stage. This circuit, *Fig.* 3, takes into account the necessity for complete stability under feedback conditions so as to eliminate tendencies toward transient instability under any type of load, including the varying impedance of loudspeaker systems.

This complete circuit offers linearity of operation of a very high order. It is based around a special output transformer, the Acrosound TO-300, which is 6600 ohms primary impedance and has taps at the optimum point indicated in *Fig.* 2. A special seven-section symmetrical winding arrangement placed on a substantial grain-oriented lamination of unique shape permits a ratio of primary inductance to leakage reactance in excess of 15,000 to 1. The response of the transformer alone is within ±1 db from 10 to 100,000 cps with extremely low phase shift and no resonances within this band.

The complete amplifier circuit is relatively simple, inexpensive, and efficient. With a 370-volt power transformer at 130-ma peak requirement, power output is almost as high as for a tetrode amplifier and twice that of a triode amplifier with cathode bias and the same power supply. No adjustments are necessary for balance either of the phase inverter or of the output-stage plate current, and there are no critical values of capacitors or resistors required. The amplifier is driven to 20 watts of output with an input of only 0.7 volt.

Feedback is carried around the complete circuit in an external loop. There are 20 db of feedback in this loop as measured under load conditions (about 30 db based on open circuit gain), and a safe margin is maintained. A small capacitor across the feedback resistor increases the feedback in the region above 100 kc to smooth the high-frequency response. This capacitor is not required to keep the amplifier stable though it does add to the stability margin.

Performance of the Amplifier

All stages of the amplifier have been adjusted for minimum intermodulation, and the IM curves based on sine-wave power output are shown in *Fig.* 4. These curves were run using frequencies of 40 and 2,000; 40 and 7,000; 40 and 12,000; 100 and 2,000; and 60 and 7,000, all mixed four to one. The IM is almost identical under all conditions of test indicating that it is completely independent of frequency, at least up to 20 watts output. This factor possibly accounts for the superlative listening quality of the amplifier.

Undistorted power, less than 2 per cent IM, is in excess of 20 watts. This power is delivered *undistorted* within 1 db over the range from 20 to 20,000 cps. This power curve (*Fig.* 5) is not a response curve run at high power level. Instead it represents *clean* power available at these frequencies. This is particularly important with today's program sources. The dynamic range of some of the best LP's is reputed to be in excess of 100 db. It is necessary to have power to handle this range, and this power is required over a wide frequency band. New standards of audio fidelity are rapidly making obsolete the five or ten watt amplifier which cannot even deliver its rated power at frequency extremes.

Another factor of considerable importance in evaluating amplifier performance cannot be seen from the curves. This is *overload characteristic.* The amplifier has been given listening tests under overload conditions with a pad on the output so as not to deafen the participants. Peaks which would require a 40-watt amplifier are transmitted without irritation even though the output can be seen to clip on the 'scope. The overload recovery is rapid and has no noticeable hangover, so a clipped peak has no time to penetrate the ear. Some amplifiers

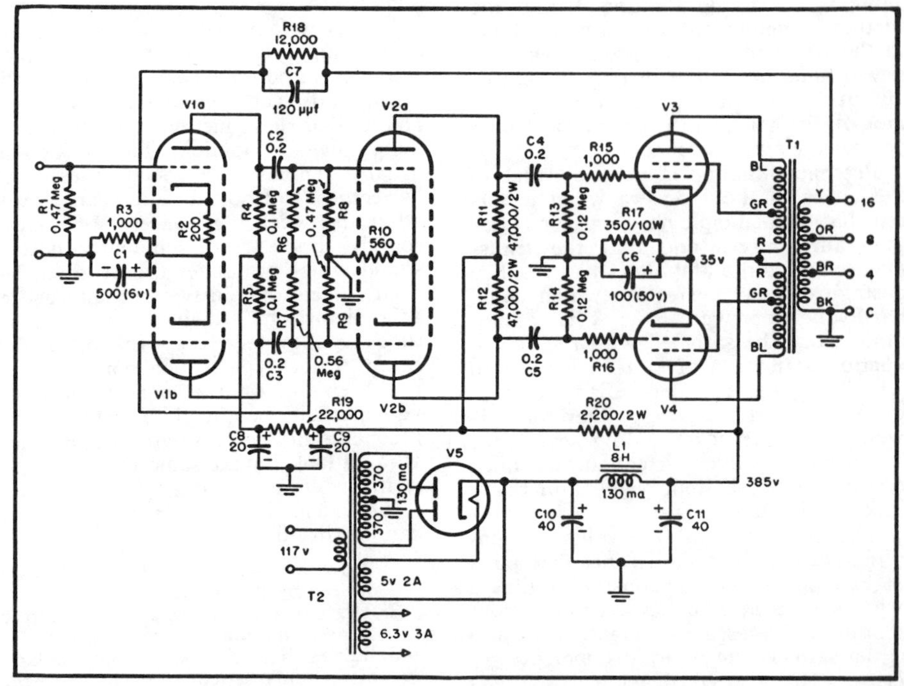

Fig. 3. Overall schematic of the Ultra-Linear amplifier and associated power supply.

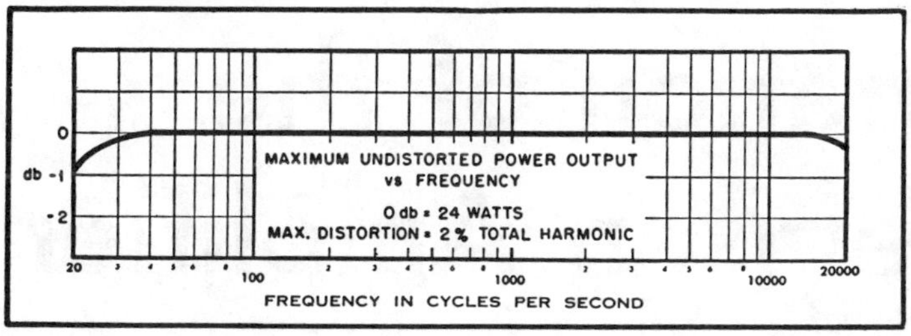

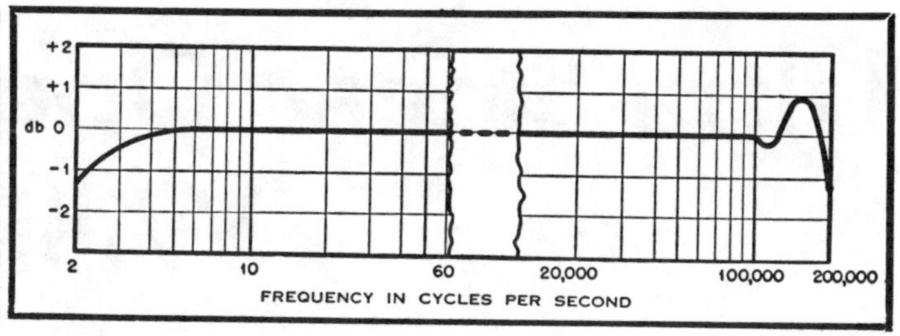

Fig. 5 (left). Undistorted power output vs. frequency. Fig. 6 (right). Frequency response, showing effects above and below the audible frequency range.

break up on a peak, and for seconds thereafter the sound is distorted badly because of poor recovery. In the ultra-linear amplifier transient instability has been eliminated—changes in amplifier characteristics caused by overload do not make the circuit unstable; and, therefore, recovery is almost instantaneous. Most feedback amplifiers fail miserably under overload listening tests.

Figure 6 shows the voltage gain versus frequency. Obviously, most present day amplifiers are flat through the audio band. However, it is the band outside of the audible region which makes some of the difference between one amplifier and another. In this circuit it is evident that smooth flat bandpass extends more than two octaves on each side of the traditional 20 to 20,000 band. This enormous band width is necessary to eliminate phase shift over the customary region and to provide good transient response.

The square wave performance of *Fig.* 7 testifies to the transient response. Evidently, a circuit with response flat within 2 db for a decade on each side of the audio band should show a presentable square wave at most frequencies. However, the low phase shift, fast rise time, and insignificance of ringing in this circuit as indicated by the square waves shows that more than just the frequency response is excellent. In addition, square waves were checked on a speaker load with practically identical results, thus demonstrating that performance of the amplifier is unaffected by a load of varying impedance.

Other circuit configurations can be used with this ultra-linear output stage. However, they should have a phase characteristic permitting substantial feedback, and they should have the lowest possible distortion for the early stages. The popular Williamson circuit has been converted to this output arrangement with gratifying results. This conversion permits 30 watts of output plus the other benefits inherent in the increased linearity of the output stage.

Listening Tests

The majority of listeners agree readily to the superiority of this circuit. None felt that other equipment was better although some could not recognize differences on the program sources used. However, during the course of the tests, certain recordings were found which demonstrated differences vividly; and after this finding, even the less discriminating listeners could identify the ultra linear amplifier on "blind" tests and could recognize its superiority.

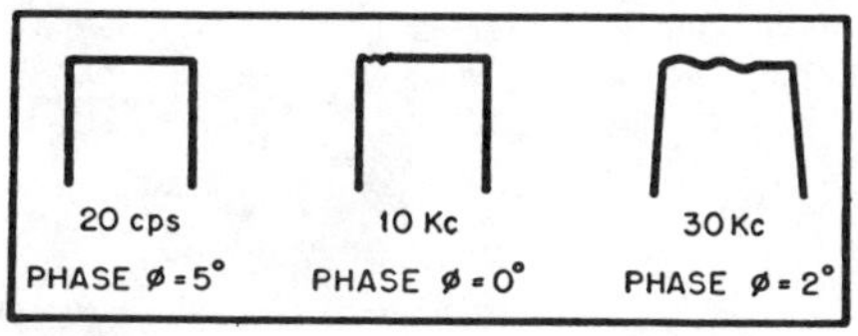

Fig. 7. Square-wave performance of the Ultra-Linear amplifier.

Listeners agree that the bass region is more articulate, better defined, and better damped than in other amplifiers. This damping is not a function of internal impedance alone but also relates to stability under dynamic conditions. For example, no low-frequency cutoff is required in the preamplifier as no ill effects are audible due to momentary overloads from turntable rumble, switching thumps, and similar disturbances. Certain types of signals such as organ pedal tones combined with rumble will cause other amplifiers to break up even at levels as low as a few watts in the mid-frequencies.

Another audible feature in the bass range is that the amplifier does not have *more* bass, but it has *lower* bass. Other amplifiers, of good quality in terms of measurements, by comparison were generating harmonics and intermodulation products. This was also apparent on scratchy "dirty" recordings which cleaned up on the ultra-linear amplifier while remaining mushy and irritating on others.

In the treble region the consensus of opinion is that the amplifier sounds "smoother." The scratch level of shellac records is less irritating while the high frequency sounds, particularly of a percussive type, cut through the scratch and seem far more prominent. This seems due to the fact that intermodulation between scratch and music is diminished, and the two assume much more pleasant proportions.

The authors believe that for sheer listening pleasure the ultra-linear amplifier represents the best that can be achieved at the present state of the art. Others who have had an opportunity to hear and try the circuit agree with this; and these beliefs will not be shaken until something comes along which *sounds* better, or at least sounds as good and can be built for lower cost.

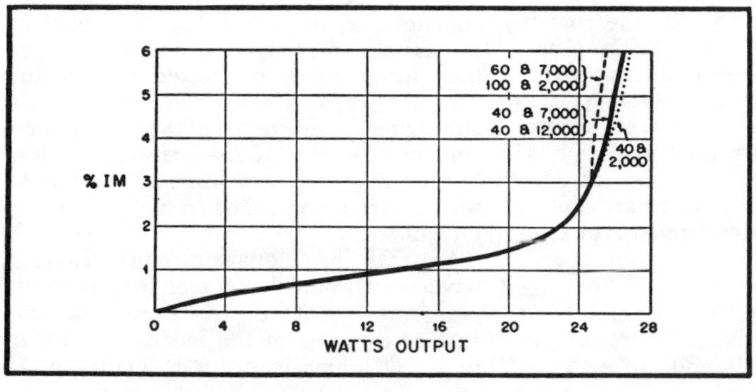

Fig. 4. Intermodulation distortion, using several test frequencies with a constant ratio of 4:1.

Parts List

C_1	500 μf, 6 v. electrolytic
C_2, C_3, C_4, C_5	0.2 μf, 600 v. paper
C_6	100 μf, 50 v. electrolytic
C_7	120 μμf, mica
C_8, C_9	20–20 μf, 450 v. electrolytic
C_{10}, C_{11}	40–40 μf, 450 v. electrolytic
R_1, R_6, R_8, R_9	0.47 meg, ½ watt
R_2	200 ohms, ½ watt
R_3, R_{15}, R_{16}	1000 ohms, ½ watt
R_4, R_5	0.1 meg, 1 watt, 5%
R_7	0.56 meg, ½ watt
R_{10}	560 ohms, ½ watt
R_{11}, R_{12}	47,000 ohms, 2 watt, 5%
R_{13}, R_{14}	0.12 meg, ½ watt
R_{17}	350 ohms, 10 watt
R_{18}	12,000 ohms, ½ watt
R_{19}	22,000 ohms, 1 watt
R_{20}	2200 ohms, 2 watt
T_1	Acro TO-300 output transformer. Primary: 6600 ohms plate-to-plate, tapped for screen; secondary: 16, 8, and 4 ohms.
T_2	Power transformer: 370–0–370 v. at 130 ma; 5 v. at 2 a; 6.3 v. at 3 a.
L_1	Filter choke, 8 Hy at 130 ma.
V_1	6SL7
V_2	6SN7
V_3, V_4	6L6
V_5	5V4

Ultra-Linear Operation of the Williamson Amplifier

DAVID HAFLER and HERBERT I. KEROES

The Famous "Williamson" can be improved simply by replacing the output transformer and making a few minor changes in other components. The results are well worth the effort and expense.

FOLLOWING the original appearance of the preceding article on Ultra-Linear operation of power output tubes, considerable interest has been evidenced in the application of this new circuit improvement to the famous Williamson amplifier. The Williamson circuit has been publicized in several arrangements including at least one commercial one, and the configuration is undoubtedly the most popular high-quality audio circuit ever developed. For many people there is little necessity to attempt to improve this basic amplifier circuit. Its listening quality is excellent; it is easy to construct; and it provides top quality at a cost comparable with units which cannot measure up to its capabilities.

The one category in which the Williamson amplifier is significantly deficient is with regard to efficiency and power-output capabilities. Peak power output is less than 15 watts, and it takes a 450-volt supply at approximately 130 ma to achieve this power output. If this limitation can be overcome without deterioration of quality, a change in the original design is justified. If simultaneously it is possible to improve the amplifier both in measureable aspects and in listening quality, then a change is not only justifiable, it is mandatory.

It is difficult to improve on something which is really good. There are some audio enthusiasts who will scoff at the idea that the Williamson circuit can be improved. However, it has been five years since Mr. Williamson published his circuit; and in the course of five years, there is little which can maintain supremacy without change or renovation. When a basic circuit improvement—the Ultra-Linear output stage arrangement—came along, it was natural to see how it could fit in with the basic Williamson circuit.

The Ultra-Linear output stage is not a triode stage as is used with the Williamson circuit—nor is it a tetrode or pentode stage. It combines the advantages of both triode and tetrode by using an arrangement in which the screen grids of tetrodes are energized from a tap on the primary of the output transformer. This connection, on which patents are pending, modifies the operating characteristics of the tube.

The authors' Ultra-Linear amplifier combined with the power supply on a single chassis.

Proper location of the tap results in optimum input-output linearity simultaneously with efficient operation, power capabilities approximately double those of a triode connection, and low-impedance output such as is offered by triodes. In short, it permits better performance than either triode or tetrode connection of the tubes, and this is substantiated in comparative listening tests and distortion measurements.

The unique merits of the Ultra-Linear stage are particularly applicable to the Williamson circuit. The mating of the two seems to have been inevitable. The simple substitution of an output transformer with primary taps for Ultra-Linear operation and a few minor changes in circuitry, which will be discussed below, combined the basic circuits into an amplifier which practically everybody agrees is an improved version in all respects. Obviously, we must gain improvement if we substitute a more linear output tube and use a transformer which exceeds the originator's stipulations for performance.

The original Ultra-Linear circuit utilizes a transformer, the Acrosound TO-300, which was designed for use with tubes of the 6L6 type. Its 6600 ohms primary impedance therefore, is also correct for 5881's and 807's in the Ultra-Linear hook-up. In addition, KT-66's can be used without deterioration of quality as the slight mismatch is in a favorable direction with respect to distortion characteristics. Therefore, this transformer can be used with the tube types normally used in Williamson amplifiers without compromise of characteristics. It is of interest to note that the change in impedance to 6600 does not violate Mr. Williamson's design considerations. The modified tube characteristics of the Ultra-Linear connection require this impedance if we wish to preserve operating conditions similar to those of the original amplifier. In other words, the tubes are still matched for minimum distortion rather than for maximum power output. The transformer, therefore, can be placed in the circuit directly and the screens of the output tubes connected to the appropriate taps as shown in *Fig.* 1. This eliminates the two 100-ohm screen stopper resistors of the original circuit. The plate and screen leads of the transformer are color coded to avoid phasing difficulties.

Several additional circuit changes have been found beneficial for optimum performance. One of these is the change in value of the feedback resistor to 10,000 ohms in order to maintain 20

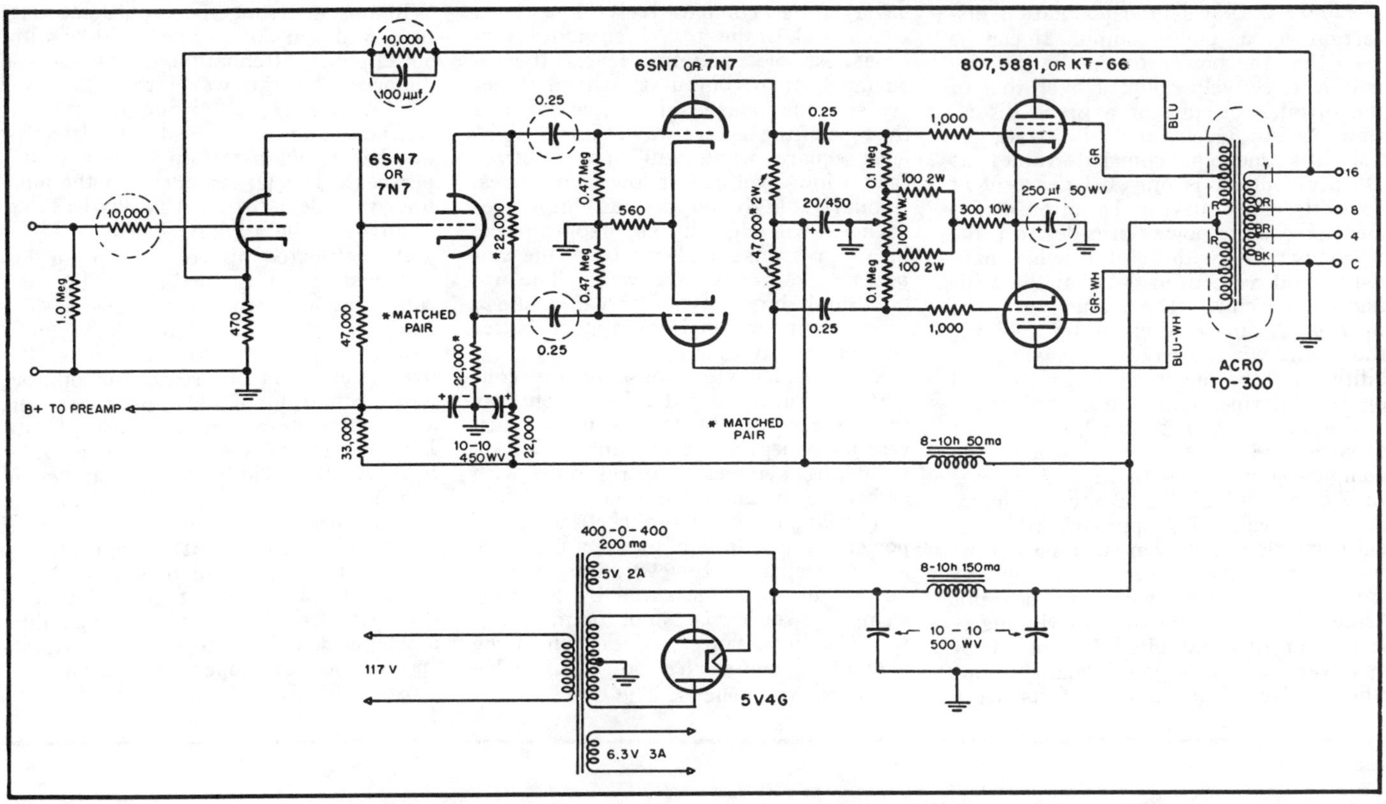

Fig. 1. Schematic of the Ultra-Linear Williamson amplifier. The components in dotted circles are those which are changed from the original circuit in making the conversion of an existing amplifier to Ultra-Linear operation.

db of feedback. In the Ultra-Linear stage the gain of the stage is greater than for a triode stage. In addition, the change in primary impedance changes the proportion of voltage fed back. Thus the feedback is increased unless the feedback resistor is changed to compensate. The readjustment of this resistor to the desired value then permits the added gain of the Ultra-Linear output stage to increase the amplifier sensitivity. It can now be driven with a little over 1 volt as compared to almost 2 volts required for the original amplifier.

The feedback is taken from the 16-ohm tap regardless of the speaker connection. This tapped secondary arrangement is extremely convenient when shifting to speakers of different impedance as it does not require a change in the value of the feedback resistor. It is made possible by special transformer design (on which patents are pending) which permits equivalent response on all taps of a tapped secondary winding.

The amplifier, as converted, now surpasses the original with respect to response, distortion, and transient characteristics. In addition, it was considered desirable to make certain other slight changes which primarily increase the stability under feedback conditions.

The low-frequency time constants of the original circuit's interstage coupling networks were the same for both such networks. This is not particularly desirable in a feedback amplifier since a given frequency loss is accompanied by maximum phase shift. Separation of the time constants permits less phase shift for the same frequency loss. Increasing one pair of coupling capacitors from .05 µf to .25 µf gives a five-to-one ratio of time constants for the two pairs of networks and increases the low-frequency stability margin at nominal increase in cost.

The insertion of a 10,000-ohm parasitic suppressor in the input grid and a 100-µµf capacitor across the feedback resistor adds to the high-frequency stability margin and eliminates a slight ringing in the vicinity of 200 kc.

One last optional difference from Mr. Williamson's original circuit lies in the use of a bypass capacitor across the cathodes of the output stage. This has been found beneficial in both the Ultra Linear conversion and in the triode Williamson at high levels of operation as distortion at the overload point is diminished.

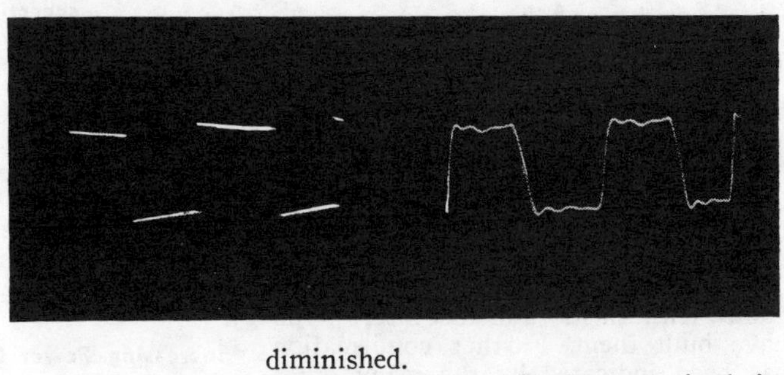

Fig. 3. Square-wave performance at 20 cps (left) and at 50 kc (right).

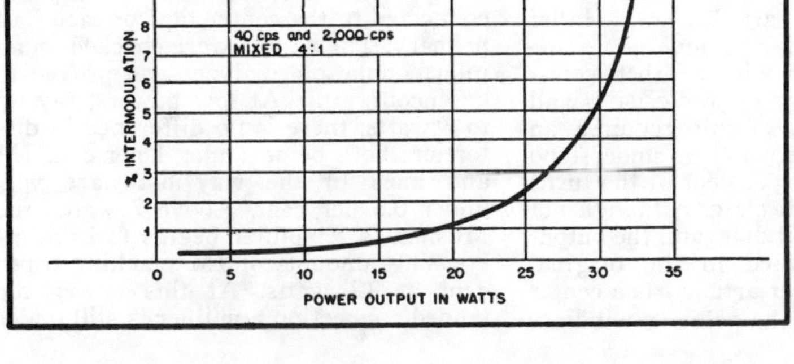

Fig. 2. Curve of intermodulation distortion vs. effective sine-wave-power output.

There are no changes required in the remaining stages nor in the power supply. Most of the publicized versions of the circuit utilize power transformers which furnish 400 volts at 200 ma. Since the drain of the circuit does not exceed 130 to 140 ma, the voltage obtained out of a capacitor input filter and 5V4 rectifier is about 450 volts. This is the correct value for the circuit as converted. Lower voltage will limit the power output capabilities.

Figure 2 shows intermodulation distortion versus power output. It can be seen that the power output of the circuit is effectively doubled over that of the original circuit for a given distortion. At low levels, around 1 watt, the IM hits such phenomenal values as .06 per cent. It is only 0.3 per cent at 13 watts. This curve is based on equivalent sine wave power in order to make it comparable with all the other published and advertised data on the Williamson circuit. The values graphed in *Fig.* 2 can be divided by 1.47 for those who wish to have direct comparability with the meter readings obtained on the intermodulation test equipment.

Figure 3 shows oscillograms of square-wave traces taken through the complete amplifier with repetition rates of 20 cps and 50 kc. Traces at intermediate frequencies approach theoretical perfection, and even such a rigorous test as the 50-kc wave shows up extremely well. The waveform has not "sined off," and the extent of ringing is less than that exhibited by the 5000-cps wave of many good quality amplifiers. These square-wave tests were made at a comparatively low level which makes the test even more rigorous. At low excitation levels, the inductance of an output transformer decreases, the phase shift increases and the tops of the square wave tilts. A high-level square wave will appear better than a low-level one at low frequencies. Similarly, high powers at high frequencies will clip any supersonic peaks in the response and improve the appearance of the square wave. The use of a high level of power can make a relatively poor amplifier appear better on square wave tests.

The frequency response of the converted amplifier is flat ± 1 db from less than 5 cps to 200 kc. Its phase shift reaches 3 deg. at 20 cps and at 20 kc, indicating symmetry of response with respect to the audio band.

The amplifier puts out 30 watts of power over a range greater than the audio spectrum. However, this type of power curve, as measured by response at high power levels, is not too meaningful. The important consideration is the amount of *undistorted* power available at various frequencies. The Ultra Linear Williamson arrangement puts out close to 25 watts at 20 cps and at 30 kc without clipping, attenuation, or other visual distortion of the waveform as viewed on a 'scope. By observing the transfer characteristic, it is possible to detect by eye harmonic distortion of less than 2 per cent. The power curve of the amplifier thus deviates from flat by less than 1 db over the range 20 cps to 30 kc.

As intimated above, the circuit has excellent listening qualities. This is a confirmation of the measurements. The additional power available shows up in cleaner and better articulated bass. The overall effect is of greater smoothness, more definition of detail in the sound, and better transient response. Ultra-Linear circuits seem to have a wider transient bandwidth—an audible benefit which is not readily susceptible to measurement. The combined effect of the Williamson circuit configuration—a wide-band, low-distortion arrangement—plus an output stage of decreased distortion and higher power capability, a stage which exceeds the original specification and operating parameters, must be heard to be appreciated.

Gilding The Lily

DAVID SARSER and MELVIN C. SPRINKLE

Details of a few simple changes in the Musician's Amplifier to improve performance and listening quality.

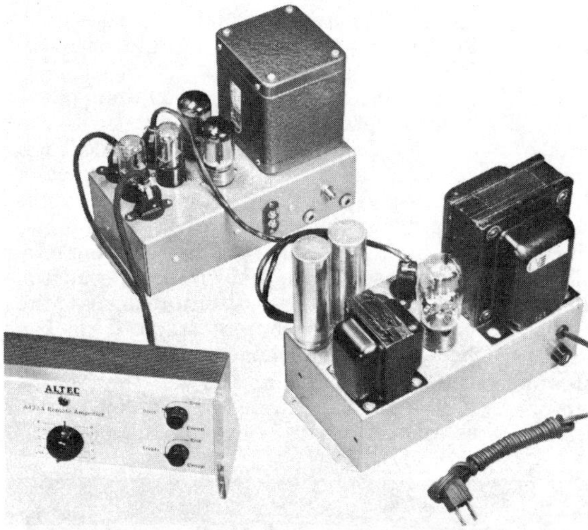

Fig. 1. The converted Musician's amplifier, using the Ultra-Linear connection of the output stage, which employs 5881's instead of the 807's previously specified. The Altec Lansing A-433-A "front end" is shown with the main amplifier and power supply.

The wise man who said that "imitation is the sincerest form of flattery" certainly must have had the Musician's Amplifier in mind. Since its introduction to the American Audio scene, the opinions of the authors have been confirmed by literally thousands of audio enthusiasts and engineers who have built them. Further confirmation has been indicated by the many other versions of "The Williamson" amplifier that have appeared both in kit and in wired form. It is conservative to say that no other audio amplifier has ever had such a wide publicity, so many unanimously enthusiastic users, and so many imitators.

As it does to all things, time has brought some changes to the Musician's Amplifier, and it is felt that the authors should bring to the attention of others certain improvements which can be made in the Musician's Amplifier. All of these changes have been field-tested and are recommended to those who have built the amplifier as per the original article. They cannot be made if the original circuit and components were not followed.

Increasing Power Output

The preceding articles by Hafler and Keroes describe a unique new power amplifier circuit which is between a tetrode and triode in characteristics and performance. This circuit requires an output transformer which is understood to have a tap at 43 per cent of the turns from center to each plate. Although not shown on the circuit diagram, the output transformer specified in the original Musician's Amplifier article has a center tap in each half of the primary winding, which is at 50 per cent of the winding, not too far from 43 per cent. As the circuit has certain features of interest, we investigated the possibility of using the taps to adapt the Musician's Amplifier and improve its performance. An amplifier was built with an A-B switch, arranged so that in one position the circuit was the conventional Musician's while in the other position the screens were connected to the center taps of each half primary. The results were checked on an intermodulation analyzer and proved to be encouraging. At low powers, say up to 7 watts, there is no difference in distortion, both being under 1 per cent IM and most of the way both are way under 0.5 per cent. Above 7 watts, the Musician's Amplifier begins to have increasing amounts of IM reaching 8 per cent at 12 watts. At this power the tapped connection amplifier is still under

1 per cent, and its IM distortion does not begin to climb until the power output is 16 watts reaching 8 per cent at 19 watts. These results are summarized in *Figs.* 2 and 3. It must be emphasized that the above power figures are those as read on the IM meter and are not equivalent sine-wave power. If the figures are converted to equivalent sine-wave power by multiplying by the factor 1.47, then the power output at 8 per cent is 27.9 watts, while the equivalent sine wave power at 1.5 per cent IM is 22 watts. Effectively, the power output has been increased to 158 per cent of its previous value. This is certainly a worthwhile improvement—particularly when it costs no more than two pieces of wire and eliminates the two 100-ohm resistors which tie the screens to the plates.

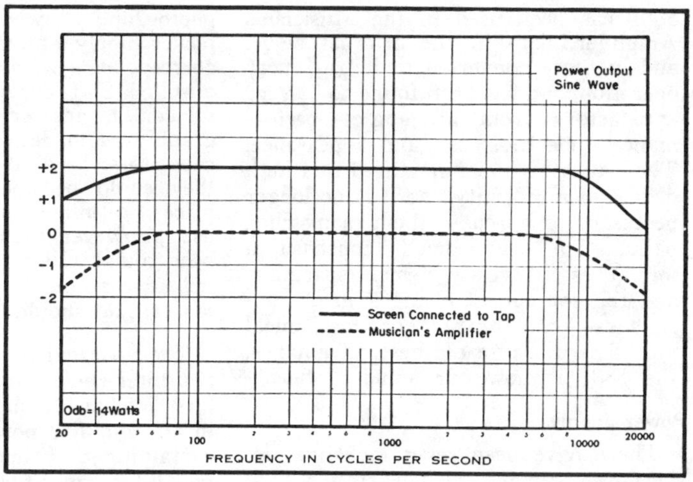

Fig. 3. Power output vs. frequency curves for original and converted amplifiers.

Operation

Checks were made on the effect of the change on the plate and screen currents and on dissipation at both full-signal and quiescent conditions. It was found that the tubes were operating within ratings so that satisfactory tube life may be expected. Checks also were made on the response, square-wave performance, and source impedance; these were found to be affected very little. One item of importance was found: as originally described, the circuit is very nearly Class A and the power amplifier is operated toward the upper regions of plate dissipation ratings, but well within ratings. The bias on the final stage was increased so as to go toward Class AB operation. It was found that any move toward higher bias caused the IM distortion to climb, even at relatively low power levels. The original bias resistor of 250 ohms gives optimum results with the new connection.

By going to the tapped connection for the screens, the gain of the amplifier without feedback is increased by about 4 db. With a 4700-ohm resistor supplying feedback voltage from the 16-ohm output connection, the gain increase by using the taps is around 0.5 db or less.

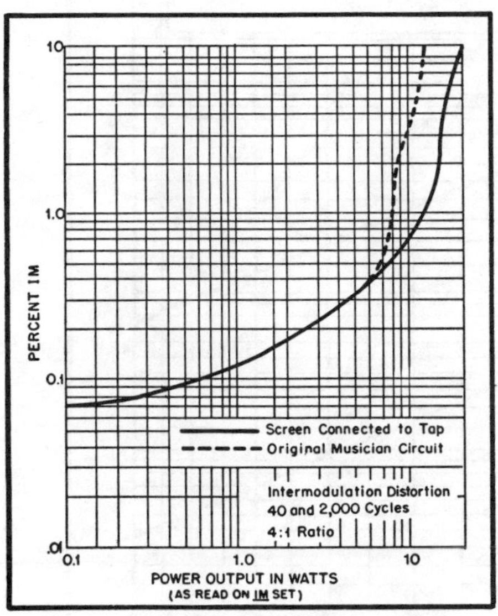

Fig. 2. Intermodulation distortion curves for the original Musician's amplifier (dotted line) and for the converted model (solid line).

Thus the amount of feedback is increased to around 24 db. With most of the amplifiers converted by the authors, there is no tendency toward instability either at sub-audible or supersonic frequencies. Depending on the capacitance and condition of decoupling capacitors, it is possible that a tendency of the loudspeaker to "breathe" or oscillate slowly at 1 cps or less may be encountered. It is recommended that when the change to tapped operation is made, the feedback resistor be increased to 6800 ohms. This value will provide 20 db of feedback, and tests have shown that no appreciable increase in distortion results. For other secondary connections, the feedback resistor may be figured as 1700 times the square root of the nominal secondary impedance.

One of the most interesting and important features of the Musician's Amplifier is the way in which it overloads. Sine-wave power output tests can be made conveniently, quickly, and more accurately than might be supposed, by feeding in a sine-wave signal and increasing the input level until the waveform of the output as seen on a cathode ray oscilloscope begins to clip at the tops and bottoms, or begins to get "bumps" on the sides at low frequencies. The original amplifier overloads so smoothly that it is often difficult to tell just when the beginning of overload is reached. Furthermore, when the tops and bottoms of the waves are being clipped, after overload really is evident there is no ringing or fuzz, but only a clean clip. With the change to the screen tap connection, it was found that the overload was just as smooth as with the conventional triode connection.

Some inquiries have been made as to whether or not a large capacitor should be connected across the self-bias resistor of the output stage. It is well known that a large bypass capacitor should be connected across the bias resistor in Class AB stages, as this improves operation at the higher power levels. During the original work, the bypass capacitor was tried and was abandoned because it produced no significant effect. This is because the power amplifier is practically pure Class A. With the tapped arrangement the capacitor was found to have an improving effect at higher power outputs. For maximum power output connect a 50-μf 50-volt capacitor across the bias resistor. However, it can be omitted with the assurance that no noticeable difference will be heard at lower levels.

Listening Tests

Of course the final test of the merits of an audio circuit is now and probably ever shall be the listening test. In music, listening quality is everything. Having an amplifier with an AB switch is an advantage in listening tests, and after considerable listening it is our opinion that the change *does* improve the sound, particularly on fortissimo musical passages when played at concert hall level. At the usual apartment house living room loudness, operation of the switch produces very little noticeable change. Several users tell us that after living with modified Musician's Amplifiers for several weeks, they are convinced that they sound better at all loudness levels.

For those who have built the Musician's Amplifier as originally written up, with the specified output transformer, here are the details for making the conversion:

1. Remove both 100-ohm resistors (R_{22} and R_{23} on the schematic) that tie screens to plates of the output tubes.
2. Connect a wire from the screen of the tube whose plate connects to terminal 1 of the transformer to the adjacent terminal 2.
3. Connect a wire from the screen of the other output tube (its plate connects to terminal 6 of the transformer) to terminal 5.
4. Change the feedback resistor from 4700 ohms to 6800 ohms (or to a value equal to $1700\sqrt{Z_{vc}}$ if an output impedance other than 16 ohms is being used).

Output Tubes

In the original paper, the authors used the 807 as an output tube in place of the KT-66 valve used in Williamson's design. At that time the KT-66 was not available in America, although it is now.

Recently Tung-Sol Electric, Inc. introduced the 5881 tube which is, in effect, a single-ended 807. The total plate and screen dissipation in the triode connection is 26 watts with a plate-to-cathode voltage of 400. It has the further advantage of single-end construction and the now almost standard octal base. The

5881 has been used in the Musician's Amplifier, both in the original model and in those converted to Ultra-Linear operation, and has been found to be very satisfactory from all angles—performance, tube life, cost, and appearance. These tubes are manufactured to a high degree of uniformity, so it is no longer necessary to purchase them in matched pairs. The ruggedized construction minimizes changes in element spacing—and the consequent changes in characteristics—with heating or mechanical vibration. Because of these advantages, the 5881 is now our standard tube.

Power Supply

There have been several changes in the power supply which warrant a discussion. In the original paper, the editor inserted the words "oil-filled capacitors" in the text material.[1] The accompanying photograph showed round cans in the power supply, and the authors had much correspondence as to where round can oil-filled 8-μf capacitors could be obtained. The answer is simple: the photograph was made with 8-μf electrolytic capacitors. In a number of cases when they could be obtained at reasonable prices, oil-filled capacitors have been used; however, the cans have not always been round. Oil capacitors of 6 or 8-μf. will give a hum-free amplifier. The voltage rating should be at least 600 volts.

The original power supply showed two filter chokes and three filter capacitors. We have found that there is no hum in an amplifier powered from a supply containing only one choke and having two filter capacitors of 6 or 8-μf. The reason for the use of only one choke is to cut down on the d.c. voltage drop in the power supply filter.

Another change in the power supply is in the rectifier tube. The original paper recommended a type 5U4G rectifier. The 5V4G or the older 83V were considered, and their advantage in having a lower internal tube voltage drop was fully recognized, but they were not used because of some past experience with internal tube leakage or shorts. The 5VAG tubes have become readily available because of their wide use in TV receivers as dampers, and it appears that modern construction has made them quite reliable. Thus, we now recommend that the 5V4G be used as a rectifier for improved results. The voltage surge during warmup is practically eliminated with this tube.

With these changes, the output voltage under full load is around 440 volts measured from B plus to ground. With a cathode bias on the output stage of 40 volts, the d.c. plate voltage as measured from plate to cathode on the 5881 tubes is just about 400 volts. With these voltages on the tubes, the cathode current is around 63 ma, and no trouble should be encountered in obtaining the power levels or the low distortion of the Musician's Amplifier. Hum and noise in the power amplifier are inaudible.

[1] We *still* prefer oil-filled capacitors. Most electrolytics rated at 600 volts or more are built-up units using two lower-voltage electrolytics in series. Ed.

Wavelength vs. Frequency Chart

Calculations in the audio field are often encountered in which the length of a sound wave in air is required. While this value is readily calculated in feet by dividing the frequency into 1130, or in inches by dividing the frequency into 13,500, it is much easier to obtain the required information by consulting a chart. The chart below gives this information in two forms—wavelength in inches or in centimeters for a given frequency at a temperature of 20° C. and a barometric pressure of 760 mm. The chart covers the entire audio spectrum by extrapolation.

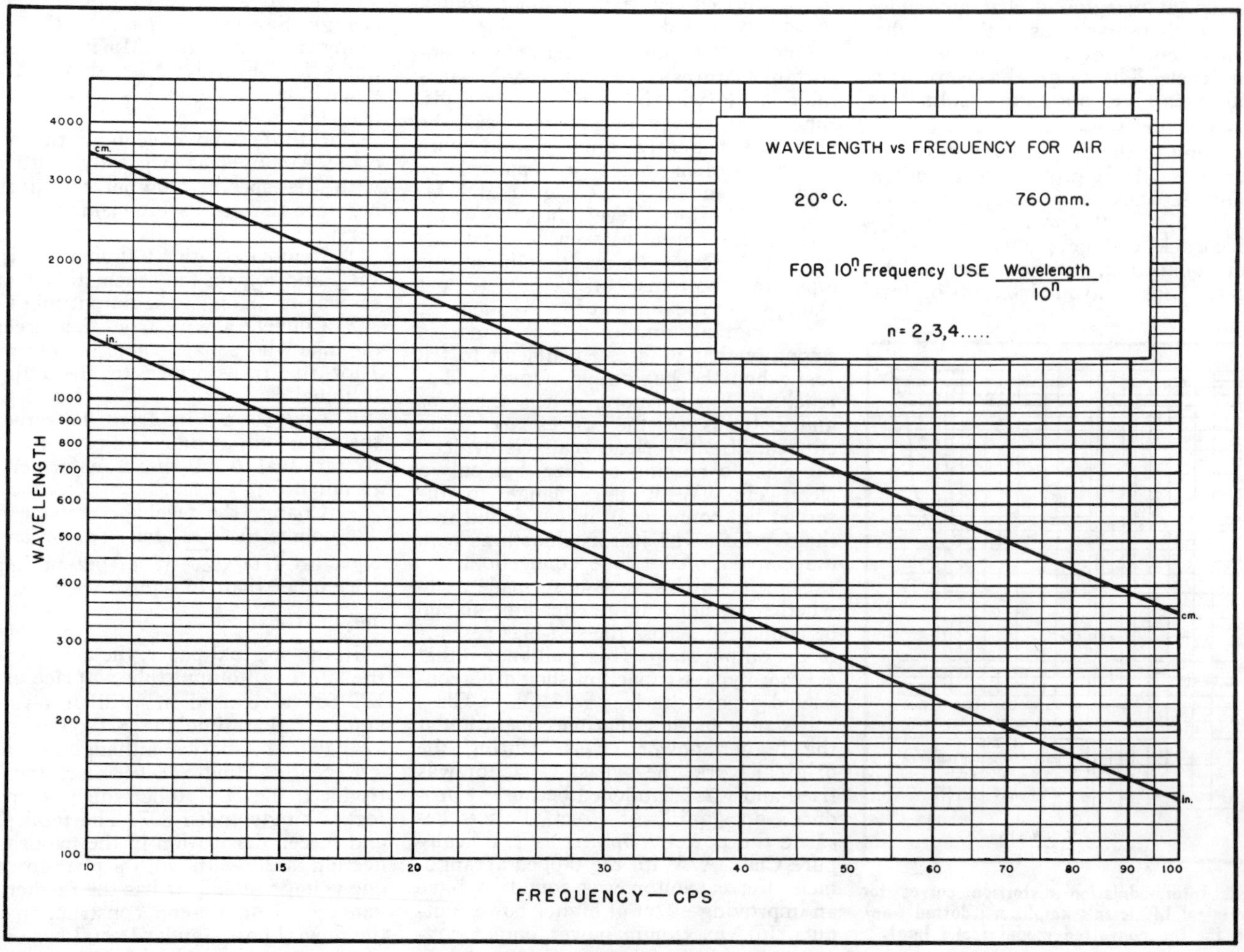

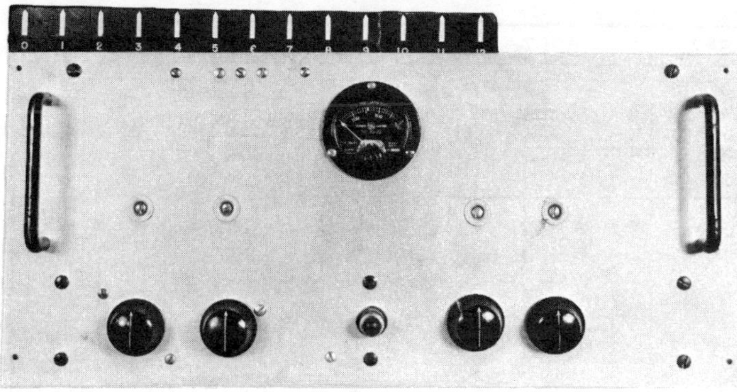

A 15-Watt Direct-Coupled Amplifier

WILLIAM B. FRASER

Describing a stable, well designed amplifier suitable for high quality music reproduction or for small commercial program distribution systems.

SOME TIME ago the author commenced the design and construction of a high quality audio amplifier for his personal use at home. Complexity of circuit design or difficulty of adjustment were considered unimportant, for it was not intended to publish the circuit. The amplifier was finally completed and gave satisfactory results. Only then did it occur to the author that perhaps others would be interested in the design finally adopted.

The circuit is not complex, though it may appear so because of unconventional circuit arrangements. The unorthodox features include a duplex thermostatically controlled power supply, a unique form of loudness control, direct coupling throughout (except preamplifier), push-pull throughout (except preamplifier), and an input circuit permitting the use of either an unbalanced or push-pull signal. At the least, the design is an interesting study in wire. At most, in the author's opinion, it is an excellent amplifier.

Design specifications are used to define certain objectives. In this case, we wanted an amplifier that sounded as we thought it ought to sound, had no hum or tube noise, and had output power sufficient for home use. How are these requirements expressed in figures? It is difficult to say. Experts argue the problem interminably.

But there must be something more specific to aim at than the generalities just mentioned, so the following specifications were set up:

Power output: 15 watts maximum
10 watts below 1% distortion from 30 to 10,000 cps

Frequency range: 20–20,000 ± 0.5 db

Hum and noise: inaudible at all volume levels

Gain: full output with 0.5 volts or less rms input. A preamplifier permitting the use of magnetic phonograph pickups is to be incorporated.

Circuit Details

The design program commenced with a study of the better known commercial circuits and a number of published diagrams. Most of these designs were more or less conventional. By great refinement, a high degree of excellence had been attained in many of them. Nevertheless, there appeared to be two general ways in which conventional design might be improved somewhat. First, almost all of these circuits employed either transformers or capacitor-resistor networks for interstage coupling, and it appeared that a part of the overall distortion of the amplifier originated in these coupling devices. Obviously, then, the elimination of coupling circuits would result in an improvement of the quality of amplification, provided the system used in lieu of conventional coupling was itself distortion free. Secondly, most of the circuits employed single-ended stages for part of the circuit rather than push-pull arrangements. It was thought that a fully push-pull circuit, if feasible, would assist in reducing the second harmonic distortion produced in most equipment.

With these preliminaries in mind, design was commenced. Low-mu triodes were tentatively decided upon for the output stage. 6A5G's were attractive, for they produced the desired power output at small distortion values: they did not require nearly as much driving voltage as the 6AS7G; they had reasonably low plate current and voltage requirements; and they were almost completely hum free.

After design and construction had been completed, it was found that the drivers were capable of providing a peak-to-peak potential of about 210 volts. This is sufficient to drive almost any output tube. Consequently, with appropriate changes, an experimenter may substitute his favorite tube for the 6A5G's shown in the schematic. The author tried 6L6's (tetrode connected), 807's (triode connected), and 6B4G's. 6A5G's seemed to give better results than any of the others, though this is difficult to prove.

Glass enclosed triodes are used for voltage amplifiers. Both 6SN7's and 6SL7's are rugged and non-microphonic. The glass envelopes facilitate trouble shooting. In addition, glass tubes are somewhat less gassy than their metal counterparts. The use of dual triodes cuts down on the total number of tubes required and is also desirable because the two triode sections are more likely to have similar characteristics than separate tubes.

To eliminate conventional coupling devices, direct coupling is used throughout. Direct coupling is inherently free of all forms of distortion. Its principal disadvantages are the high plate supply voltage required, critical balancing, and the possibility of operating tubes at incorrect potentials. Of these problems, maintenance of balance of the circuit was found to be the most difficult to overcome. Balance was finally secured by the use of direct-coupled inverse feedback from the cathodes of the drivers. This arrangement not only corrects for tendencies of the tubes to shift their operating potentials and currents, but also maintains signal balance between the two halves of the push-pull voltage amplifier circuit.

It can be shown that plate supply resistors common to both tubes of a push-pull arrangement assist in stabilizing the d.c. potentials of a direct coupled circuit. Resistors R_{25}, R_{30}, R_{37}, and R_{38} have such an effect.

This feedback does not, of course, correct distortions which may arise in the output tubes and output transformer.

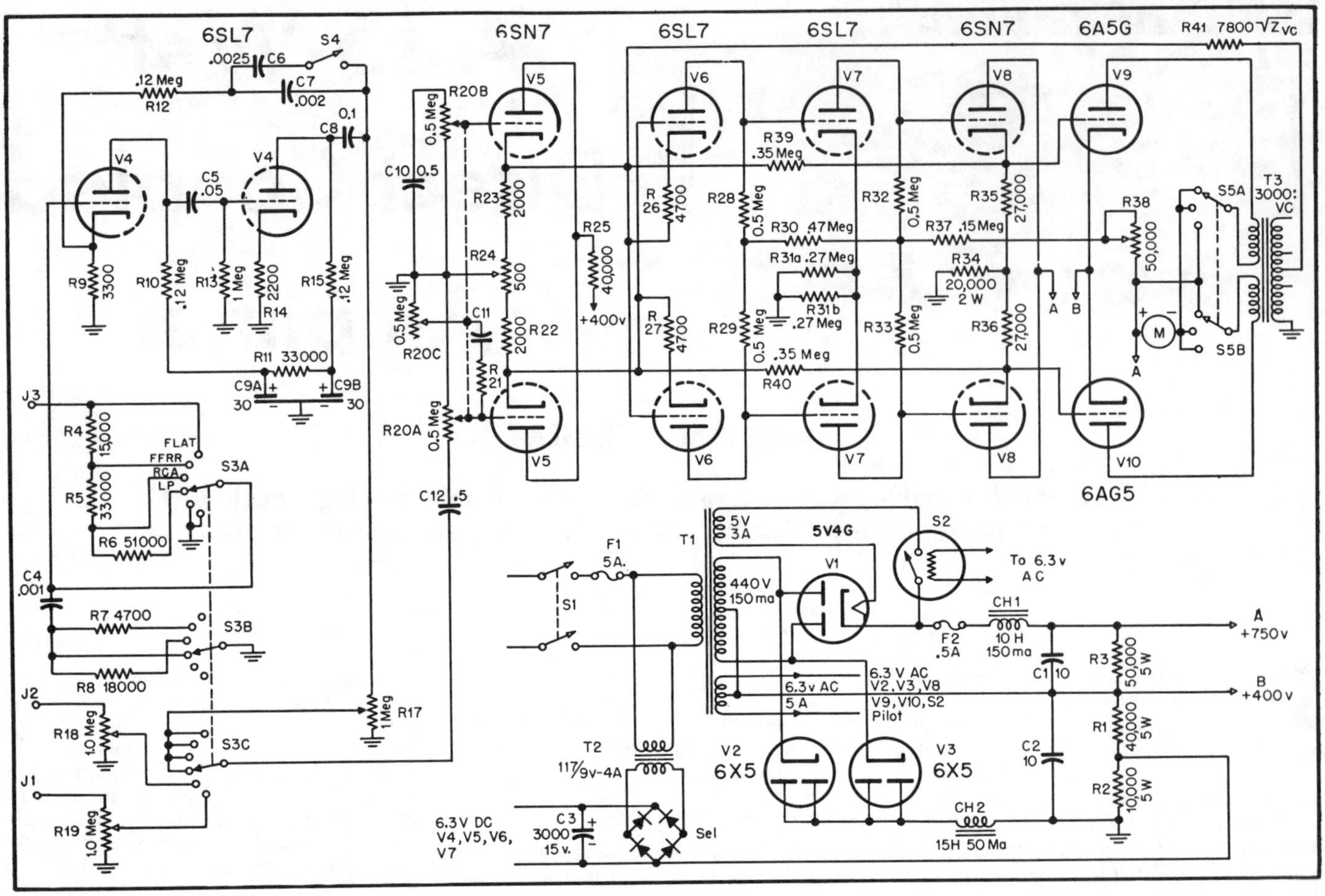

Fig. 2. Over-all schematic of the direct-coupled amplifier.

Such distortions are cancelled by inverse feedback from the output transformer secondary tap to which the loud-speaker voice coil is attached. This arrangement effectively feeds back an accurate sampling of the voltage supplied to the loudspeaker. The feedback resistor used in this circuit should have a resistance equal to 7800 times the square root of the voice coil impedance.

The values of the feedback resistors were selected after considerable investigation of the effect of feedback on signal waveform. Both sine and square wave inputs were used and the resultant oscilloscope patterns were carefully studied. Because of the small phase shift inherent in this direct-coupled circuit, unusually large amounts of feedback can be used. However, feedback in excess of that recommended will result in reduced amplification and may cause high-frequency oscillation.

Signal input can be either single ended or push-pull. Referring to *Fig. 2*, phase inversion is accomplished by V_5 and V_6 in case a single-ended signal is used. This type of phase inverter has no frequency discrimination and produces a perfectly balanced push-pull signal, provided the corresponding parts of the circuit are matched. The design was adapted from similar circuits which have been published recently. A unique form of loudness control is shown as the three-section ganged potentiometer R_{20}, capacitor C_{11} and resistor R_{21}. It will be observed that the high frequencies fed to the grid of V_5 are automatically attenuated as volume is decreased, thus giving an increased proportion of lower frequencies at reduced volume levels. The values of R_{21} and C_{11} can be varied to suit the individual. The author used a value of .04 µf for C_{11} and a value of 0 ohms for R_{21}. After values have been established for the elements of this loudness control, it will work with greatest effectiveness for input signals which have the same average strength as the signal for which the loudness control was originally designed. Therefore, individual semi-adjustable volume controls (R_{17}, R_{18}, R_{19}) have been provided for each signal device.

In case a balanced signal is to be used instead of a single ended signal, the signal should be fed to the ungrounded ends of R_{20A} and R_{20B}. C_{10} is then attached to R_{20B} in the same way as C_{12} is attached to R_{20A}. A fourth potentiometer should be ganged to the loudness control and wired similarly to R_{20A} and R_{20C}.

The slightest trace of d.c. appearing on the grid of V_5 will upset the balance of the entire amplifier. Therefore, C_{12} is used to insure that d.c. from the signal sources is eliminated.

It will be noted that V_5 is a cathode follower and hence produces no amplification. V_6 and V_7 amplify in the normal manner. The common cathode resistor of V_7 tends to correct for any signal unbalance which may occur. Finally, the balanced feedback from V_8 to V_5 corrects for any small residual signal unbalance. Oscilloscope tests show that the signals supplied to the output tubes are balanced under all conditions. This is an important requirement in push-pull circuits.

It will be noted that V_6 has an unbypassed cathode resistor. The resultant degeneration improves frequency response and stability.

V_8 is a cathode-follower driver. Since the 6A5G's are to be operated Class AB_1, and presumably draw no grid current, it may be wondered why V_8 is used. The principal reason for the presence of this tube is that the grids of the 6A5G's do draw current, even though they are not driven positive. This characteristic is typical of many triode-output tubes. V_7 cannot supply current from its plate to the grid of the following tube without suffering serious distortion in its output. However, a cathode follower can supply the small amounts of power required without ill effects, and so this arrangement is used for the driver. An inspection of the circuit diagram will show that R_{38} controls the total plate current of the output tubes and that R_{34} balances the plate current.

A number of excellent preamplifier designs are available. The one shown has been described previously.

Power Supply

At first glance, the power supply may appear to be unusual. Actually, the high voltage secondary of the power transformer merely employs a bridge type rectifier (V_1, V_2, V_3) so arranged that the center tap of the winding is +400 volts. This type of power supply is sometimes referred to as a "duplex" power supply.

The thermostatic delay relay is included to prevent the application of plate voltage to the output tubes before the indirectly heated cathodes of the voltage amplifiers have warmed up sufficently to provide correct bias.

The 6.3 volt a.c. heater winding is baised at +400 volts. The cathodes of V_2 and V_3 are biased at +400 volts, so the same heater winding that supplies the output tubes can be used for the 6X5 heaters. Also, it will be noted that the cathode of V_8 operates at approximately +327 volts which permits this same source of 6.3-volt a.c. to be used for the heater of V_8.

A full-wave selenium rectifier and associated transformer are used to provide d.c. heater current for V_4 to V_7, inclusive. The use of d.c. heaters in these tubes reduces hum disturbances.

Construction

The entire amplifier can be mounted on a 15×19 chassis, but it is recommended that the power supply be mounted on a separate chassis. If only one chassis is used, the parts must be arranged so compactly that a cooling fan is almost a necessity, especially if the amplifier is to be placed in a confined box. If a single chassis is used, the parts should be laid out so that the power supply is at the opposite end of the chassis from the low level stages. Since the circuit is completely push-pull (except for the preamplifier), hum is minimized and shielded wire need not be used. However, the preamplifier must be carefully shielded.

The use of a single ground and grounding bus is recommended to avoid hum which sometimes results from multiple grounds. In this case of the preamplifier, an insulated input jack should be used. The grounded side of this jack should be attached to the grounded end of R_9. If this precaution is not observed, a high hum level will almost invariably result.

In the interest of good construction, filter capacitors C_1 and C_2 should be oil filled. Because of the push-pull arrangement with its inherent hum cancellation characteristics, no large capacitance electrolytic capacitors are required except in the case of the preamplifier and d.c. heater supply.

It is not absolutely essential to match resistors, capacitors, tubes, etc., of the two halves of the push-pull circuit, because cross-coupling, cathode degeneration, inverse feedback, and balancing potentiometers provide for a reasonably well balanced output, even if exact push-pull symmetry is not maintained. Nevertheless, accurate balance and superior performance of the amplifier can be attained only by electrical and mechanical symmetry. Furthermore, changing line voltage will result in unbalanced operation if parts are not fairly carefully matched. Therefore, matching of corresponding parts is recommended insofar as possible.

Wiring of those portions of the circuit operating at +400 volts or less should follow conventional procedure. For the higher voltages, wire with fibre glass insulation is recommended.

Potentiometers R_{24} and R_{38} should be so located that they can be reached easily with a screwdriver while the amplifier is in operation.

Transformers and chokes should be of good quality. The output transformer is especially important. The quality of the entire amplifier will depend largely on this item. This circuit was designed, among other things, to eliminate expensive interstage audio transformers, and the money so saved can be invested in the output transformer. A number of excellent makes are available. The author used a UTC linear standard LS55, and found it very satisfactory.

Tubes V_4, V_5, and V_6 should be mounted in non-microphonic tube sockets. Switches S_3 and S_5 should be the shorting type to prevent noisy switching. The feedback loop attached to the secondary of the output transformer should not be finally soldered in place until adjustments of the amplifier are completed.

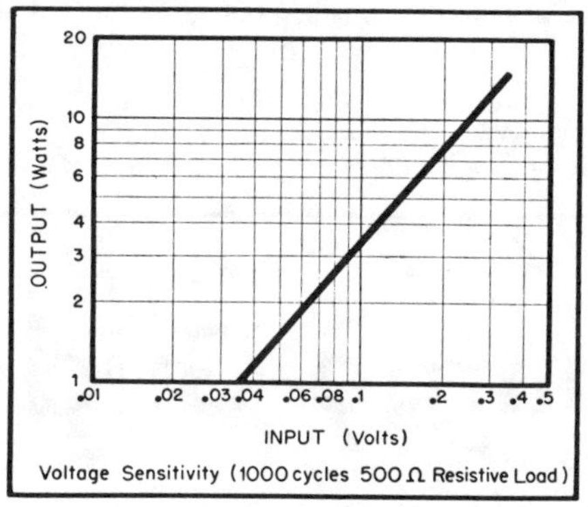

Fig. 3. Curve showing voltage sensitivity of the amplifier when feeding a 500-ohm resistive load.

Adjustment

After completing the construction, insert all tubes heated by d.c. Turn on the amplifier and check the heater voltage to make sure it is 6.0 to 6.3 volts.

Next, insert all tubes heated by a.c. The thermostatic switch should not be inserted until later. Turn on set and measure a.c. heater voltage.

If everything is operating normally, place the thermostatic switch in its socket and turn on the amplifier. Adjust R_{24} and R_{38} to produce correct operating current and voltages for the output tubes. If parts have been well balanced, minimum hum will be obtained when plate currents are balanced. If parts have not been carefully selected, minimum hum may occur when plate currents are slightly *un*balanced. Adjust R_{24} for minimum hum.

Finally, attach feedback loop from output transformer secondary to one of the cathodes of V_5. If noise increases when feedback loop is attached, the loop has been reversed and should be attached to the other cathode.

A final check of voltages should be made; Table I shows typical values for plate, grid, and cathode potentials referred to ground. If everything is in correct working order, hum and noise will be inaudible when the ear is held more than three or four inches from the speaker. When the preamplifier is switched into the circuit, a small amount of noise will become apparent, though this noise should be so slight as to cause no objection.

In order to maintain balance of the amplifier, potentiometers R_{24} and R_{38} will have to be adjusted periodically as the tubes age. The frequency of these adjustments will decrease after the first few weeks of operation, during which time the tubes' characteristics are changing quite rapidly. Since line voltages will vary throughout the day, it is suggested that balancing be done when the line voltage is at its average value. The amplifier should be adjusted only after it has warmed up at least half an hour.

Performance

Full output of 15 watts is attained with an input voltage of 0.35 volts rms. An output of over 20 watts can be attained, though the amplifier begins to produce appreciable distortion over 15 watts.

Frequency response is flat to approximately 20,000 cps, with a gradual droop above that point. Voltage sensitivity is shown in *Fig* 3. Hum and noise voltages were so low that they could not be measuered with equipment available.

Conclusions

In the past, direct-coupled amplifiers have proved unpopular probably because of several problems associated with this type of design. The circuit described herein overcomes all these difficulties by unconventional arrangements, except that periodic adjustment of the current of output tubes will be required.

However, for those who require unusually good performance, and enjoy the work of achieving it, it is believed that this circuit will provide satisfactory results. Its superiority to most typical designs can be shown either by instruments or by listener tests.

TABLE I

Tube	Plate	Grid	Cathode
V5	170	0	6.5
V6	120	6.5	8
V7	309	120	122.7
V8	750	309	327
V9	745 @ 42 ma	327	400
V10	745 @ 42 ma	327	400

NOTE 1. Actual voltages may vary as much as 15 per cent from figures shown without detrimental effect. However, output tube plate current and voltage should be adjusted as accurately as possible. The exact voltage of output tube grids is unimportant, provided plate current and voltage are correct.

NOTE 2. Voltage measurements should be made with a vacuum tube voltmeter.

PARTS LIST

R_1	40,000 ohms, 5-watt, wire wound
R_2	10.000 ohms, 5-watt, wire wound
R_3	50,000 ohms, 5-watt, wire wound
R_4	15,000 ohms, ½-watt
R_5	33,000 ohms, ½-watt
R_6	51,000 ohms, ½-watt
R_7	4700 ohms, ½-watt
R_8	18,000 ohms, ½-watt
R_9	3300 ohms, ½-watt
R_{10}, R_{15}	0.12 meg, 1-watt
R_{11}	33,000 ohms, 1-watt
R_{12}	0.12 meg, ½-watt
R_{13}	1.0 meg, ½-watt
R_{14}	2200 ohms, ½-watt
R_{16}	75,000 ohms, 1-watt (Not shown, but connects C_{9b} to 400-volt supply at "B")
R_{17}, R_{18}, R_{19}	1-meg potentiometer
R_{20}	3-sect. ganged potentiometer, 0.5 meg. each section
R_{21}	See text
R_{22}, R_{23}	2000 ohms, 1-watt
R_{24}	500-ohm wire-wound potentiometer
R_{25}	40,000 ohms, 10-watt, wire wound
R_{26}, R_{27}	4700 ohms, 1-watt
R_{28}, R_{29}, R_{32}, R_{33}	0.5 meg, 1-watt
R_{30}	0.47 meg, 1-watt
R_{31}	0.135 meg (2 0.27-meg 1-watt resistors in parallel)
R_{34}	20,000 ohms, 2-watt
R_{35}, R_{36}	27,000 ohms, 2-watt
R_{37}	0.15 meg, 1-watt
R_{38}	50,000-ohm wirewound potentiometer
R_{39}, R_{40}	0.35 meg, 2-watt
R_{41}	7800 $\sqrt{\text{V.C. impedance}}$
C_1, C_2	10 μf, 600-volt, oil filled
C_3	3000 μf, 15-volt electrolytic
C_4	.001 μf, mica
C_5	.05 μf, 600-volt, tubular
C_6	.0025 μf, mica
C_7	.002 μf, mica
C_8	0.1 μf, 600-volt, tubular
C_9	30-30 μf, 450-volt, electrolytic
C_{10}, C_{12}	0.5 μf, 600-volt, oil filled
C_{11}	See text
V_1	5V4G
V_2, V_3	6X5
V_4, V_6, V_7	6SL7
V_5, V_8	6SN7
V_9, V_{10}	6A5G
S_1	DPST toggle switch
S_2	Amperite 20-sec delay relay
S_3	3-gang, 6-position, shorting type
S_4	SPST toggle switch
S_5	2-gang, 3-position, shorting type
T_1	440-0-440v at 150 ma; 5v at 3a; 6.3v at 5a
T_2	9v at 4a filament transformer
T_3	20-watt output transformer, 3000 ohms plate-to-plate
F_1	5-amp fuse
F_2	0.5-amp fuse
SEL	15-volt, 4-amp selenium rectifier
M	Milliammeter, 0-100 ma
Ch_1	10-henry, 150-ma choke
Ch_2	15-henry, 50-ma choke
J_1, J_2, J_3	Input jacks

Space-Charge-Grid Amplifier

MELVIN C. SPRINKLE

A new low-power amplifier using the space-charge-grid tubes first publicized as audio output tubes almost four years ago.

FROM TIME IMMEMORIAL, it seems, the audio enthusiast has been searching for wider frequency range and lower distortion in his amplifiers. These desirable features are expensive, and cost is an important consideration to the average audio fan. So, the effort has gone on to get the most for the least. The amplifier to be described represents what is believed to be the highest quality yet attained for the required financial investment. The remarkable performance is made possible by two important components: a new and radically different tube type and a high-quality low-cost output transformer.

In spite of its well known limitations which increase cost, the triode tube has for years been the standby of the audio crowd. The beam-power tube overcame a number of the limitations of the triode, but many builders never accepted the beam power tube as the equal of the triode. Thus the argument has raged for some years on the beam tube vs. the triode, with good points on both sides. Now, the National Union Radio Corp. has developed for commercial use a new tube type which combines the best features of the triode and the beam tube, and which opens a new era in high-quality amplifiers. This tube is known as the NU 2160 and is a space charge tube. Its plate family of curves resembles those of triodes, but its efficiency and drive requirements are like a beam power tube.

Fig. 1. Separate chassis for the amplifier and power supply simplify mounting in many applications.

The space charge tube has been described in the literature, but for the benefit of those who do not have access to the references, its characteristics are briefly summarized. The space charge tube is a tetrode, with a cathode, two grids, and a plate. The grid nearest the cathode is operated at a positive d.c. voltage with respect to the cathode, the effect being to counteract the negative space charge and produce a larger cloud of electrons at the plane of the first grid. This cloud acts as a virtual cathode. The second grid is then the control electrode, being operated with a negative bias in the conventional manner. The operation of the space charge tube is similar to a triode with a large effective cathode. It should be pointed out that it is not possible to connect an ordinary tube as a space charge tube because so doing would cause excessive current in the first grid. In the NU 2160 the current in the first grid has been lowered by mounting a pair of side rods between the space charge grid and the cathode, and connecting these rods to the control grid. The side rods thus shield the space charge (No. 1) grid from the cathode except in the vicinity of the lateral wires of the grid. They also have an effect in reducing harmonic distortion.

The positive voltage on the space-charge grid is supplied from the plate supply through a 10,000-ohm resistor. The space charge grid is *not* bypassed, and the omission of the bypass capacitor produces three beneficial effects: the power output is increased slightly for a given average cathode current, the output is less affected by changes in external load impedance, and the odd harmonic distortion is reduced. It has been mentioned that the plate current family resembles triode curves. To this must be added the fact that the plate resistance is low—2,500 ohms—the μ of 16 is medium, and the transconductance is 6,500 micromhos. The uniformity and regularity of the plate family indicate that these parameters are fairly constant and therefore the distortion is low. The tube is operated with a bias of about 18 volts on the control (No. 2) grid so that operation out of a resistance-coupled stage or phase inverter is very practical.

Through the courtesy of National Union, the writer was provided with a pair of NU 2160's for experimental use. At the present time, National Union is marketing this tube through radio parts distributors under the new designation 6BL5GT/2160 at a price of $18.00. Because of this high price, most of the discussion of this tube for audio amplifier applications is purely academic, since better amplifiers could be built at much lower cost.

Design of Amplifier

Examination of the data sheets provided with the tubes, showed that the tube was ideal for a low-cost amplifier, the power output as calculated being of the order of 6 watts for two tubes in push-pull. The Peerless S-508-A output transformer, originally designed for type 6V6 tubes, has ratings as to plate current and primary impedance which suit it to the 2160. Its design, based on the 6V6 plate resistance, means that it will perform even better at low frequencies with the low-plate-resistance space-charge tube. Its physical size is small, and its price—scarcely higher than a large replacement type transformer—puts it within the reach of the most modest budget. This transformer has low insertion loss, wide frequency response, and ability to deliver power at frequency extremes. The performance data on the complete amplifier to be given later attests to this.

The amplifier itself is simple and with the possible exception of the space-charge stage is conventional. As will be seen from *Fig. 2* it consists of a pentode amplifier stage using a type 6SJ7, direct coupled to a 6J5 split-load phase inverter which drives the output

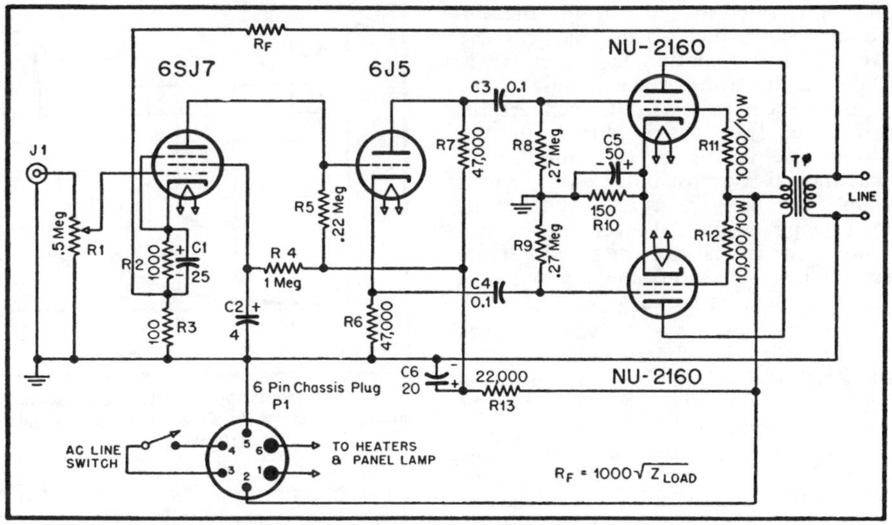

Fig. 2. Schematic of the space-charge-grid amplifier chassis.

tubes. The circuitry of the pentode stage is conventional, the only precaution being to use an adequate bypass capacitor on the screen. The split-load phase inverter is also well known and conventional. The writer has found it to be about as good a circuit as is available, having none of the drawbacks sometimes ascribed to it. One of the big advantages is that it lends itself so well to direct coupling from a preceding stage. Use of direct coupling extends the frequency response of the stage to d.c. and eliminates coupling networks which can cause phase shift and low-frequency attenuation. There are only two coupling capacitors in the amplifier, from the phase inverter plate and cathode to the power amplifier grids. The low-frequency response in the entire amplifier is very good, much better than the transformer curves alone might lead one to believe. It should also be pointed out that the direct coupling system used here is almost automatic in its action to provide optimum operation for both stages once the plate voltage of the preceding stage is brought close to the positive voltage on the phase inverter cathode. There is plenty of good clean signal from the phase inverter to work directly into the space charge control grids and give full power output.

Output Stage

The space-charge-tube circuit differs from the usual push-pull beam-power circuit in one respect. Two 10,000-ohm 10-watt resistors are used to provide the space charge grid voltage. It is necessary to use a separate resistor for each tube. As has been pointed out previously, these space-charge grids are *not* bypassed. The cathodes are tied together and go to ground through a 150-ohm resistor, which is heavily bypassed. The tubes are operated more Class AB_1 than recommended by the manufacturer, as tests with intermodulation equipment showed that reducing the bias increased the distortion slightly at the higher output levels.

Inverse feedback around the amplifier is sufficient to reduce hum, noise, and distortion almost to the vanishing point at the usual levels. The source impedance is lowered to 2.5 ohms on the 16-ohm tap. Connections are available on the output transformer to match loads of from 4 to 16 ohms. Values for the feedback resistor for various load impedances are given in *Fig. 2*.

Construction

The amplifier is constructed on an aluminum chassis 5×7×3 in. There is plenty of room for all parts, most of which are mounted on two terminal boards. One board is used for the pentode and phase-inverter components, while the other holds the coupling capacitors and resistors for the space charge tubes. A ground bus is used with connection to the chassis at the input so as to avoid chassis currents and hum. The controls are simple—a volume control, a pilot lamp and an a.c. line switch.

The power supply, *Fig. 3*, is simple and conventional. The power trans-

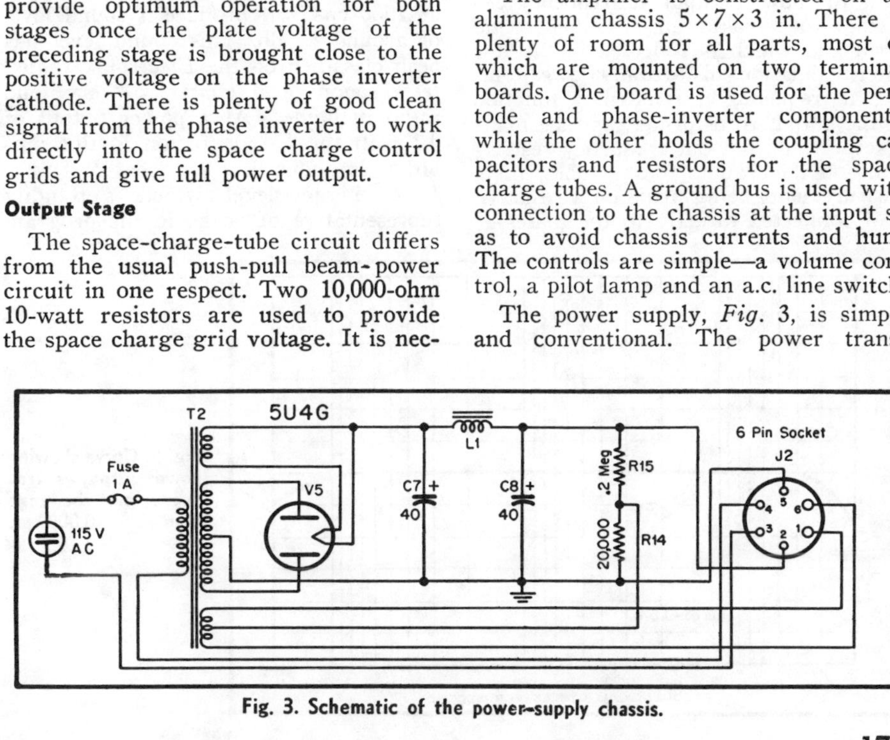

Fig. 3. Schematic of the power-supply chassis.

former is rated at 120 ma, and the amplifier draws just about that current. After 12 hours continuous use in 75° F. ambient, the transformer is just warm. A 6-wire cable is used to connect the power supply to the amplifier. The a.c. power line is run to the amplifier so that there is no possibility of the power supply becoming energized, without its load being connected even if the a.c. power switch is not installed. If no switch is desired, terminals 3 and 4 of the plug must be wired together.

The two-chassis system was employed for this unit—one chassis being used for the amplifier and another for the power supply—so that the amplifier can be placed wherever desired and the power supply can be located at some remote point. Furthermore, they can be placed side by side, end to end, or even stacked if required. One large chassis can be used if the amplifier and power supply are to be constructed as a single unit.

Performance

At the beginning it was stated that this amplifier was intended as a high-quality, low-cost unit. It is believed that this has been accomplished, as will be seen from the performance data. The NU 2160 was not intended as a high-power tube, and hence the power output is limited when compared with 6L6 tubes. The useful power output of the amplifier in a load resistor connected to output transformer secondary is about 5.5 watts, based upon the point where the intermodulation begins to climb rapidly. This power rating compares very favorably with the data sheet which gives the power in a resistive load connected from plate to plate as 6 watts for 2 per cent harmonic distortion, as shown in *Fig.* 4. The single-frequency sine-wave output power before the waveform begins to be distorted (about 5 per cent) is 7.5 watts. The power output as a function of frequency is given in *Fig.* 5. It is the power output at frequency extremes that makes an amplifier sound good and which is reflected in the low intermodulation distortion. Power output at frequency extremes is a function of the output transformer, as tube manufacturers do not put frequency restrictions on their audio power tubes, their measurements being made on a resistive load connected directly to the plates.

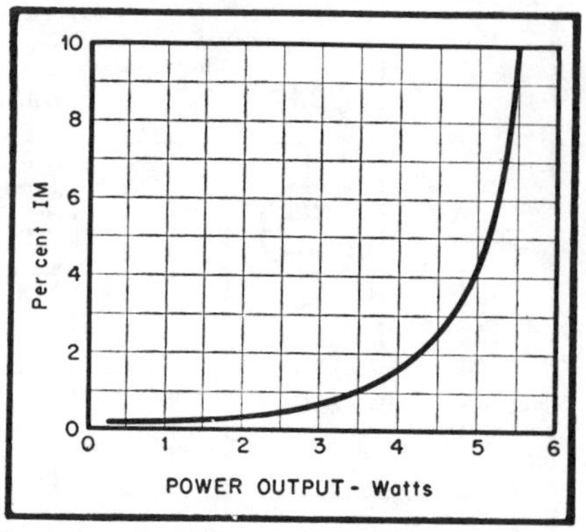

Fig. 4. Intermodulation distortion curves for the amplifier indicate its excellence for low-power applications.

The absolute gain of the amplifier is 81 db, and is practically constant from 20 to 20,000 cps., the maximum departure being about 0.2 db at the limits. It must be pointed out again that frequency response is not nearly as important in the "ear appeal" of an amplifier as power output as a function of frequency. The output transformer used has a rated frequency response of no more than 1 db down at 30 and 15,000 cps, yet the amplifier frequency response is much better. The answer lies in the fact that inverse feedback improves the frequency curve. However, inverse feedback will not help power output measurements as a function of frequency to a great extent. Hence the power curve is more important.

The total noise level, including hum, is – 51.2 dbm which is 88.7 db below the rated output of 5.5 watts. The distortion was measured with an intermodulation set using frequencies of 40 and 2000 cps. Up to within 1 db of the rated power the IM distortion is no more than 2 per cent, and at 2 db below rated power the IM is 1 per cent or less. These are entirely negligible amounts. An efficient speaker with 0.1 watt input at 1000 cps will produce a sound level of around 83 db. This sound level has been classified on the *Electronic Industries* sound level chart as "very loud radio in home." At a power output of 0.1 watt the IM distortion in this amplifier is about 0.15 per cent and at a 10 db higher level, which is roughly representative of peaks in the program material, the IM distortion is 0.25 per cent. Thus at all usable levels, the IM distortion is too small to worry about.

The total cost of the parts to build this amplifier and power supply, using the transformers specified for its performance and other parts of first grade is about $35.00. This does not include the space-charge-grid tubes, which add up to another $36.00, making the entire amplifier rather more costly than many amplifiers of considerably better performance. However, if the demand for the tube should ever increase to the point where its cost became comparable to that of a 6V6, the performance of this amplifier would recommend it.

It was not intended that this amplifier take the place of a deluxe amplifier, such as the Musician's Amplifier, where the ultimate in quality is desired and where the performance requirements are rigorous. This amplifier "folds up" at power levels where the Musician's Amplifier is still below 2 per cent IM. However, this amplifier is especially commended for the average home, especially when cost is an important consideration. At the 1950 Audio Fair it was compared on A-B test with larger amplifiers of equal quality and until the level became so loud that it was deafening, there was no audible difference. In the author's home it has done everything that a home amplifier is called upon to do. There is not enough gain to operate directly from a magnetic pick-up cartridge but with a preamplifier, many of which have been described in these pages, it reproduces all that is on a record. There is, however, plenty of gain for an FM tuner.

Parts List

C_1	25 µf, 25 v. electrolytic
C_2	4 µf, 450 v. electrolytic
C_3, C_4	0.1 µf, 600 v. paper
C_5	50 µf, 50 v. electrolytic
C_6	20 µf, 450 v. electrolytic (FP type)
C_7, C_8	40–40 µf, 450 v. electrolytic (FP type)
L_1	Peerless C-325-A filter choke. 10 H. at 120 ma; d.c. resistance, 240 ohms
R_1	0.5 meg, audio taper vol. control
R_2	1000 ohms, 1 watt
R_3	100 ohms, ½ watt
R_4	1.0 meg, 1 watt
R_5	0.22 meg, 1 watt
R_6, R_7	47,000 ohms, 1 watt (5% or matched)
R_8, R_9	0.27 meg, ½ watt
R_{10}	150 ohms, 10 watt
R_{11}, R_{12}	10,000 ohms, 10 watt
R_{13}, R_{14}	22,000 ohms, 1 watt
R_{15}	0.22 meg, 1 watt
T_1	Peerless S-508-A output transformer. Pri. Z: 8000 ohms plate-to-plate; Sec. Z: 4, 8, 12, 16 ohms. Frequency response ±1 db from 30 to 15,000 cps.
T_2	Peerless R-480-A power transformer. 350–0–350 v. at 120 ma; 5 v. at 3 amps; 6.3 v. at 5 amps.
V_1	6SJ7
V_2	6J5
V_3, V_4	National Union 6BL5GT/2160
V_5	5U4G

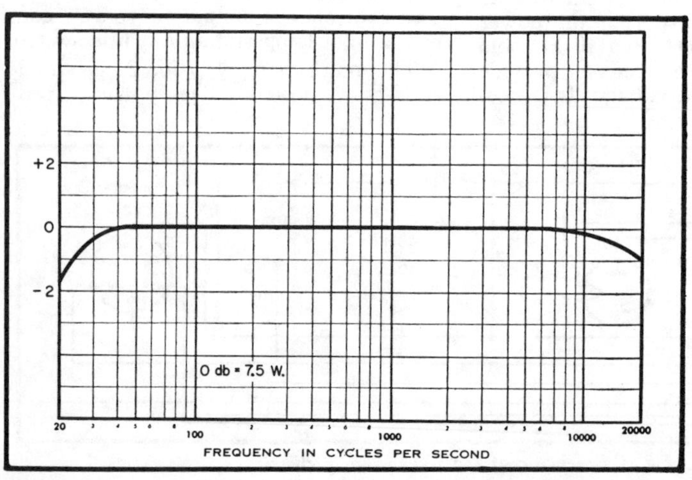

Fig. 5. Curve showing power output vs. frequency over the range from 20 to 20,000 cps.

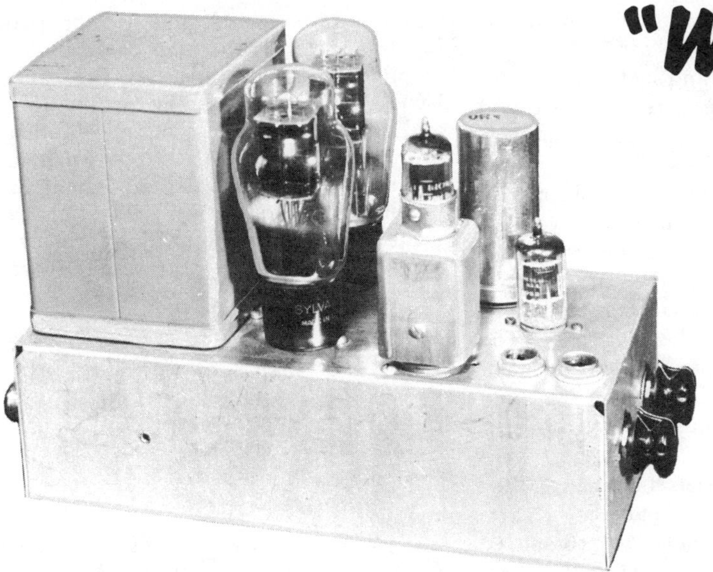

Figure 1

"Williamson" Type Amplifier Using 6A5's

J. H. BEAUMONT

A simple and effective amplifier based on the now-familiar circuit originally appearing in "Wireless World."

THE CONTINUALLY INCREASING POPULARITY of the Williamson amplifier among music lovers and technicians alike prompts another version of this circuit, the third to appear in this country. Designed for 6A5's, it will not deliver as much power to the speaker as either of the others; but use shows its nominal rating of six watts to be ample for home use. The arguments for and against large reserve power have never been resolved among engineers, and no attempt will be made to do so here. The low-power amplifier is regaining its popularity after having been nearly lost in the welter of claims that grew up around amplifiers using the 6L6 and the attendant fanfare about 25- and 30-watt outputs. There is no particular need for arguing that there should *not* be as large a power reserve as is reasonably possible since most circuits exhibit greatly improved distortion and linearity characteristics when they operate well below their ratings, but it is worth considering that the average power delivered to the speaker for "normal" room volume in the usual living room is of the order of 6 milliwatts, as has been established by tests. On this basis, the thousand-to-one reserve ratio of this amplifier seems adequate. If an efficient speaker is used, the amplifier will produce more volume than most listeners can tolerate, with sufficient reserve to maintain clean reproduction of any music.

The original plan was to provide a fixed gain package which would operate from a preamplifier containing all the necessary controls. However, the idea of a self contained preamplifier seemed so attractive that it was added to the design, with consequent modifiations in the circuit. To avoid an additional stage of gain, high-mu triodes were used for inverter and driver stages. The direct-coupled inverter was changed slightly to conform to the configuration of the direct-coupled preamplifier in order to preserve uniformity and to ensure the excellent characteristics which have been observed with this arrangement in several circuits. Direct-coupled pairs, which are a characteristic of the Williamson, are seldom seen in amplifier service, though their use for this purpose was established in this country as long ago as 1940.[1] The configuration shown here has been used for some five types of stages with considerable satisfaction. The cathode circuit of the 6A5's had to be modified because the cathodes in these tubes are tied to the filaments. There are no other major changes from the original circuit. The output transformer chosen was a Freed F-1951, which gives excellent results. Miniature tubes were used in place of octals because of the higher mu available in this series. *Figure* 1 shows the external appearance of the complete amplifier.

Preamplifier Description

Starting at the input plug, let us consider those parts of the circuit that merit discussion. The preamplifier, which is constructed in a Vector plugin can, is a familiar circuit which has been modified slightly for direct coupling. Its operating characteristics remain essentially the same as the capacitance-coupled version. Referring to the schematic, *Fig.* 2, the input grid resistor, R_{11}, was set at 0.1 meg because this value works well with the General Electric cartridges which were used with this amplifier. When other cartridges are to be used, it is preferable to use the values specified by the manufacturer for flat response.

R_1 and C_5 form a compensating network in the feedback loop to eliminate the tendency of this preamplifier circuit to droop at frequencies above the crossover point, a condition which is inherent to the circuit since the crossover point is located on the knee of the compensation curve produced by the feedback-loop components. When this droop is added to that often found in cartridges and present even in amplifiers which are rated flat, the result can be as much as a 5-db drop at frequencies between crossover and 5,000 cps. This is sufficient to be noticeable, and has an effect of reducing the brilliance of music. The compensating network tends to decrease the feedback as frequency increases, thus offsetting the droop. Values for these components may have to be varied to suit an individual amplifier, but normally the capacitor should have a reactance of about ten times R_3 at the crossover frequency. If the response rises, the value of the capacitor should be decreased; if response drops off, C_5 should be increased. R_1 may vary between the value of R_3 and one-half that value, increasing if there is a rise at 10,000 cps, and decreasing if there is a droop at this point.

Adjustment of the network was made with a Cook Series 10 frequency record so that the cartridge would be included. The pickup used with this amplifier exhibits excellent response characteristics with the Cook record, showing no more

[1] C. G. McProud and R. T. Wildermuth, "Direct-coupled input stage for phase inversion in audio amplifiers," *Electronics*, October 1940.

than plus or minus 1-db variation through 10,000 cps.

The preamplifier may be built on a regular Vector socket if the plug-in feature is not desired. However, the preamplifier plug-in socket may be used as a source of power for an external preamplifier if at any time it is desired to use one, and plug-in preamplifiers allow changes to be made without disturbing the internal wiring. The circuit shown here is the third used with this amplifier since its construction was started. It is worth noting that if the octal socket is used as a source of external power there can be no heater grounds in the external equipment because the heater string is connected to the 6A5 cathodes and is "off ground" by the amount of their bias. If a separate heater pair is supplied on the power supply used with the amplifier, it may be wired to the preamplifier socket, though in general, the d-c bias on the heater string is considered desirable.

Parts mounting on the post of the Vector unit will be much simplified if the screw joining the parts is taken out and the octal plug removed entirely. This allows the post to be rotated so that its lugs are conveniently located with respect to the novel socket lugs. Components which pass into the octal base, such as C_4, should be fastened at one end and plenty of lead length left on the other. Leads from the lugs to the socket are fastened to their lugs and left projecting below the end of the post far enough to pass through the octal pins. Fiber-glass sleeving should be used on leads only where one crosses another, or comes near a lug; the rest of the wiring may be done with bare wire. *Figure* 3 shows the appearance of the preamplifier, both cased and uncased.

When all wiring and soldering is complete, the octal plug is re-assembled and all leads passed through their respective pins. When the screw has been tightened, the lead ends are cut off flush with the pins and soldered. If a layout is made before assembly is undertaken, it is possible to arrange the parts so that all the leads will pass almost straight from the post into the plug pins. The ground lug on the base of the can should be connected to the common ground point of the preamplifier.

The remainder of the stages were wired on Vector sockets because they offered an interesting approach to parts mounting which could be completed stage by stage and inserted in the chassis with all leads attached. This simplifies assembly greatly, and the resulting job will be both compact and neat, as shown in *Fig.* 4. This type of mounting avoids the troubles often found in getting strip mountings to operate satisfactorily due to the necessarily greater length of critical leads. The only critical leads leaving the Vectors are the grid leads of the following stage, and the capacitors may be hung from one socket to the next if these give trouble.

Further ease in assembly can be obtained by using a 6 x 10 Bud Minibox in place of the chassis indicated here. These boxes come in two sections, one of which forms the top and ends of a chassis, and the other the sides and bottom. Work on the top section can be done, unhampered by sides, and the completed unit will be housed in a neatly finished box.

Moving into the amplifier proper, R_{15} and R_{16} provide isolation for the two inputs and R_{36} is the gain control. The amplifier-inverter stage has the same basic configuration as the preamplifier and requires no comment, except that R_{18} and R_{19} should be a matched pair to obtain as nearly a balanced output as possible. C_6 and C_7 were intentionally made small in value in order to preserve a smooth bass response down to the low-

PARTS LIST

R_1, R_2	2200 ohms, ½ watt
R_3, R_4	3900 ohms, ½ watt
R_5, R_6	10,000 ohms, 1 watt
R_7, R_8	47,000 ohms, 1 watt
R_9, R_{10}	68,000 ohms, 1 watt
R_{11}, R_{15}, R_{16}	0.1 meg, ½ watt
R_{12}, R_{13}, R_{14}, R_{17}, R_{20}, R_{21}, R_{22}, R_{23}	0.1 meg, 1 watt
R_{18}, R_{19}	0.1 meg, 1 watt, 5%
R_{24}, R_{25}	0.27 meg, ½ watt
R_{26}, R_{27}	0.47 meg, ½ watt
R_{28}, R_{29}	47 ohms, 2 watt, 5%
R_{30}, R_{31}	100 ohms, 2 watt
R_{32}	350 ohms, 10 watt wire-wound
R_{33}	1.0 meg, audio taper
R_{34}	50,000 ohms, linear
R_{35}	100 ohms, wire wound
R_{36}	0.25 meg pot, Ohmite
$C_{1a, b, c, d}$	20–20–20–20/450 v, elect.
C_2, C_3	0.1 µf, 400 v, Aerolite
C_4, C_5	.01 µf, 400 v, Aerolite
C_6, C_7	.004 µf, 400 v, Aerolite
C_8, C_9	.002 µf, 400 v, Aerolite
T_1	Output transformer, Freed F-1951
V_1, V_2, V_3	12AX7
V_4, V_5	6A5
P_1, P_2	Amphenol 80-C
P_3	Jones P-202
P_4	Jones P-306-AB
P_5	Octal socket, Amphenol
	Vector B8-N socket ass'y
	Vector 10–O–9T
	Vector 8–N–9T

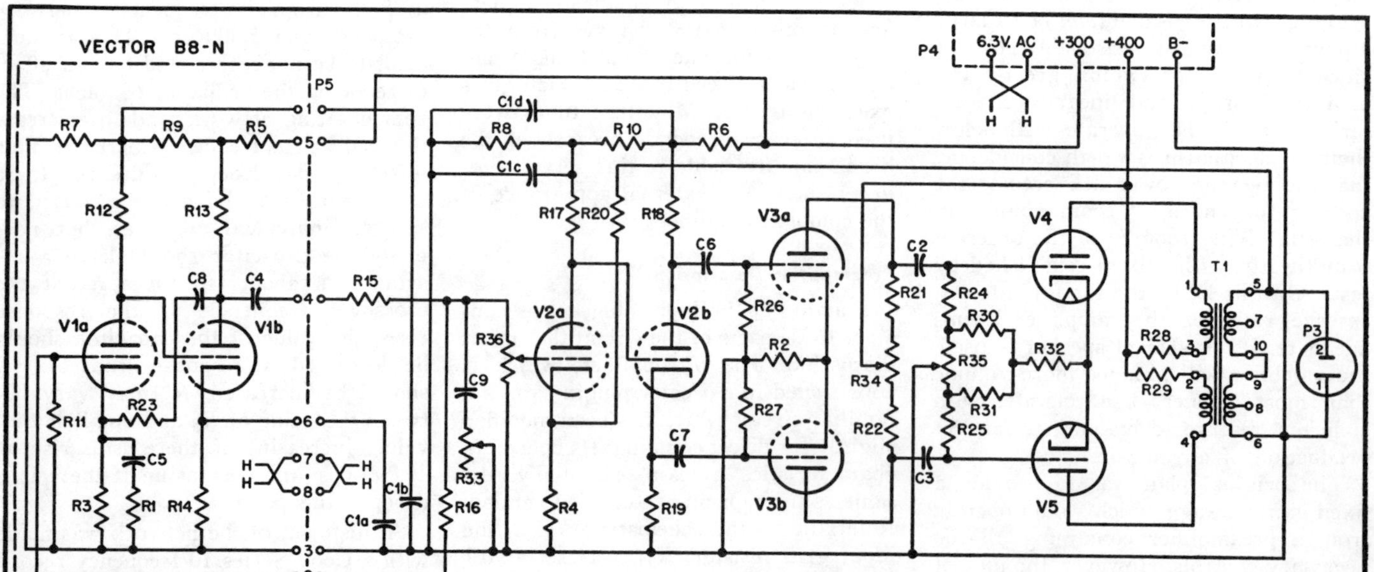

Fig. 2 Complete schematic of the 6-watt amplifier described.

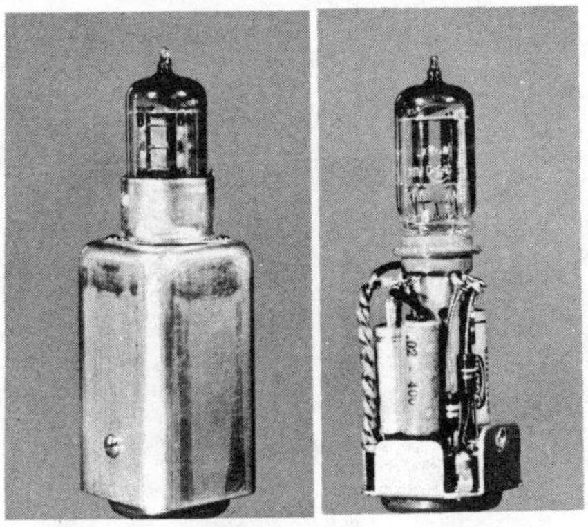

Fig. 3. The plug-in preamplifier, cased (left) and with the case removed (right).

est frequency because feedback loops around two or more stages usually require that some adjustment be made within the loop to avoid response peaks at the frequency extremes, with consequent oscillation. A simple means of achieving this is to make the response of one stage poor with respect to the others.[2] There seemed to be no need of special components to reduce the high-frequency response, and the amplifier exhibits excellent stability throughout its range.

C_9 and R_{35} provide a controllable high-frequency roll-off. These values provide approximately 20-db loss at 10,000 cps. This is ample to provide proper equalization for long-playing records. No high boost was desired, or considered advisable. When the response of a system can be made truly flat to at least 20,000 cps there should be no need of high boost. All long playing records require de-emphasis; most shellac records of recent vintage have pre-emphasis added, and will require a certain amount of de-emphasis. In other cases, surface noise becomes excessive when boost is used to capture highs on a record where they are deficient.

No bass control is provided in this design at all. If one is desired it may be added either internally or as part of an external preamplifier. It can be shown that a satisfactory bass compensation can be worked out for virtually all records so far as their recording characteristic is concerned,[3] and that the crossover frequency may be adjusted by listening preference. Beyond this there is a legitimate doubt that control is needed, except to compensate for low listening levels which may best be done in any case with a loudness control. While it is true that records vary in bass response, so do live concerts, and this variation should be considered as a normal part of listening. An amplifier should have a clean and flat response down to its lowest frequency to assure the *clarity* of bass reproduction, which is more important than any amount of volume.

R_{34} in the plate circuit of the driver serves to balance the signals to the grids of the 6A5's. Should any serious unbalance occur when the control is at its mid-point, preceding circuits and tubes should be checked. R_{33} equalizes the plate currents of the 6A5's by adjusting a portion of the bias voltage to which the grid returns (R_{24}, R_{25}) are connected. The balance is read by connecting a milliammeter at the points marked *2* and *3* on the output transformer. Balance within a few microamperes can be obtained when the circuit is warm. R_{28} and R_{29} must be matched within 1 per cent to ensure an accurate reading, and 10 per cent resistors may be matched on a bridge to find a pair. This method was used because cathode metering cannot be used with 6A5's—their cathodes being effectively tied together by the filament wiring.

No power supply is shown inasmuch as none was built specifically for the amplifier. The supply voltages shown on the schematic may be obtained from any well designed supply. With these voltages the amplifier is in a somewhat overbiased Class A condition. If the 6A5 supply voltage is reduced to 300 volts and R_{32} adjusted to give 45 volts bias from cathodes to ground, the tubes will then be operating at their rated Class A point.

The strapping of the output transformer shown in *Fig. 2* is for a speaker load of 16 ohms. Other strappings may be determined from the instructions accompanying the transformer. The feedback, using the component values indicated, will be approximately 5 db. The amplifier is stable under all conditions observed, and the driving conditions are such that more feedback would have required more gain elsewhere in the circuit for full output, which did not seem desirable. The over-all response of the equipment is essentially flat from 35 to 20,000 cps, using the Cook record, and the listening quality of the amplifier fully validates the superiority of the Williamson circuit.

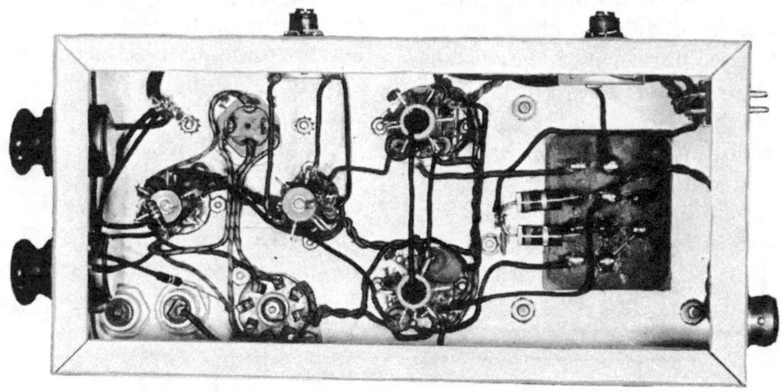

Fig. 4. Under-chassis view of the amplifier. Note the neatness resulting from the use of Vector sockets.

[2] F. E. Terman, "Radio Engineer's Handbook," p. 397 ff, 1st Edition. New York: McGraw-Hill Book Co., 1943.

[3] Howard T. Sterling, "Simplified Preamplifier Design," AUDIO ENGINEERING, November 1949.

Fig. 1. Rear view of rack-mounted amplifier, showing simplicity of construction.

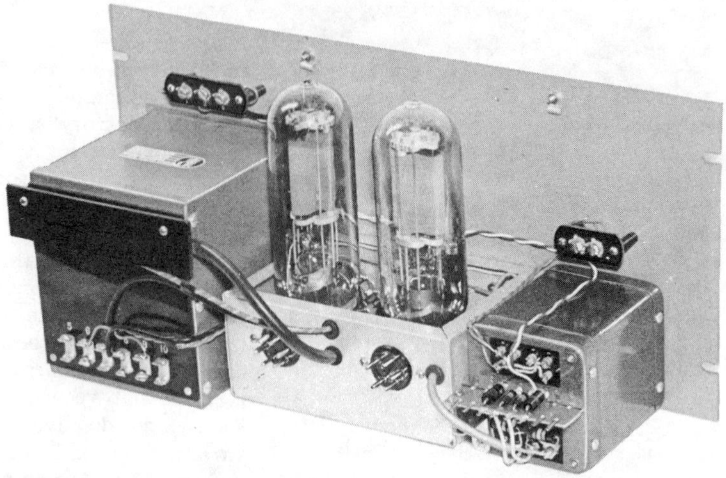

The Musician's Amplifier Senior

DAVID SARSER and MELVIN C. SPRINKLE

Construction details on a "Big Brother" to the original Musician's Amplifier. This model has adequate power for recording or similar applications.

THE MUSICIAN'S AMPLIFIER, which was brought to the attention of the American audio world by the authors, has gained an enviable reputation in Europe and Australia for its excellent fidelity. In an Americanized, and now fully naturalized version, the Musician's Amplifier has caused a sensation in this country. Literally thousands of these amplifiers have been built and all those who did not cheat on the quality of the parts used have been hearty in their praise. It is not amiss to mention at this time that "The Musician's Amplifier" has been installed in the homes of some of the world's great names in music and they have been just as impressed with its performance as the audio enthusiast whose mouth drops open when he measures one. As a matter of fact, one world famous musical figure insisted on having a Musician's Amplifier with him on his travels, so that at no time would he be without his music.

While the Musician's Amplifier has been setting new standards for performance, the authors in their constant search for perfection have unearthed the one application in which it is a little deficient—cutting disc recordings.

While we like to attend the live concerts and recitals in New York, time and financial considerations prevent us from attending them all. The next best thing to live concerts is a live FM broadcast or a good recording. However, there is an increasing paucity of live concerts on the networks, and in spite of all the professional skill of recordists, many commercial recordings leave much to be desired in the way of fidelity and interpretation. In view of the above circumstances we have spent much time and money in making disc recordings for our own use in reliving the performances we have heard, either of FM broadcasts or in recital halls.

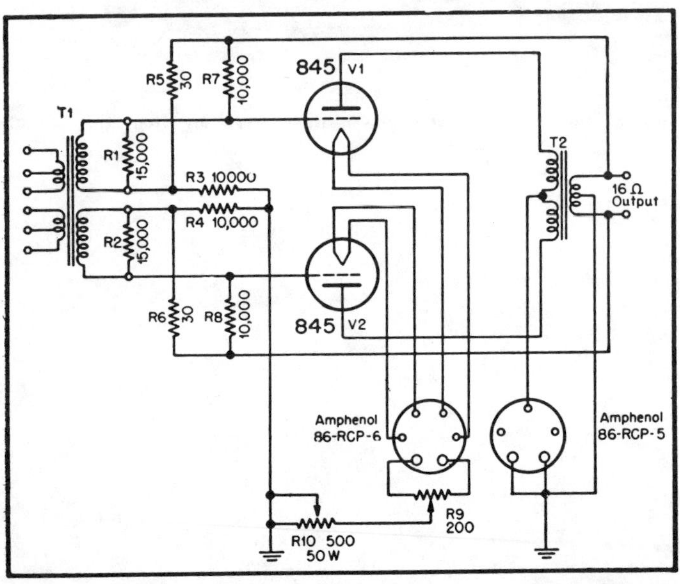

Fig. 2. Schematic of amplifier section.

Someone will raise the question of why we are interested in disc recording when fine tape machines are available. Of course we are familiar with tape and we often use it, but for our purposes the disc is still supreme. There are several reasons: (1) the cost of disc recordings is less than an equivalent time on tape; (2) the storage space for microgroove disc recordings is less than an equivalent playing time on tape; (3) not all of our friends are equipped with tape machines but all of them do have microgroove disc reproducing equipment so that sharing the recordings does not become a problem; (4) with the hot stylus technique and the amplifying system to be described, we have made discs which cannot be distinguished from tape. As a matter of fact many of the visitors at the Audio Fair who heard the Musician's Amplifier Senior playing one of its own recordings swore that we were using tape. Therefore, we make disc recordings.

Need for More Power

Naturally, one of the first things done was to use the Musician's Amplifier to drive a magnetic cutting head. Here the output power of the amplifier was just a little inadequate to give the equalization that makes a disc recording sound good to musical ears. There have arisen also a few cases where someone wanted to have his music at ear shattering levels (these individuals for some reason live away out in the country) and then there have been certain installations where it was necessary to drive a system of speakers throughout an entire home. Thus there is a need for a power amplifier which will deliver more power than the Musician's Amplifier but yet maintain the same or possibly even higher standards of naturalness in reproduction.

The power level desired was 40 watts, *clean.* This figure was arrived at from disc recording considerations as follows: Most modern cutting heads make excellent microgroove recordings when fed with ¼ watt average level as read on a VU meter. We feel that the pre-emphasis at 10,000 cps should be no more than 10 db but preferably 6 db as mentioned in Æ April 1948, page 15. In addition, we wanted at least 10-db reserve power to handle the peaks in the music and speech. The total power required comes to 25 watts on a 10-db reserve power basis, while 40 watts gives 12 db of reserve power. For cutting 78-r.p.m. records, little or no pre-emphasis is required so that when an average level of +31 dbm is fed to the cutter in cutting 96 lines per inch, there is a reserve of 15 db to handle peaks.

Having arrived at the desired power output, we examined the several possible ways in which the power output of the Musician's Amplifier could be increased. We came to the conclusion that the simplest and best way was to use a push-pull power amplifier stage of more than adequate capacity and drive it from the Musician's Amplifier as is. The possibility of push-pull-parallel 807 stages was considered and several models were built but the results were not too satisfactory. An advantage of the method decided on—the booster stage—was that it does not in any way make obsolete one's investment in equipment. This is an important consideration in these days of high prices. Conventional receiving tubes were ruled out by the power level desired, and from the list of transmitting tubes available to our pocketbook the 845 was selected. This old standby can give 75 watts in push-pull Class AB_1 and so would be coasting at 40 watts Class A. Operation in Class A was arrived at by the fact that an increase in bias to Class AB_1 conditions caused the IM distortion to jump to values considered excessive for recording work. Class B operation would, of course, produce even higher IM distortion.

Fig. 3. Rear view of power supply section, with space provided for mounting the power supply for the driver amplifier.

Transformer Selection

Selection of the output transformer was relatively easy, since there are only a few types available. The one chosen—Peerless S-275S—has a gain-frequency response within 1-db limits from 20 to 20,000 cps, and the power delivery is no more than 3 db down from its rated power of 80 watts at these frequencies. Since the 3-db power drop-off point at low frequencies is determined by the magnitude of the a.c. exciting current in the primary, we were assured of 40 watts clean at 20 cps. Leakage reactance and shunt capacitances are controlled so as to give 40 watts at 20,000 cps. Subsequent measurements proved that this transformer would provide us with a flat power-frequency characteristic from 20 to 20,000 cps.

The input coupling could have been resistance-capacitance, but this would have involved high-impedance leads with consequent frequency errors. Thus the elimination of an input transformer would have been poor economy. The Peerless S-281Q input transformer is made to operate at 30 db above 6 milliwatts, or at a level of 6 watts, and so can drive a 250-watt booster. The drive for 40 watts just "tickles" the transformer. Primary impedances are provided for 14 ohms as well as for 250 and 500 ohms, and the 14-ohm primary impedance is used with the 16-ohm output of the conventional Musician's Amplifier. A 500-ohm input is available for those who have a 500-ohm output on their amplifiers. The secondary of the input transformer is loaded with two 15,000-ohm, 5-watt resistors to absorb the drive power and to terminate the Musician's Amplifier properly. Non-inductive resistors should be used if at all available. The only other circuit components required for the amplifier proper are resistors, as shown in the schematic, *Fig. 2.*

It will be noted that inverse feedback is used, in two loops. The loop from the output transformer secondary to grids is used to wipe out just a trace of overshoot which appears on 10,000-cps square waves. The second loop, to the low ends of the input transformer secondaries, is used to lower the source impedance slightly and, as a by-product, to improve the IM distortion at lower power levels. Only 4 db feedback is used—including the effects of both loops—and there is absolutely no trace of oscillation in the combination of amplifiers. The source impedance on the 16-ohm output winding is 10 ohms, which is sufficient to give adequate damping to a good speaker.

The bias circuit for the 845's provides for balancing plate currents between the two tubes in addition to permitting a wide range of adjustment of bias voltage. The potentiometer R_9 adjusts the balance between the two tubes, while R_{10} sets the average bias, serving as the self-bias resistor. Bias is set normally, under the operating conditions selected, at 85 volts for a plate supply of 1000 volts. This gives an effective plate voltage of 915, and results in minimum intermodulation distortion.

Analysis of Operating Conditions

The tubes are operated at a condition which is slightly in excess of normal rating in order to keep distortion as low as possible at the desired power output. Thus, with a 1000-volt supply and

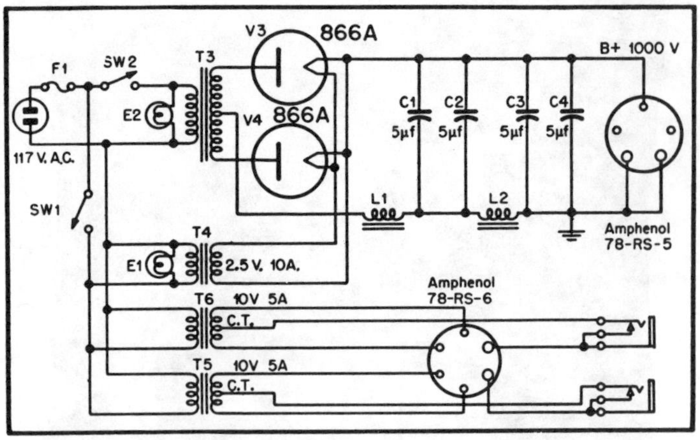

Fig. 4. Schematic of power supply section. Connections between amplifier and power supply are made with cables which plug into the two units.

an effective plate voltage of 915, the plate current is 125 ma per tube, or a quiescent plate dissipation of 114 watts per tube. This is 14 per cent higher than the rated plate dissipation of the 845, but in view of the improved operation the excess was felt to be justified. Tests were made of the amplifier with 99 watts plate dissipation (900 volts plate, 100 volts bias, resulting in a current of 110 ma per tube) but the intermodulation distortion was approximately doubled. For example, at an output power of 38 watts, the IM distortion for 99-watt dissipation is 8.2 per cent, while for the 114-watt condition the IM distortion is 4.4 per cent. Increased tube failures, if any, will be a small price to pay for the lower distortion on records. Those who want to operate their tubes within the ratings may do so with the assurance that their amplifier will have no more distortion then most commercial amplifiers, and in all probability the distortion will be less. The 8.2 per cent IM at 38 watts in the 99-watt condition is the distortion rating of commercial high-quality amplifier manufacturers, Roys[1] states that IM distortion in excess of 10 per cent is evident to trained observers, when using test frequencies of 400 and 4000 cps. The 10-per cent IM point, using 40 and 2000 cps (a much more severe test) occurs in the 99-watt condition at about 45 watts output and in the 114-watt condition at 50 watts. Our own opinion is that the IM distortion in a recording amplifier should not be more than 2 per cent at operating levels in order that the distortion in the recordings be as low as possible. The 2-per cent IM occurs in the 114-watt operating condition at 25 watts output. Thus the limiting factor on quality in the recordings made with this amplifier is the cutting head.

Power Supply

The power supply for the Musician's Amplifier Senior resembles that of an amateur transmitter in that it is required to produce high voltage. USE EXTREME CARE WHEN WORKING ON THIS POWER SUPPLY. THE HIGH VOLTAGE PRESENT IS LETHAL. YOUR FIRST SHOCK MAY BE YOUR LAST, AND DEATH IS SO PERMANENT. These cautions may be redundant, but the builder must be made fully aware of the danger involved before attempting work with high-voltage units.

The 845 filaments are fed from two filament transformers so that the plate currents may be balanced. The rectifiers are type 866A mercury vapor tubes, and their filaments are supplied from a third transformer. The plate transformer furnishes a.c. voltages of 880 or 1175 each side of center tap. With choke input, the d.c. output voltage is 1000 at the current drain required. The plate leads are connected to the high-voltage tap, as shown on the schematic, *Fig.* 4. Separate power switches are used in filament and plate transformer primaries, and pilot lamps are arranged to indicate when the separate circuits are energized. In equipment of this type it is customary to delay the application of the plate voltage for 30 to 60 seconds after turning the filaments on in order for the amplifier and rectifier filaments to be thoroughly heated.

The filter is of the brute-force type, using 1500-volt oil-filled capacitors and two chokes. The latter are placed in the negative lead where the filtering is just as effective and there is less danger of breakdown to ground. The measured noise and hum level with both amplifiers connected normally and with an open grid in the driver amplifier is −29 dbm. or 75 db below 40 watts output. No trouble was experienced with mercury vapor "hash" in the output, and no r.f. chokes were required in the rectifier circuit.

Construction

Generally when one begins to build amplifiers of the power of this one, chassis-type construction is abandoned for the more efficient relay rack. For this unit, two sections are utilized, the amplifier proper being 8¾ in. high and the power supply 14 in. The total rack space occupied by the complete system—including the driver amplifier, the power amplifier, and the power supplies—is but 29¾ in. All equipment for a complete disc recording system may be placed on a single six-foot rack, with microphone inputs, preamplifiers, equalizers, mixers, and FM tuner, and a VU meter panel.

As will be seen from *Fig.* 1, the amplifier proper is quite simple in layout. Viewed from the rear, the output transformer is on the left, the input transformer on the right. The upper left terminal strip is the output connection, while the input terminal strip is at the upper right. The 845's are mounted in an inverted 3 × 5 × 7 chassis, with a 6-prong male plug being used for the filament connections (3 leads for each tube because of the center tap). The plug at the right is for the high-voltage plate supply. The plate current balancing potentiometer R_9 and the bias resistor R_{10} are mounted within the inverted chassis, although not visible in the photograph. Access to the balance control is through a hole in the front panel. It should be pointed out that accurate plate-current balance does not affect bass response as much as it does the hum and IM distortion. The feedback resistors are mounted on a strip attached to the input transformer.

Layout of the power supply is equally simple. A chassis 3 × 7 × 15 is fastened to a 14-in. rack panel, as shown in *Fig.* 3. The plate transformer, two 845 filament transformers, and the filter capacitors are mounted on what is normally the top of the chassis, while the filter chokes and the filament transformer for the 866A's are mounted inside. The mounting of the rectifier tubes is so arranged as to leave space for the driver amplifier power supply. The two jacks shown are for measuring plate currents in the amplifier tubes.

The small junction box at the lower left of *Fig.* 3 mounts the cutter and

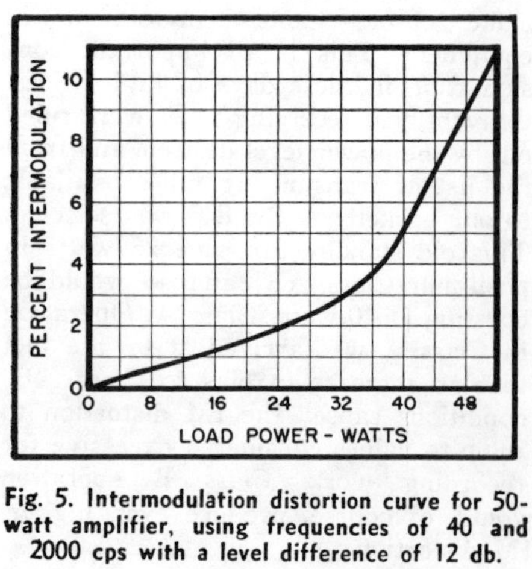

Fig. 5. Intermodulation distortion curve for 50-watt amplifier, using frequencies of 40 and 2000 cps with a level difference of 12 db.

[1] "Recording and fine-groove technique," H. E. Roys; AUDIO ENGINEERING, Sept. 1950

A Survey of Audio-Frequency Power-Amplifier Circuits

PETER G. SULZER

A discussion of a number of amplifier circuits—both conventional and otherwise—along with the presentation of a "single-ended push-pull" amplifier.

IT IS THE PURPOSE of this paper to describe some circuits that are useful as audio-frequency power amplifiers. Although the use of several of the circuits in this application is uncommon, they appear to offer interesting possibilities of obtaining special characteristics. The discussion will be restricted to amplifiers employing output transformers since with present tube types it is economically impractical to obtain sufficient current to excite the voice coil of a loud speaker.

One of the first circuits employed as a power amplifier is that of *Fig.* 1(a), which is the plate-loaded, single-ended stage. The undistorted power output is comparatively low because class-A operation is required. The presence of d.c. flux in the output transformer is a disadvantage, as is the generation of even-harmonic distortion.

The push-pull circuit of *Fig.* 1(b) is an improvement, since the d.c. flux in the output transformer can be canceled. At the same time, class-AB or class-B operation becomes possible, as the tube characteristics are complementary. If symmetry has been attained, even-harmonic distortion will be very low.

These two circuits have been modified by the use of cathode loading, as shown in *Figs.* 1(c) and 1(d). Although the output impedance is decreased, no real advantage is gained that could not be obtained by employing inverse feedback.[1] In practice, distortion in an amplifier employing a cathode-follower output stage may be severe because of driver distortion which is the result of the large driving voltage required.

Two-Tube Arrangements

In devising other two-tube power amplifiers, the following considerations

[1] Howard T. Sterling, "The cathode follower as an audio amplifier," AUDIO ENGINEERING, p. 14, Dec. 1949.

---

speaker outputs, and a panel-mounted switch transfers the 40-watt output from the cutting head to a speaker, or terminates the output on the resistors shown along the base panel. All high-voltage terminals on both the plate and output transformers are covered by a Bakelite strip, and safety caps are used on the rectifier tubes. These safety precautions are essential.

Performance

The performance of the complete system indicates that the "big brother" is a worthy companion to the Musician's Amplifier. All tests were made using both amplifiers, and the frequency response is flat within 0.5 db from 20 to 35,000 cps. The amplifiers together will pass square waves with no ring, distortion, or roughness on the top up to a 10,000-cps fundamental. This means that the frequency response is reasonably flat and the phase shift is linear up to at least 200,000 cps. Furthermore, there is no transient oscillation. The absolute gain of both amplifiers is 84 db.

The IM distortion, using 40 and 2000 cps, is shown in *Fig.* 5. These values are based on power as read on the IM-Set meter, and not upon equivalent sine-wave power. It will be evident that at normal listening levels—say up to 2 watts—that the IM distortion is too low to measure. It is no more than 2 per cent up to 25 watts output, while at 40 watts the IM distortion is about 5 per cent. The 8-per cent point appears at about 47 watts, and distortion does not climb rapidly until about 60 watts.

While the Senior amplifier was intended for making disc recordings, the authors were pleased by its performance as a playback amplifier when playing recordings which had just been cut. Visitors to the Audio Fair confirmed those opinions, and many made mention of the "cleanness" of sound. This is the result of two factors: the tremendous reserve of power, and the low intermodulation distortion even at high power levels. *Crescendo* and *fortissimo* passages are handled with effortless ease. The user should be cautioned, however, not to turn up the gain unless the load is adequate to absorb the power. Ordinary speakers will not handle the full output of this amplifier.

Parts List for Amplifier and Power Supply

Amplifiers

R_1, R_2	15,000 ohms, 5 watts
R_3, R_4, R_7, R_8	10,000 ohms, 1 watt
R_5, R_6	30 ohms, 1 watt
R_9	200-ohm potentiometer, 4-watt Mallory M200P
R_{10}	500-ohm adjustable, 50 watts
T_1	Input transformer, line to p-p grids, high level; Peerless K-281Q
T_2	80-watt output transformer, 4000-ohm pri., speaker impedances 2, 4, 8, or 16 ohms; Peerless S-275S
V_1, V_2	845's

Power Supply

C_1, C_2, C_3, C_4	5-µf, 1500-volt, oil filled
E_1, E_2	110-v pilot lights, with sockets and jewels
L_1, L_2	3-H, 225-ma chokes, Peerless C-315X or equivalent
T_3	Plate transformer, 1180 v. each side of c.t., 300 ma; Peerless P-330K
T_4	2.5-v. 10-a. filament transformer; Peerless F-096X or equivalent
T_5, T_6	10-v. 5-a. filament transformer; Peerless F-140E or equivalent
V_1, V_2	866A's

While certain transformers have been specified by the authors, and were used in the construction of the amplifier described, it is possible that equivalent performance could be obtained by the substitution of certain other types for those listed, provided that components of similar quality were chosen. Power components are readily replaceable, but the output transformer presents somewhat more of a problem of selection.

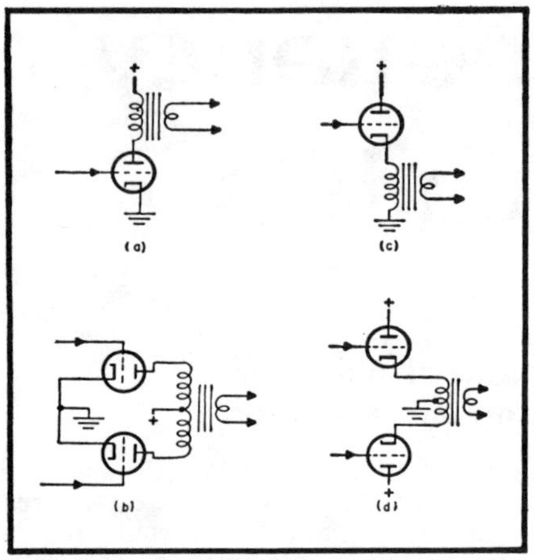

Fig. 1. Conventional single-ended and push-pull output stages.

should be kept in mind: (1) d.c. magnetization of the output transformer should be avoided; (2) the tubes should be connected in such a manner that the production of even-harmonic distortion is minimized; and (3) the lowest practicable output-transformer turns-ratio should be employed; this may be accomplished by adding plate *currents* rather than plate *voltages* in the output transformer. The last requirement permits the use of a lower turns ratio, which is highly desirable from the standpoint of output-transformer design.

Considering a two-tube power amplifier, nine combinations are possible, as follows:

(1) Plates in push-pull, with grids driven in push-pull; (2) plates in push-pull, with cathodes driven in push-pull; (3) the push-pull cathode follower; (4) parallel operation with plate loading; (5) parallel, grounded-grid operation; (6) the parallel cathode follower; (7) plate loading of one tube and cathode loading of the other, with both grids driven; (8) plate loading of one tube and cathode loading of the other, with one grid and the free cathode driven; and (9) loading both plates and both cathodes, with the grids driven in push-pull.

The first six can be rejected with the aid of the above considerations, leaving but three circuits for analysis.

The seventh can be connected as shown in *Fig.* 2(a), with the tubes connected in series across a split power supply. If the circuit is balanced, direct current does not flow through the output-transformer primary. Push-pull operation is obtained, since the plate current of one tube increases as the plate current on the other tube decreases. To satisfy the third requirement it will be noted that the alternating components of the plate currents of both tubes flow through the single output-transformer primary.

Disadvantages of the scheme are that twice the normal power-supply voltage is required, and that a large driving voltage is required by the upper tube.

It is of interest to note that the output impedance is very low—essentially that of the upper tube operating as a cathode follower—while the plate load is computed as for parallel operation. If a split-primary output transformer is available, the two halves of the primary can be connected in parallel, with improved high-frequency response.

The method of obtaining grid drive shown is one possibility; it may not be the best.

The need for a double-voltage power supply can be eliminated by rearranging the circuit as shown in *Fig.* 2(b), and employing a transformer with two primaries, P_1 and P_2. The other advantages are retained, as is the disadvantage of large driving voltage for one tube. The coupling capacitor between plate and cathode will improve high-frequency response if the coupling between the two halves of the output-transformer primary is not perfect.

The eighth circuit appears to be of no particular advantage, as grounded-grid operation of one tube will require additional driving power. The remaining grid, which is associated with a cathode follower, will still require a large voltage. However, an interesting combination of (7) and (8) exists,[2] as shown in *Fig.* 2(c).

It will be noted that V_1 is driven in the normal manner, while the drive for V_2 results from the voltage across R_k. Since this voltage will increase as the load impedance is decreased, one would expect the circuit to have a low output impedance. A practical advantage is that push-pull driving voltage is not required.

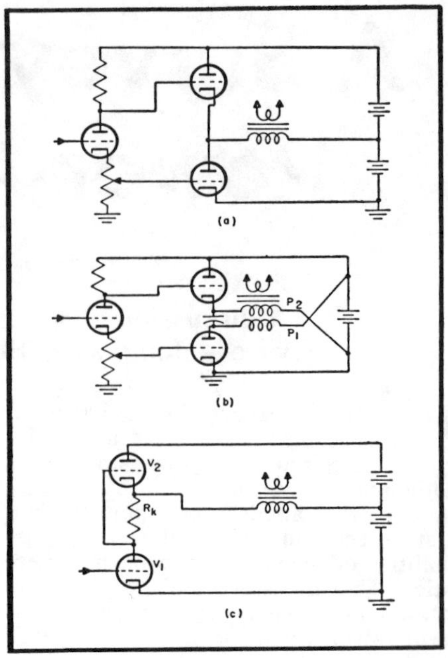

Fig. 2. Series-balanced output stage arrangements.

Gain and Impedance

It is of interest to derive the equations for the open-circuit voltage gain and the output impedance of the circuit. To obtain the voltage gain, the amplifier is split, as shown in *Fig.* 3(a). The equivalent circuits are shown in (b) and (c). Here V_2 is considered as a cathode follower with a resistance R_k connected between its grid and cathode, while V_1 functions as a grounded-cathode amplifier, with a load consisting entirely of V_2. Assuming the tubes to be identical, and considering *Fig.* 3(b),

$$\frac{e_3}{e_2} = \frac{R_p + \mu R_k}{R_p + (1+\mu) R_k}; \quad (1)$$

considering *Fig.* 3(c),

$$\frac{e_2}{e_1} = -\frac{\mu\,[R_p + (1+\mu)\,R_k]}{2\,R_p + (1+\mu)\,R_k}. \quad (2)$$

Forming the product of *(1)* and *(2)*, it is found that the overall voltage gain

$$\frac{e_3}{e_1} = -\frac{\mu\,R_p + \mu^2\,R_k}{2\,R_p + (1+\mu)\,R_k}. \quad (3)$$

If $\mu \gg 1$,

$$\frac{e_3}{e_1} \approx -\frac{R_p + \mu\,R_k}{\dfrac{2\,R_p}{\mu} + R_k} \quad (4)$$

Figure 3(d) shows that the output impedance Z of the amplifier consists of $R_p + R_k$ in parallel with the output im-

[2] Maurice Artzt, "Survey of direct-current amplifiers," *Electronics,* p. 112, Aug. 1945.

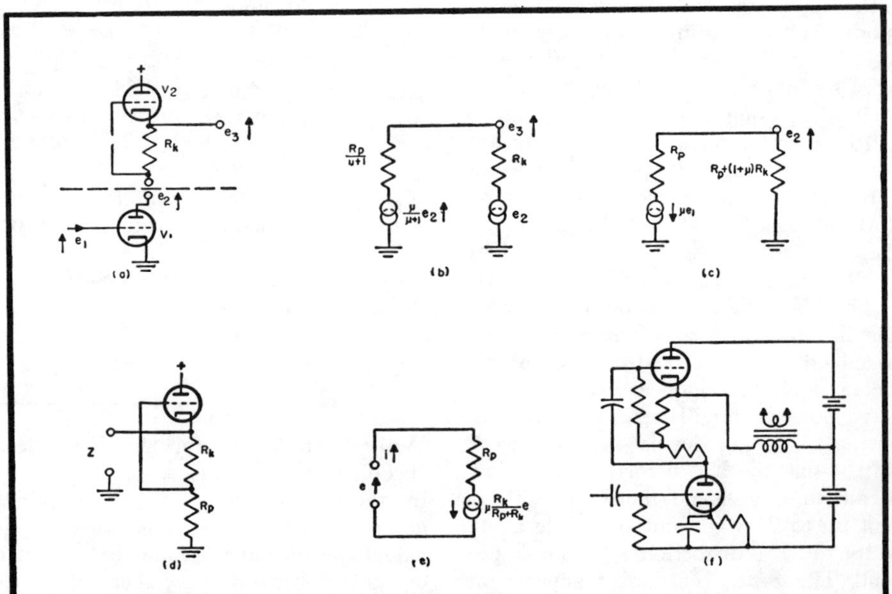

Fig. 3. Equivalent circuits for analyzing the series-balanced output stage.

pedance of V_2, which must be evaluated separately. *Figure* 3(e) is the equivalent circuit for V_2 alone. Applying a voltage e as shown,

$$\frac{e}{i} = \frac{R_p}{1 + \dfrac{\mu R_k}{R_p + R_k}} = Z_2 \qquad (5)$$

Where Z_2 is the impedance of V_2 alone. Paralleling Z_2 with $R_p + R_k$ to obtain the output impedance of the amplifier,

$$Z = \frac{R_p (R_p + R_k)}{2 R_p + (1 + \mu) R_k} \qquad (6)$$

To study a practical example, one might consider the use of two 6L6 tubes with a 500-volt plate supply. If $R_k = 170$ ohms, $R_p = 22{,}000$ ohms, and $\mu = 135$. Substituting in *(4)* and *(6)*, $e_3/e_1 \approx -90$, and $Z = 7300$ ohms. Inspection of a tube handbook[3] shows the optimum load impedance to be 1250 ohms, with an output of approximately 14 watts. The required r.m.s. grid drive is 10 volts, while the r.m.s. signal current per tube is approximately 50 ma. Since this current flows through R_k, it is necessary to increase R_k to 500 ohms to obtain sufficient driving voltage for V_2. The more elaborate circuit of *Fig.* 3(f) was employed to obtain the proper bias for V_2 with the increased value of cathode resistor. Substituting the new value of R_k in *(4)* and *(6)*, $e_3/e_1 \approx -110$, and $Z = 4{,}500$ ohms. The output impedance is lower than that normally obtained with a pentode amplifier, which permits a good damping factor to be obtained with a moderate amount of feedback.

The amplifier of *Fig.* 3(f) gave good results; however, it was found that clipping occurred at the comparatively low level of 15 watts. This resulted from the fact that the drive from V_2 comes from V_1; and therefore, when V_1 overloads, there is a rapid deterioration in the output of the amplifier.

A disadvantage of the circuit is that the value of R_k for proper a.c. balance is a function of load impedance, with the possibility of even-harmonic distortion production when operating into a fluctuating load such as a loudspeaker.

Plate- and Cathode-Loaded Circuit

The last scheme to be considered employs plate and cathode loading of both

[3] RCA Tube Handbook HB-3, vol. 3–4.

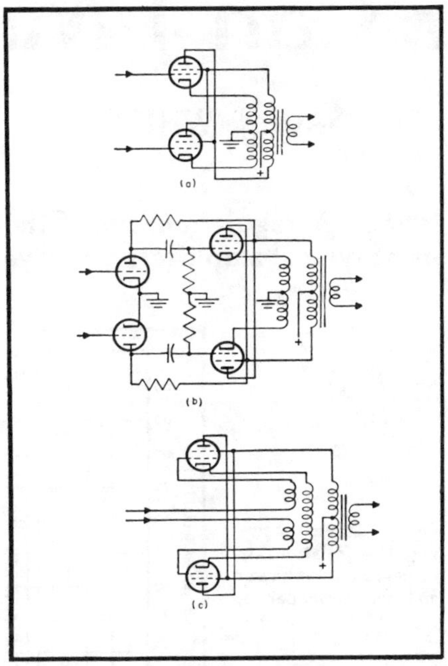

Fig. 4. Plate-and-cathode-loaded output stage circuits.

tubes,[4] as shown in *Fig.* 4(a). By using a bifilar output-transformer primary, excellent coupling is obtained between the two tubes, permitting class-B operation with very low distortion. Signal currents add in the transformer windings, permitting the use of a low load impedance, while the d.c. flux is canceled. However, because of the cathode loading, degeneration occurs, and a large driving voltage is required.

A resistance-coupled driver was tried with this amplifier; results were disappointing until the driver plate supply was increased to 500 volts. This disadvantage was eliminated by use of the "bootstrap" driver shown in *Fig.* 4(b). Here each of the driver plate-load resistors is returned to the plate of the opposite output tube which, to an approximation, has the effect of doubling the driver plate-supply voltage. An amplifier and special output transformer were constructed and excellent results were obtained provided that the output stage was operated without grid current. The effect of "bootstrapping" is to raise the equivalent driver impedance, which produces severe clipping when grid current flows.

[4] Frank H. McIntosh and Gordon J. Gow, "Description and analysis of a new 50-watt amplifier," AUDIO ENGINEERING, p. 9, Dec. 1949.

A third modification of the plate-cathode output stage is shown in *Fig.* 4(c). It is seen to be similar to that of (a), with the addition of a grid winding for each tube. Driving voltage is applied between the grid and cathode windings, avoiding degeneration, while the other advantages of the original circuit are retained. This driver impedance can be made low if the transformer resistance and leakage reactance are small. Satisfactory results can be obtained by using a trifilar winding. Driving voltage is obtained conveniently from a push-pull cathode follower, which eliminates the need for an interstage transformer. The principal disadvantage of the circuit is that a special output transformer is required. An experimental amplifier using this scheme was constructed, and tests showed that a total harmonic distortion of less than 0.5 per cent was obtained over the frequency range from 20 to 20,000 cps when employing 6L6 tubes. The power output was maintained constant at 25 watts.

Conclusions

After considering the several circuits discussed above, one is left with some doubt concerning the best application of each. Assuming that low distortion is essential, the circuit of *Fig.* 1(b) is satisfactory where the low-power output associated with class-A operation is sufficient. The special circuits of *Fig.* 2 are also subject to this limitation; however, they have the advantage of operating the tubes in parallel rather than in series, permitting a less critical output-transformer design. The circuit of *Fig.* 4(a) permits class-B operation with low distortion, and is to be recommended where the maximum power output is required. The circuit of *Fig.* 4(b) will give good results with class-AB operation, but is not suitable for a class-B output stage because of the high apparent driver impedance. The capabilities of *Fig.* 4(c) have not been fully explored; however, with a good output-transformer design it should be similar in performance to (a), with the advantage of requiring a smaller driving voltage.

How Far Can I Mismatch?

SAUL J. WHITE

A discussion of output stage regulation, and of the load characteristic and damping factors of typical amplifiers employed in audio systems.

THE QUESTIONS often arise, "Can I connect the 8-ohm output on my amplifier to a 16-ohm loudspeaker?" or "How would my performance be affected if I connect my 8-ohm amplifier output to two 8-ohm speakers in parallel?" Thus the problem of matching amplifier output to speaker has always seemed to be present with most audio enthusiasts. At the outset it can be stated that under certain conditions mismatching up to several hundred per cent can be tolerated. Under certain other conditions an almost perfect match is essential.

The extent of permissable mismatch depends upon what is known as the "load characteristic" of the amplifier output stage. This expresses the relationship between output power (watts) and impedance of the load. Such curves are shown in *Fig.* 1. Studies of several commercial amplifiers show, as a rough average, that a 100 per cent mismatch results in a reduction in output power of 25 to 50 per cent. Curves A, B, and C were made on popular priced phono-amplifiers. Curve D shows the regulation on a Langevin 101-D amplifier. Curve E represents the relationship obtained on a McIntosh 50-watt amplifier.

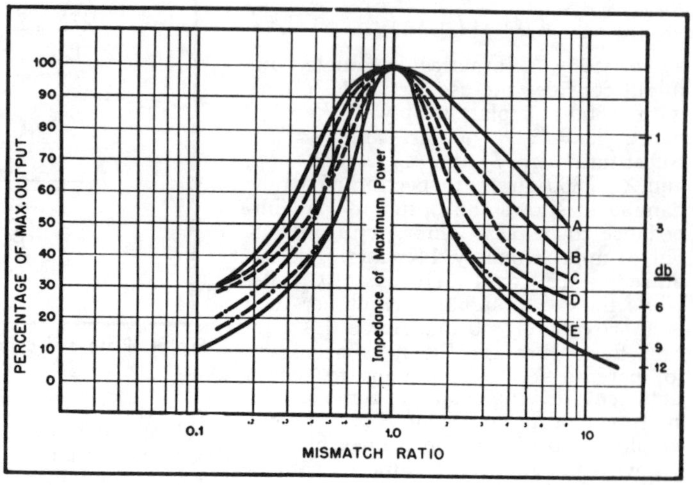

Fig. 1. Curves showing the relationship between output power and load impedance.

Expensive amplifiers with low internal impedance will fall between curves C and E. Curve F is the computed load characteristic from a true constant voltage amplifier, having a theoretical internal impedance of zero ohms.

Thus, an 8-ohm output may be connected, on an average medium-priced amplifier, to a 16-ohm speaker whereby the *maximum* available power between the ideal match and the mismatch will be reduced by 25 per cent (1¼ db). Where the loudspeaker is intended to be operated at only a fraction of the available power inherent in the amplifier, then wider mismatches can be tolerated, and there is no noticeable effect upon the performance. According to data published on a Stromberg-Carlson amplifier,[1] it is shown that a mismatch of 100 per cent dropped the maximum available power from 32 watts to 25 watts. However, if the application requires that this particular amplifier operate below 25 watts, then there is no apparent effect upon the performance with 100 per cent mismatch.

Actually, the degree of mismatch that can be tolerated, as stated above, depends upon how much of the amplifier power output it is desired to use. If one must utilize the *absolute maximum* of power of which the amplifier is capable, then an accurate match is necessary. A load impedance which does not vary by more than 25 per cent from the rated output impedance may be considered a close match for all practical purposes and will absorb practically 100 per cent of the amplifier energy when fully driven.

On the other hand, the less power to be utilized, the greater may be the mismatch. In the case of most quality phonograph systems, these usually have an output power of 10 to 15 watts. For operating under normal living room conditions it is seldom that more than 2 watts have to be utilized with perhaps

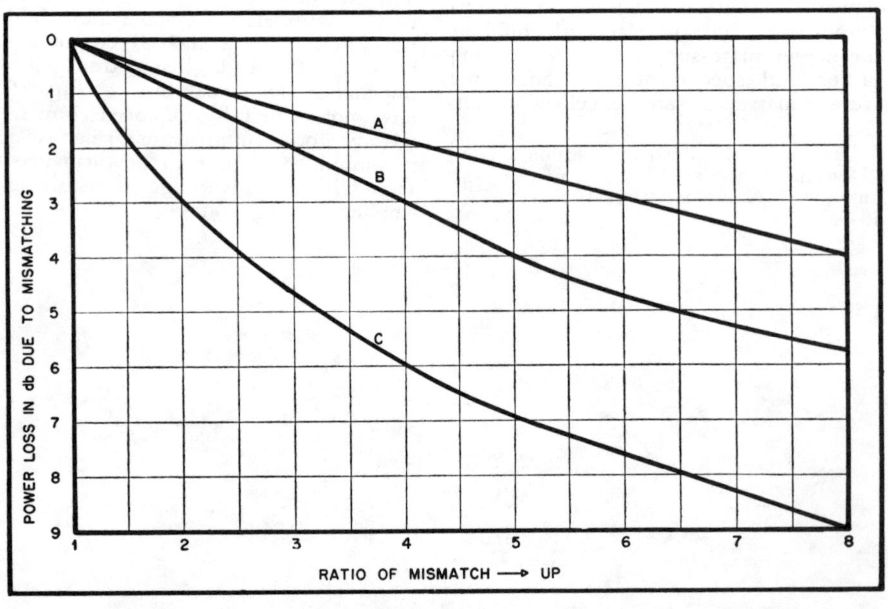

Fig. 2. Chart showing mismatching losses on a conventional 6L6 amplifier without feedback. (A) Unregulated output stage at ¼ of full power; (B) same amplifier at full power; (C) theoretical constant-voltage output.

[1] "Impedance Matching" by O. L. Angevine, Jr., AUDIO ENGINEERING, December, 1947.

6 watts allowed for unusual high level peaks. Thus it can be seen that only 50 per cent of the amplifier power will be utilized. Where this situation prevails, mismatches up to 200 per cent can be tolerated. Where ¾ of the amplifier output must be preserved for high-level operation, then it is permissable on many amplifiers to have 100 per cent variation in matching without affecting in any way the frequency range of the reproducing system.

However, this is on the assumption that the mismatch is upward, that is, the load impedance must be greater than the amplifier impedance. Mismatching upward preserves the optimum performance from the point of view of frequency range. Mismatching downward to a lower load impedance is not recommended since it may cause loss of low frequencies in addition to power. Low-frequency losses are due to loss of magnetization inductance when the output transformer is abnormally shunted down.

The type of output-stage regulation sought above is that wherein the output *power* tends to stay up in spite of load changes, as in *Fig.* 1, Curves A and B. This is of value in maintaining maximum high-frequency output because of the increased impedance of the voice coil at high frequencies. Constant power output permits maximum mismatching of load impedance, as shown in *Fig. 2*, Curves A and B. Such amplifiers have output which may be characterized as "poor regulation," or "high internal impedance." With such amplifiers it is wasted effort to attempt to achieve a precision match.

Constant Voltage Amplifiers

There are, however, some disadvantages in such "constant power" regulation which are manifest at the low-frequency resonant point of the loudspeaker. At resonance, the speaker impedance rises by two or three times its rated impedance. If the power remains constant at resonance as it does at other frequencies, the speaker will overshoot and produce exaggerated resonant boom.

Another type of regulation is that wherein the output *voltage* and not power, remains constant over a wide range of load impedance. Thus the power will drop as the impedance rises. The power into the load will fall proportionately with the ratio of mis-match. See Curve F, *Fig.* 1 or Curve C, *Fig. 2.*

It has been claimed that a speaker system which has overpronounced bass resonance and a poor transient characteristic may be improved by employing it with an amplifier where the *voltage* output is constant and independent of the load impedance. At the low-frequency resonance, the voice-coil impedance rises greatly. If, however, the voltage from the amplifier remains constant for a given signal, then less power is delivered to the speaker at the resonant frequency, tending to restrict the extreme amplitude which causes boom. If the output voltage is sensitive to load changes as in a poorly regulated output stage, then the voltage will rise as the speaker impedance rises, thus contributing unnecessary watts to a condition which is already over-sensitized. Constant output voltage is achieved by incorporating large degrees of inverse feedback, and the use of efficient output transformers. The output impedance of the amplifier *acts* as if it were considerably lower than its *rated* impedance. This is known as a low internal impedance. The ratio between the rated impedance and the internal impedance establishes the damping factor of the amplifier. Many amplifiers today have a damping factor of 10 or more.

Resonance in a speaker is the point of maximum acoustic sensitivity and may frequently cause a peaked and distorted output.

If the regulation were such that the voltage rises with an increase in load impedance, then the cone might be overdriven at its resonant frequency, introducing undesirable distortion and hangover. Furthermore, amplifiers which have good voltage regulation are characterized by a low internal impedance. This acts as a sort of electrical brake, absorbing the counter e. m. f. which a cone generates when the signal terminates abruptly as in a transient impulse. Therefore, it may be said that well-regulated amplifiers with a good damping factor (a low internal impedance) will provide (*a*) reduction of the resonant effect and (*b*) improved transient response. This effect is greatest when the loudspeaker has a high conversion efficiency. With insensitive speakers, there is no improvement on the transient response. To realize the above advantages, a loudspeaker must further possess a flat or smooth response without sharp dips or peaks, and its cutoff at either end of the spectrum should be gradual rather than abrupt. This reduces phase-shift distortion, an illusive and debatable form of distortion not readily measurable, but which definitely contributes to poor transient response.[2]

On the other hand, the disadvantage of constant-voltage output lies in the fact that the full high-frequency capabilities of the loudspeaker may not show up. The voice coil impedance rises with frequency. At 10,000 cps it may be two to four times its value at 1000 cps. Thus, the power delivered from the regulated amplifier reduces with frequency and the speaker is under-driven. An increase up to 6 db at high frequencies may result by substituting an unregulated amplifier for a well-regulated one. Or, for a regulated amplifier, an increase in highs is possible by assuming a high load impedance and selecting a corresponding output tap.

While the constant-voltage type of amplifier appears to be the most sought after, it must be borne in mind that for applications where changes in load may occur frequently, the unregulated type of amplifier will prove more satisfactory. A study of the chart of *Fig.* 2 will reveal that a fairly large range of load impedances can be connected to this amplifier without reducing the power output too seriously. The amplifier used to develop this chart was a straightforward public-address unit built some years ago and without inverse feedbacks. The output stage uses a pair of 6L6s and is rated at 20 watts. At low output levels, extremely large mismatch is permissible, Curve A, *Fig.* 2.

The so-called "rated" output impedance of an amplifier is not necessarily the "internal" output impedance. These may be wide apart, depending upon the damping factor of the output stage. This relationship between internal impedance and rated load impedance indicates the voltage regulation characteristic or damping factor of the amplifier. This is expressed as the number of db the output voltage will increase when the proper load impedance is removed and the output is left on open-circuit. Where precise and accurate information must be had as to permissable mismatch, the data on load characteristics of the amplifier must be obtained from the manufacturer.

The rated "load impedance" is that impedance into which the amplifier will deliver its maximum power for a given distortion. The "internal impedance" is the effective impedance of the output winding. Its value is considerably lower than the rated load impedance, and is established by the design and degenerative feedback system of the amplifier as well as the coupling efficiency of the output transformer. In true constant-voltage amplifiers, the power into the load will be proportional to the load impedance. Therefore, the power loss will be proportional to the mismatch ratio, and for full utilization of the amplifier capacity, a perfect match is required.

In large sound distribution systems using many speakers, loudspeakers are

[2] "Phase Shift in Loudspeakers," Ewaskio and Mawardi, *J. Acous. Soc. Am.*, July, 1950.

The Cathode Follower Output Stage

ROBERT M. MITCHELL

Summarizing the advantages and disadvantages of a much discussed circuit arrangement which has many zealous adherents.

Although the advantages and method of operation of the cathode follower have been known for years and covered thoroughly in the literature, there has been relatively little material devoted to the use of the circuit as an audio power output stage.[1] As a result, there are widespread misconceptions regarding this use of the cathode follower, even among otherwise well-informed engineers. These misconceptions involve: (1) the relation between the static and the dynamic characteristics, i.e., between d.c. and a.c. conditions; (2) impedance matching; and (3) damping effects. These three concepts are those usually misunderstood in treating negative voltage feedback amplifiers in general, and, consequently, the conclusions obtained here for the cathode follower will be directly applicable to other voltage feedback circuits.

The simple general negative voltage feedback circuit, its equivalent, and the simple cathode-follower circuit are shown in *Fig.* 1 at (A), (B), and (C) respectively. The fraction of the output voltage that is fed back out of phase with the input voltage is denoted by β. For the cathode follower, β is 1, and consequently there is 100 per cent inverse

[1] C. J. Mitchell, "The cathode-follower output stage," *Wireless World,* April 1944, pp. 108–110.

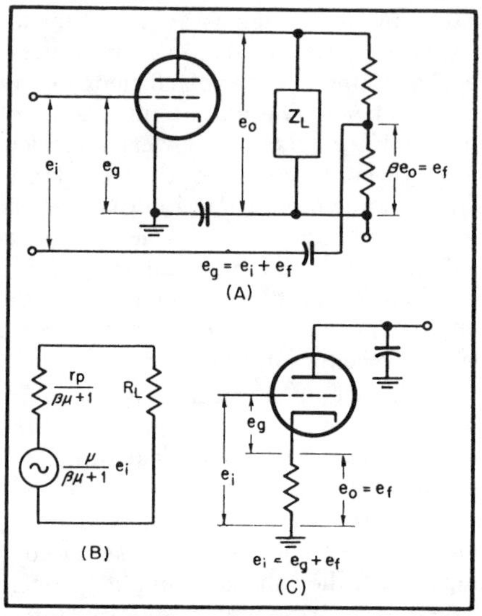

Fig. 1. (A) General schematic of voltage-feedback circuit, and (B) its equivalent. (C) Simple cathode-follower circuit.

feedback. From (B) it may be seen that since the equivalent generator magnitude is $\mu/(\mu\beta + 1)$ times the input, the cathode follower can never give a voltage gain, but must always result in a loss, the value of which depends directly on the amplification factor of the tube. There can be, however, a considerable power gain, and it is this fact that makes the circuit useful as a power output stage.

D-C versus A-C Conditions

The equation of the output voltage e_o of a negative-voltage-feedback amplifier is

$$e_o = e_i\left(\frac{\mu}{\mu\beta + 1}\right) \times \left(\frac{Z_l}{Z_l + r_p/(\mu\beta + 1)}\right) \quad (1)$$

Comparison of this formula with that for an ordinary voltage amplification stage

$$e_o' = e_i(-\mu)\left(\frac{Z_l}{r_p + Z_l}\right) \quad (2)$$

shows that the magnitudes of both the amplification factor and the plate resistance have been effectively reduced by the factor $1/(\mu\beta + 1)$. Since the amplification factor and the plate resistance are shown to have been changed when feedback is present, it is possible to construct an equivalent characteristic for the tube under feedback conditions.

Figure 2 shows the plate characteristic for a rather high r_p power triode—the 807 triode connected. For cathode-follower operation the value of β is unity. Then for any cathode voltage e_o the feedback voltage e_f equals βe_o, which is e_o. A voltage of 300 on the cathode represents a feedback voltage of 300. Reference to (C) of *Fig.* 1 will show that the voltage

How Far Can I Mismatch?

(from preceding page)

equipped with line-matching transformers which contain many taps. The purpose of these taps is to provide an intentional mismatch with particular speakers for the purpose of regulating volume. This is a common practice since volume reduction is obtained not by burning up output power into a pad, but by the simple method of not draining it from the amplifier. These mismatches are always effected upwards, that is, the transformer impedance is always greater than the impedance of the amplifier feed line. Thus, mismatches are used for a reduction of sound without distortion and there is negligible loss of quality or frequency range. This method of volume regulation requires the use of amplifiers of the constant voltage output type.

Effect of Mismatching on Crossover Networks

When a mismatch occurs on a crossover network, a shift in the crossover frequency is most likely to occur in addition to a loss of transfer power. If the mismatch is upward—that is, if the speaker impedance is larger than the impedance indicated on the network—the crossover frequency will be lower. Conversely, if the speaker impedance is lower than that for which the network is designed, then the crossover frequency is shifted to a higher value. In the case of networks, a reasonable mismatch is quite permissable since in effect a mismatch is constantly occuring in spite of the fact that a perfect match is assumed. This is because the voice coils have a rising characteristic with frequency. Actually, a so-called 600-cps crossover is never exactly a 600-cps crossover at all instants. It is varying up and down depending upon the voice coil impedance at any particular instant, which in turn depends upon the frequency content of the program at any given moment.

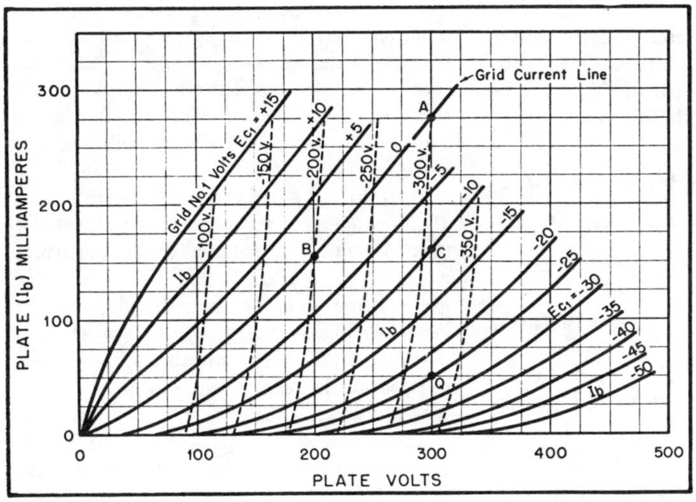

Fig. 2. Conventional plate characteristic for triode-connected 807 (solid lines) and cathode - follower characteristic (dotted lines).

polarity is defined as positive in the direction from ground to cathode and that the plate of the tube is at ground potential for a.c. voltages. The plate-to-cathode voltage is therefore negative as far as the defined voltages are concerned. If the plate voltages on the plate family characteristics (which are, of course, plate-to-cathode voltages) are taken as negative for the feedback conditions of the cathode follower, these characteristics can be used to create a new set of curves valid for cathode-follower operation.

The grid-to-cathode voltage, e_g, is the difference between the input voltage e_i and the feedback or cathode voltage $e_f = e_o$.

$$e_g = e_i - e_o$$

From this relation it is seen that at $e_g = 0$ the value of e_i must be equal to that of e_o. If points along the $e_g = 0$ curve are taken at different values of e_o (which are the plate-voltage ordinates on the graph) the value of e_i at each point will be exactly equal to the plate voltage ordinate at that point. For example, at the intersection of the $e_g = 0$ curve and "plate volts" = 300 volts on the 807 characteristic *(Point A)* the value of e_i is -300 volts (negative since e_o values were defined as negative). In like fashion, the value of e_i at the intersection of the $e_g = 0$ curve and the 200 volts ordinate is -200 volts *(Point B)*. Similarly, the intersection of $e_g = -10$ volts and the 300 volts ordinate represents -310 volts *(Point C)*, since

$$e_i = e_g + e_o$$

$$-310 = -10 + (-300)$$

If a number of points thus plotted representing a fixed value of e_i (such as $e_i = -300$ volts) are connected, a curve for this value of input voltage is constructed. Continuation of this process for different values of e_i permits the construction of the equivalent cathode-follower characteristic for the tube. Although this may appear a lengthy process, it is actually the work of only five minutes or so to construct an equivalent cathode follower characteristic, depending upon the number of curves desired. The 807 cathode - follower characteristics are shown in dotted lines on the original graph. Note that (1) the lines are much steeper, representing a tube of much lower plate resistance, and (2) the distance that represented an input of 5 volts on the original characteristic represents an input of approximately 50 volts on the cathode - follower characteristic, showing a reduction in *mu* to one tenth of the original value. Construction for other percentages of feedback in plate-loaded amplifiers can be made in a similar manner.[2] Note that the figures on the parameter curves represent values of e_i and not e_g for the feedback characteristic.

It is fairly evident that quiescent conditions will be designed on the basis of the original curves. The bias point Q is chosen, for example, at a plate voltage of 300 volts and a bias of -25 volts. This value gives a quiescent plate current of 50 ma, and the bias resistor, if cathode bias is used, is calculated in the usual manner. It is less evident that the point of grid current has *not* been changed—i.e., it is not given by the locus curve of $e_i = 0$ of the cathode-follower characteristic. It is still given by the original grid current curve, that for $e_g = 0$. This is because grid current is drawn when the grid-cathode voltage e_g is positive, and e_i does not represent this voltage in a negative-voltage-feedback amplifier. Thus it is seen that the available plate-voltage swing (or correctly, the cathode-voltage swing) without current has not been increased, as might have been expected from preliminary inspection of the new curves.

[2] Albert Preisman, "Graphical Constructions for Vacuum Tube Circuits," New York: McGraw-Hill Book Co., 1943, pp 226–231.

Impedance Matching

The method of proper impedance matching in the cathode-follower output stage is also not immediately discernible from the new curves. For any system, the maximum power output is obtained when the load impedance equals the generator impedance. For the cathode follower

$$Z_l = Z_g = r_p/(\mu + 1)$$

It is customary in vacuum tube impedance matching, however, to base the relationships on the concept of maximum power with a prescribed amount of distortion. For triode tubes, to a first-approximation, the optimum load is twice the plate resistance of the tube. For a cathode follower (or any inverse feedback amplifier), the optimum load is also twice the plate resistance, *not* of the equivalent tube, however, but of the original, unaltered tube, i.e., the optimum load is not changed by feedback. The validity of this relation between load and

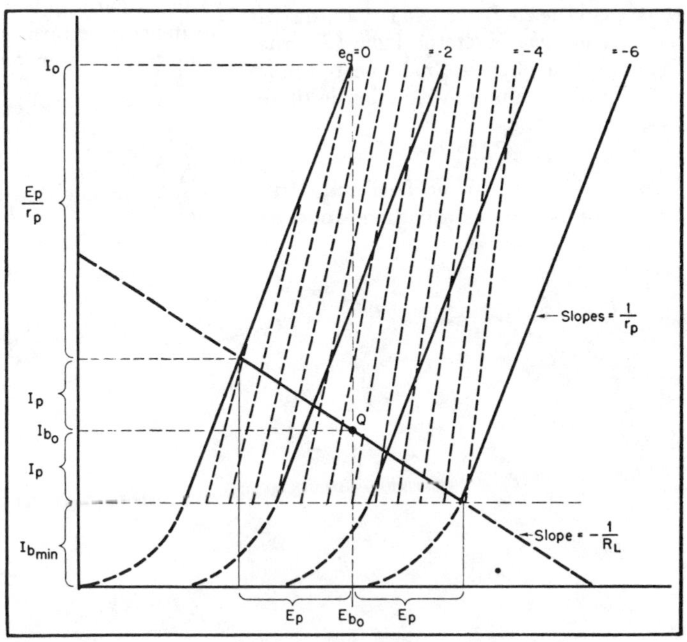

Fig. 3. Idealized plate and cathode-follower characteristics for a triode.

generator may be proven as follows:

Figure 3 shows an idealized plate characteristic for a triode tube (solid lines). Above a certain current value the characteristics are straight, parallel, and equally spaced for equal increments of voltage. Because of the absence of curvature, no distortion is assumed to take place above the Ib_{min} coordinate. The intersection of coordinates Ib_o and Eb_o determines Q_o—the quiescent operating point. The vertical projection of Q up to the $eg = O$ line gives the value marked I_o on the graph. Since the slope of the lines above Ib_{min} is rp, the equation for I_o may be written readily with the use of elementary geometry.

$$I_o = Ib_{min} + 2ip + ep/rp \quad (3)$$

In any amplifier the following relations hold

$$Rl = ep/ip \quad (4)$$

$$P_o = \tfrac{1}{2}(ep)\ ip \quad (5)$$

The above equations may be rewritten through the use of the relation

$$ep = \mu eg$$

$$I_o = Ib_{min} + 2ip + \mu eg/rp \quad (3a)$$

$$Rl = \mu eg/ip \quad (4a)$$

$$P_o = \tfrac{1}{2}(\mu eg)ip \quad (5a)$$

To solve for maximum power output under the given restrictions, equations *(3a)* and *(5a)* are differentiated and the results set equal to zero. Implicit differentiation and solution of *(3a)* gives

$$-\mu \frac{deg}{dip} = 2rp \quad (3b)$$

while differentiation with respect to ip and solution of *(5a)* gives

$$\frac{\mu eg}{ip} = -\mu \frac{deg}{dip} \quad (5b)$$

It is easily seen from *(4a)* that the left member of *(5b)* is Rl and from *(3b)* that the right member is $2rp$. The optimum load Rl for minimum distortion is therefore

$$Rl = 2rp$$

This is the result for no feedback. If the same tube is used as a cathode follower, the characteristics are changed, as shown in the graph by the dotted lines. For these curves, a new form of *(1)* may be written; *(4)* and *(5)*, being general, are unaltered.

$$Io = Ib_{min} + 2ip + \mu' eg/r'p \quad (1c)$$

For the cathode follower the effective plate resistance $r'p$ is

$$r'p = rp/(\mu + 1)$$

and the effective amplification factor μ'

$$\mu' = \mu/(\mu + 1)$$

Substitution of these values in *(1c)* gives

$$Io = Ib_{min} + 2ip + \frac{(\mu/\mu + 1)eg}{rp/(\mu + 1)}$$

$$= Ib_{min} + 2ip + \mu eg/rp \quad (3a)$$

The final equation is *(3a)* of the original conditions. Consequently, solving it and *(5a)* for the optimum load will give the same result for the cathode follower that it did for the original tube, namely

$$Rl = 2rp$$

In actual practice this value of load resistance is not strictly adhered to, and other, more complicated formulas may be used[3,4], or the optimum load may be determined by experiment.

Damping Properties

The superior damping ability of the cathode follower is generally one of the main reasons cited for its use as an output tube, and this damping quality attributed to its low effective generator impedance. It is not only the low effective plate resistance that provides the superior damping, however, but primarily the fact that the impedance matching is done on the basis of the original characteristics.

For example, the 6A5 triode has a *mu* of 4.2 and an rp of 800 ohms. The optimum load on a $2rp$ basis is therefore 1600 ohms. To match a 10-ohm speaker to this tube requires a transformer with an impedance ratio of 10: 1600 or 1 : 160. The transformer also changes the impedance that the speaker sees by the inverse of this ratio, i.e., 1/160. The effective generator reflected impedance is therefore 800/160 or 5 ohms. The same tube hooked up as a cathode follower has an effective plate resistance of $rp/(\mu + 1) = 800/5.2 = 154$ ohms. To match a 10-ohm speaker to twice this value (308 ohms) would require a transformer ratio of 10: 308 or 1: 30.8. The reflected generator impedance seen by the speaker is $154/30.8 = 5$ ohms, the same impedance that it saw in the original circuit, and therefore the damping factor has not been changed.

However, if the speaker is matched to twice the plate resistance of the original tube (the proper method), the same transformer is used as in the first case, and the speaker load looks like 10(160) = 1600 ohms to the tube. The tube looks like $154/160 = 0.96$ ohms to the speaker, and the damping factor has been increased by $5/0.96 = 5.2 = \mu + 1$ of the original tube. Since in general the optimum load given by manufacturers is greater than $2rp$, the damping factor is further increased. The manufacturers' rating of 2500 ohms for the 6A5 would give an effective generator impedance of $154/250 = 0.61$ ohms and a damping factor increase of $5/0.61 = 8.2$ for cathode-follower operation.

The principal relations in cathode-follower operations may be summarized briefly as follows:

1. Bias point unchanged
2. Grid-current point unchanged
3. Effective amplification factor reduced
4. Effective plate resistance reduced
5. Optimum load unchanged
6. Damping factor increased (if 5. is followed).

The advantage of cathode follower operation are numerous and include good frequency response, low distortion, and good damping properties. The disadvantages are few, but in some cases serious enough to preclude use of the stage. Principal among these are lack of voltage amplification (which imposes extremely severe requirements on the preceding stage if distortion is to be avoided), very low efficiency (for most designs), and low power sensitivity.

[3] K. R. Sturley, "Radio Receiver Design, Part II." New York: John Wiley and Sons, 1948, pp 56–64.

[4] W. B. Nottingham, "Optimum Conditions for Maximum Power in Class A Amplifiers," *Proc. I. R. E.*, Dec. 1941, p 620.

Loudspeaker Damping as a Function of the Plate Resistance of the Power Output Tube

DONOVAN V. GEPPERT

Logical reasoning is applied to the problem of choosing between triodes and pentodes or tetrodes to obtain low output impedance and the resulting high damping factor.

A MISCONCEPTION appears to be common with regard to the effect of the dynamic plate resistance of a tube driving a loudspeaker on the "damping" properties of the tube. It is apparently considered a general rule that the lower the plate resistance of the output tube the greater the amount of speaker damping. The 2A3, with a relatively low plate resistance of 800 ohms, has long been considered the tube *par excellence* for driving a loudspeaker. Recently the 6AS7, with a plate resistance of only 200 ohms, has made its appearance in the output stage of amplifiers designed to feed loudspeakers. Carried to its logical end, we would eventually use output tubes having zero plate resistance and, naturally, zero amplification since

$$\mu = r_p g_m$$

Is there any real justification for this apparent race towards an infinitesimal plate resistance? It will be shown that there actually does exist a practical advantage in using triodes over beam tetrodes or pentodes, as far as speaker damping is concerned, but there is practically *no* advantage of one triode over another. Also it will be shown that the advantage of triodes over beam tetrodes and pentodes does not lie in the fact that they have lower values of plate resistance as such, but that for the values of load impedance commonly chosen on the basis of allowable distortion, the impedance reflected into the secondary of the output transformer is inherently lower for triodes than for pentodes or beam tetrodes.

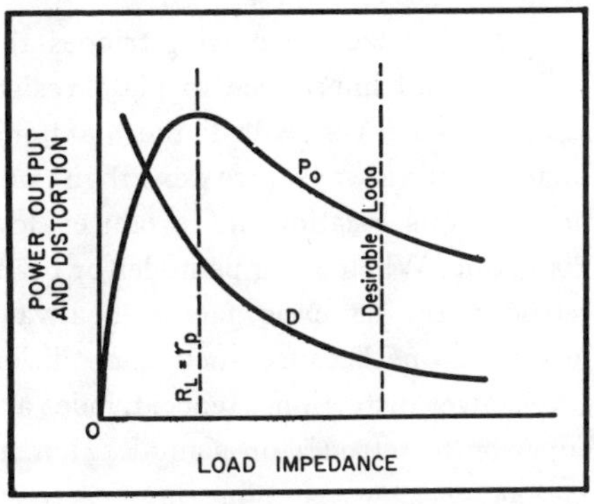

Fig. 1. Power output and distortion vs. load impedance for a typical single-ended triode power amplifier.

Consider the curves in *Fig.* 1 which show the variation in power output and harmonic distortion for a single-ended triode amplifier. For Class A linear operation, maximum power output occurs when the equivalent load impedance reflected into the primary of the output transformer equals the dynamic plate resistance of the tube. This is shown by the dotted line marked $R_L = r_P$ in *Fig.* 1. However, a more desirable load from the standpoint of distortion is a value somewhat higher than the plate resistance of the tube. A load impedance equal to twice the plate resistance of the tube is frequently stated as being an optimum value. In actual practice, values between two and four times the plate resistance are used to advantage.

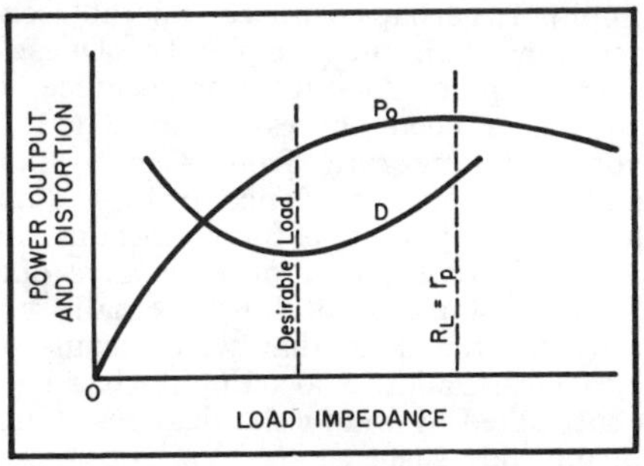

Fig. 2. Power output and distortion vs. load impedance for a typical single-ended pentode power amplifier.

Pentode Curves

The corresponding curves for a pentode or beam tetrode are given in *Fig.* 2. In this case a desirable load is one giving a minimum harmonic distortion which inevitably occurs at a load impedance lower than the plate resistance of the tube. Hence, for typical triodes, loads *greater* than the plate resistance are used, whereas for typical pentodes or beam tetrodes, loads *less* than the plate resistance are used.

Refer now to *Fig.* 3. For operation over the linear portion of the tube's characteristics, the series equivalent circuit can be used to replace the tube as far as a.c. computations are concerned.

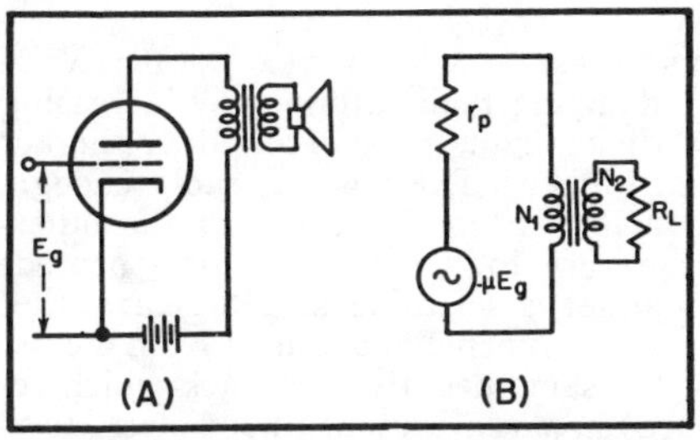

Fig. 3. (A), left, actual circuit and (B) right, equivalent circuit for Class A linear single-ended power amplifier.

If the output transformer has a turns ratio of N_1/N_2 and is considered to be perfect, the impedance reflected into the primary is $\left(\frac{N_1}{N_2}\right)^2 R_L$ and the impedance reflected into the secondary is $\left(\frac{N_2}{N_1}\right)^2 r_P$. In other words the impedances can be referred to either side. The ratio of load impedance to plate resistance is the same on either side. In the case of triodes, it has been shown that the ratio of load impedance to plate resistance is always made *greater* than one and generally two or three to one. Therefore the equivalent generator internal impedance is about one-half or one-third of the load impedance. But for pentodes, the ratio of load impedance to plate resistance is always made *less* than one, and the equivalent generator internal impedance is therefore greater than the load impedance. Consequently, triodes reflect a lower impedance into the secondary than tetrodes or beam pentodes, not because of their lower plate resistance, but because the values of load impedance dictated by distortion considerations make it so.

As to the relative merit of different triode tubes, it should be obvious that as long as the same ratio of load impedance to plate resistance is used, there is no difference whatever in the amount of speaker damping. As an example, consider a hypothetical tube having a plate resistance of 1000 ohms and a desirable load of 2500 ohms. To feed a 10-ohm loudspeaker requires a turns ratio of

A New Approach to Loudspeaker Damping

WARNER CLEMENTS

Modestly styled by the author as "the hottest thing in audio," positive feedback added to a negative-feedback amplifier is shown to reduce output impedance to zero or below, achieving a damping factor of infinity.

Is appreciably better loudspeaker damping really attained by increasing the damping ratio of amplifiers higher and higher? The answer, sadly enough, is "No!" Some lowering of the otherwise very high impedance of a pentode amplifier is desirable, and there are other concurrent benefits obtained from the use of the same negative feedback which reduces apparent output impedance. However, it appears that there is a widespread misunderstanding of the principles involved.

In this article, the author will endeavor to show that we have been deluding ourselves to a great extent about the merits of a high damping ratio, and that there is a way to achieve high damping other than by the application of more and more negative feedback. By the method to be described herein, it is possible to take the damping ratio right up to infinity and beyond. In fact, it is in the region well beyond infinite damping ratio that an amplifier must operate in order to provide theoretically perfect speaker diaphragm control. By the addition of only one inexpensive part, your own amplifier may be transformed to operate either at infinite damping ratio, or at a condition of near-perfect electrical damping.

Let us examine the principles involved in amplifier-damping of loudspeakers. Take an ohmmeter or bridge to the nearest speaker and measure the d.c. resistance of the voice coil. Ten ohms is a representative value for a 16-ohm speaker. Now measure the resistance of 16-ohm secondary of an output transformer. Another ohm or so. Now add the 11 ohms thus obtained to the tiny apparent output impedance—0.6 ohms, maybe?—you have obtained in your amplifier, perhaps with considerable expense and circuit complexity. Doesn't look so good, does it? For instance, if you have been successful in doubling your damping ratio from 4 to 8—thus cutting output impedance from 4 ohms to 2 ohms—you have, for your pains, cut *total* effective generator resistance only from 15 ohms to 13 ohms. Actually you haven't even done that well because in either case additional damping has been contributed by the mechanical resistance of the cone suspension.

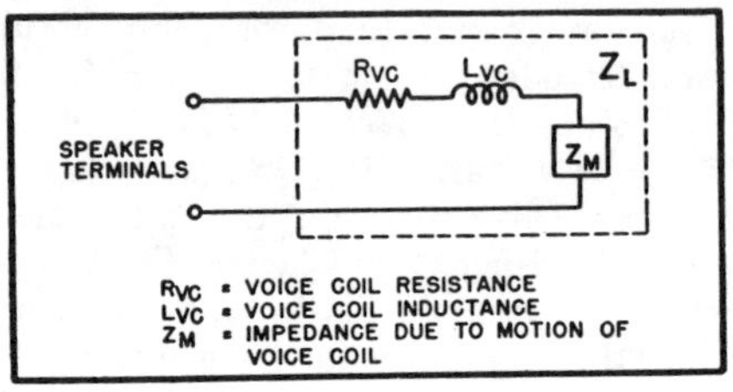

Fig. 1. Electrical equivalent of loudspeaker load.

It is justifiable to take voice-coil resistance, normally considered as part of the load, and add it to the source, for two reasons. First, as in the case of motor rotor windings, resistance and self-inductance of the voice-coil are necessary evils and contribute nothing to performance. Second, the voice-coil parameters are effectively in series with, not shunted across, the "business" part of the load. If one measures the electrical impedance of a speaker at a given frequency and then impedes the motion of the voice-coil, the impedance is seen to go *down*, being lowest with the voice-coil completely blocked. At this point the voice-coil shows the same impedance that it would if completely removed from the speaker except that its inductance is affected somewhat by the metal surrounding the gap. Conversely, if a weightless voice-coil could be suspended in the gap so that it could move freely without encountering stiffness or friction, the electrical impedance of the speaker would be infinity. The total electrical impedance of the circuit will always be greater than that of the voice-coil alone unless the speaker is inoperative. Clearly, then, the equivalent circuit is the series one of *Fig.* 1.

Transient Response

Good transient performance is often equated to the reduction of overshoot or "ringing." There is more to it than that,

---

Loudspeaker Damping as a Function of Plate Resistance

(from preceding page)

$\sqrt{\frac{2500}{10}} = 15.8$ to 1. The impedance reflected into the secondary becomes $(1/15.8)^2 \times 1000 = 4$ ohms. Consider another tube having a plate resistance of 5000 ohms and a desirable load of 12,500 ohms (2½ times the plate resistance as for the first tube). To feed a 10-ohm loudspeaker requires a turns ratio of $\sqrt{\frac{12{,}500}{10}} = 35.3$ to 1. The impedance reflected into the secondary becomes $(1/35.3)^2 \times 5000 = 4$ ohms, the same as for the first tube.

Comparisons

This does not mean of course that all triodes *as actually used* have the same damping properties. What it does show is that if the choice of load impedance is governed by allowable distortion for each tube type, one type may give better speaker damping than the other *for the particular load impedances chosen.* Clearly the higher the ratio of load impedance to plate resistance, the greater the amount of speaker damping. Fortunately, in the case of triodes, the higher the ratio of load to plate resistance the lower the distortion, so that one does not have to sacrifice speaker damping for distortion or vice versa.

To summarize, when using triodes, the ratio of load impedance to plate resistance, (sometimes called the damping factor), is always greater than one because this relationship produces low distortion. When using pentodes or beam tetrodes, the damping factor is always less than one because such a condition gives lower distortion. Hence, triodes are superior to tetrodes or pentodes, but as far as speaker damping only is concerned all triodes must be considered about equally as effective.

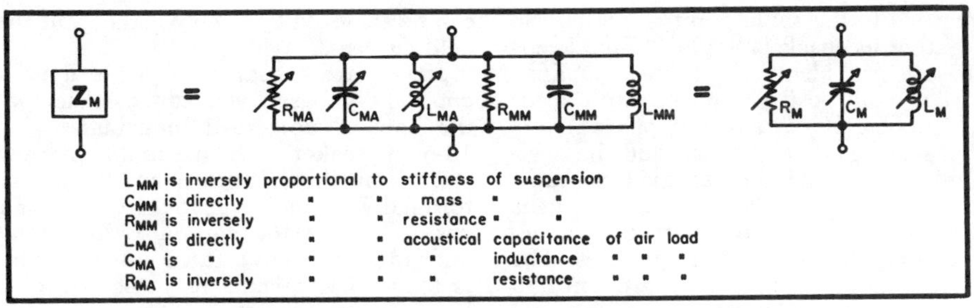

Fig. 2. Equivalent circuit for motional impedance of loudspeaker.

but the yardstick thus suggested is a useful one for underdamped systems. Certainly ringing contributes largely to the characteristic "loudspeakerish" sound from which all direct-radiator speakers seem to suffer. To induce ringing in any underdamped system it is only necessary that an impulse be introduced which contains components higher in frequency than the resonant frequency of the system. The familiar "mouth-harp" is an example of such a system. The pitch of the vibrating part of the mouth-harp does not vary. Yet when placed in the performer's mouth, tunes are produced by tuning the cavity (formed by the mouth) which the initial transient starts ringing. Circuit switching can produce widely differing sounds in different speakers. This writer once arranged a half-dozen speakers of various sizes and makes so that he could play a simple tune with the thumps created by closing toggle switches connected between the respective voice-coils and a dry-cell. Fine tuning was accomplished by changing baffling. In no case was the speaker in-

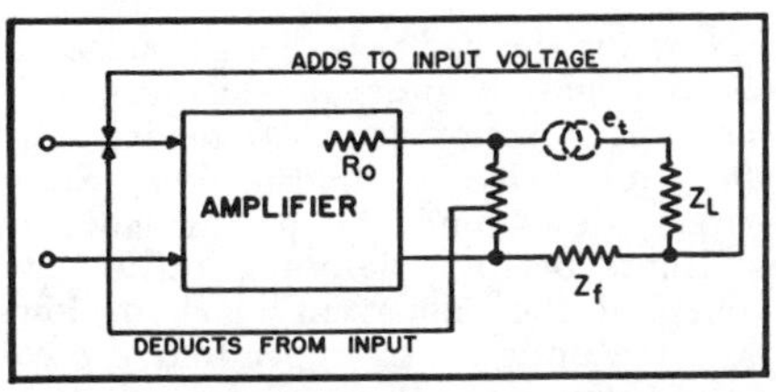

Fig. 3. Arrangement for applying voltage-proportional negative feedback simultaneously with current-proportional positive feedback.

stallation any different from one which might have been intended for sound reproduction. This experiment suggests one reason why two-way systems are likely to not *sound* good to the layman. If the pitch of the click of the high-frequency speaker bears an inharmonic relation to the pitch of the thump of the low-frequency speaker, a disagreeable noise is produced on every transient—which is to say on at least every note of music and syllable of voice reproduced.

It is clear that damping is a desirable objective, but it is also clear that beyond a certain point inverse feedback does virtually nothing toward attaining this objective. Actual tests with pulses and interrupted wave-trains show that a triode output *without* feedback represents about the point of vanishing return. A lower output impedance than that thus represented is of very little importance in restraining the ringing of a typical diaphragm assembly. Direct-radiator speakers will ring at their resonant frequency even if fed by a generator of zero internal impedance. Just how badly they will ring under that circumstance will depend on the amount of flux in the gap and upon the amount of damping inherent in the diaphragm suspension. The role of mechanical damping, which takes place mostly in the outer roll, is illustrated in *Fig.* 2. The power that

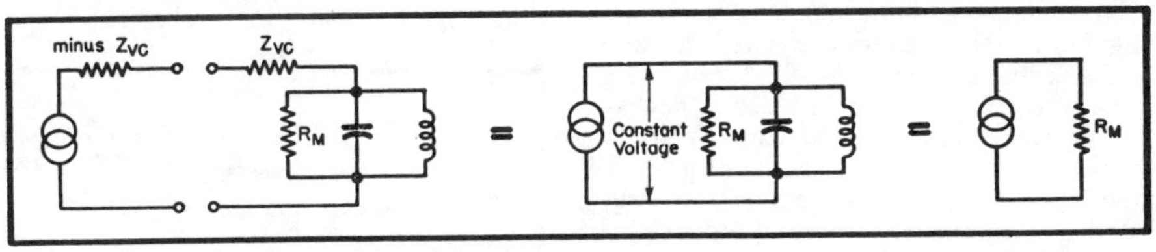

Fig. 4. Method of using negative impedance to achieve perfect damping.

goes into R_{ma}, shown in the middle circuit of that figure, is the power that counts—the power that represents acoustic output. The shunting reactances are necessary evils. Any power lost in R_{mm}, the edge damping, is power wasted from the standpoint of efficiency. Nevertheless it can be seen that a low value of R_{mm} will do much to minimize the effects of the shunting reactances.

It is impossible to get all of the way to zero apparent output impedance by means of negative feedback. Even assuming that it were possible, the speaker would still not be cured of ringing, because of the intervening impedance of the voice-coil. High flux-density in the gap can reduce the effect of the voice coil, but cannot eliminate it. But suppose that it were possible to make the amplifier exhibit *negative* impedance as viewed from the output terminals. The negative impedance would subtract from the voice-coil impedance. If the two just matched, the effect of the latter would have been eliminated entirely. This is no idle dream, but is, in fact, quite easy of attainment. The means is *positive* feedback. Readers will recall that constant-*voltage* inverse feedback from an output causes apparent output impedance to decline, while constant-*current* feedback causes it to rise. With *positive* feedback the opposite holds—voltage-proportional feedback causes the output impedance to rise and current-proportional feedback lowers it. Furthermore, current-proportional positive feedback will take the output impedance right down past zero and into the negative region if desired. The ideal setup, then, is to use voltage proportional negative feedback *plus* current-proportional positive feedback.

Direct and inverse feedback can thus be used simultaneously without cancelling each other out, although cancellation occurs insofar as the distortion-reducing and frequency-response-smoothing effects of inverse feedback are concerned. But with regard to reducing apparent output impedance, the effects of the inverse and direct feedback are additive. Referring to the block diagram of *Fig.* 3, it will be seen that any voltage developed only in the load causes the input voltage to be modified in the *same phase* by the resultant out-of-phase voltages in the two respective feedback circuits. The operation of the circuit is as follows: starting with negative feedback alone, the apparent output impedance goes down as positive feedback is applied. It requires only a slight amount of positive feedback to bring it to zero. Damping-ratio at this point is, of course, infinity. As more positive feedback is added, the apparent output impedance becomes negative and starts to "erase" the series part of the speaker load impedance. At some point the voice-coil impedance is exactly matched by the negative impedance of the amplifier and theoretically perfect damping is achieved, as illustrated in *Fig.* 4. If positive feedback is increased considerably beyond this point, oscillation will eventually occur. In actual installations it may take place before the midband gain gets up to what it would be with no feedback at all. This is due to the difficulty of securing an exact phase match between the load impedance and the negative impedance. This last consideration is of only academic interest, however. There will always be an ample safety margin before oscillation, even if the inductance of the voice-coil is ignored and the amplifier is made to exhibit a negative resistance.

Fig. 5. Waveforms across output of negative-impedance amplifier with 30-cps square-wave input. Speaker resonant frequency = 120 cps.

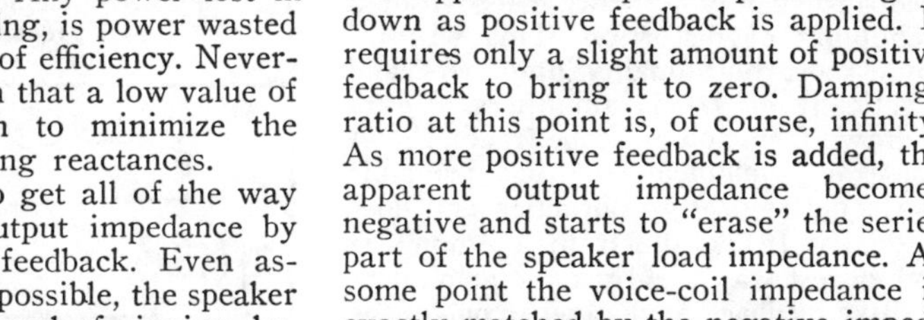

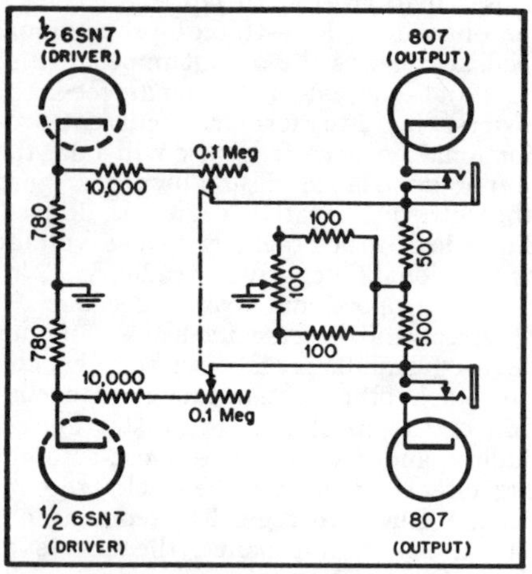

Fig. 6. One method of obtaining current-proportional positive feedback.

"Perfect" Damping

Thus perfect damping is arrived at long before the benefits of the negative feedback have been cancelled out. The exact amount of positive feedback necessary varies with the installation. With a typical high-efficiency speaker and a fairly high amount of negative feedback, the ideal amount of positive feedback to be added will be found to be that amount which just about doubles the gain. Circuit algebra shows that this means the effective negative feedback will have been cut in half. In other words, the principle is applicable to any circuit from which you can get roughly 6 db more of feedback than you really need from a standpoint of distortion. This will include most high-quality amplifiers and practically *all* triode amplifiers. *Figure* 5, drawn from actual 'scope traces, shows that the combined feedback principle works out as neatly in practice as it does in theory. It also illustrates the quickest and easiest way to adjust the amount of positive feedback applied when one of the circuits to be described is added to an amplifier. Note that if too much positive feedback is used, ringing resumes, but is now reversed in phase.

Figure 6 shows a circuit that has been recommended for applying positive feedback to the popular Williamson amplifier. It would seem that this circuit has several drawbacks. Since it applies feedback before even-order harmonics have been bucked out, it increases the magnitude of those harmonics and lessens the likelihood that they will be perfectly cancelled. Since it introduces constant-current negative feedback, it defeats its own purpose up to a point and requires high output from the driver tubes. Finally, since ordinary ganged potentiometers run well above 10 per cent "tracking error," the signal balance of the output stage is likely to be rather bad.

The circuit of *Fig. 7* was devised by the writer in 1949 and is believed to be far superior to that of *Fig.* 6. It is easy to apply to existing amplifiers using any of the popular circuits and the single adjustment has no effect on push-pull balance. It may be seen that a split-load type of phase inversion is applied to the positive feedback, permitting it to be returned to the same point at which the negative feedback is applied. This keeps phase shift identical in the amplifier part of both positive and negative feedback loops. If desired, the phase-inversion feature can be left out and the positive feedback applied to grid of same stage, as shown in *Fig.* 8. This circuit is satisfactory where the damping is desired only at the lower frequencies. The stray and interelectrode capacitance to ground from the top of the grid resistor prevents the higher frequencies from being fed back, even without the inclusion of *R* and *C*. R_g may be a volume control if desired.

Negative Inductance

An interesting refinement on the negative-output-resistance amplifier is to make it display negative *inductance* as well, thereby cancelling out the inductance of the voice-coil and improving high-frequency response and stability. This can be done by including an inductance in series with the positive-feedback rheostat, as shown dotted in *Fig. 7*. The optimum value for this inductance is given by the formula:

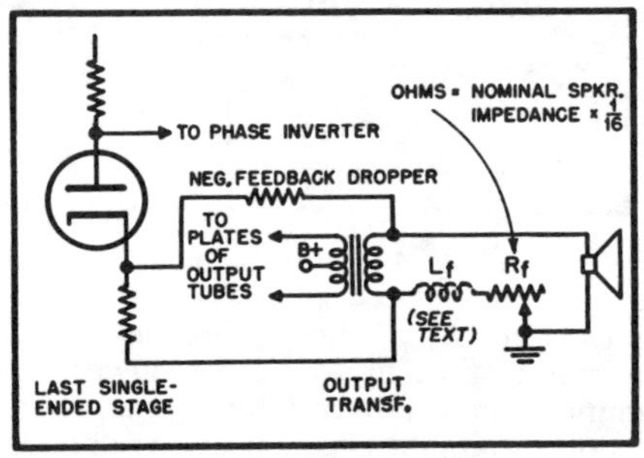

Fig. 7. Preferred method of applying positive feedback.

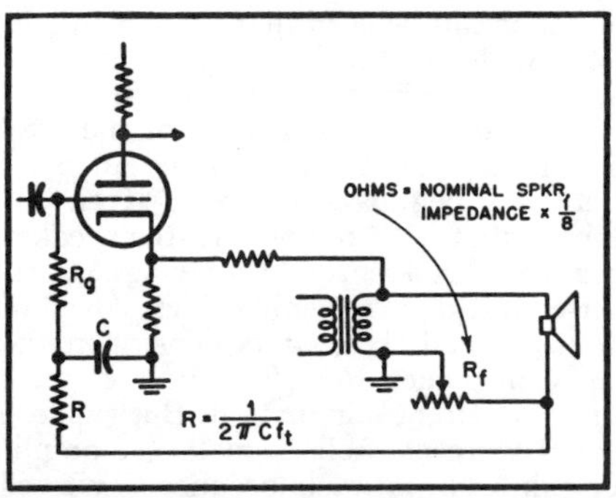

Fig. 8. Variation of Fig. 7 that produces low-frequency boost but damps speaker only at one low frequency, which may be set to be resonant frequency of the speaker.

$$L_f = \frac{L_{vc}}{A' - 1}$$

Where *A'* is gain of part of amplifier enclosed by feedback loop, with *negative feedback only* and *unloaded.* To apply this formula, you need to know L_{vc}, the blocked voice-coil inductance of your speaker (or speakers). Unfortunately, I know of no way to block the voice-coil of a speaker securely without injuring it, so even if a means of measuring inductance is available it will do no good unless the speaker has an electromagnetic field. In that case the effect of blocking can be achieved by simply leaving off the field current.

If you are prepared to tackle it on a cut-and-try basis, you can assume that the blocked voice-coil inductance of a 16-ohm speaker is about 1 millihenry and start from there. Such a coil should be toroidally wound of #16 wire on an air core. Connect the coil into the circuit and add or remove turns until a value is found that permits R_f to be advanced the furthest (greatest resistance) without oscillation taking place. Actually there is nothing critical about this inductance. If it permits R_f to be turned at all further than it could be without the coil in the circuit, it will improve the operation of the circuit. After turns on coil are adjusted, R_f can then be adjusted to its final setting by means of square waves applied to amplifier and 'scope across speaker terminals, as in *Fig.* 5. It is also possible to do a fairly good job by ear with program material consisting of male speaking voice by adjusting for the least boominess. When you are through, you will have an outfit that provides the cleanest reproduction you have ever heard from direct-radiator speakers. This is, of course, provided that other things are right. Bear in mind that no amount of damping in one circuit will dampen another resonant circuit that is but loosely coupled to it. For instance, if a room resonance is present—it is almost impossible to get away from it in small rooms—no amount of speaker damping will obviate it. Look out also for acoustic feedback to turntable or microphonic tubes.

Limitations

Now for the bad news. A generator of low or negative internal resistance is in some regards not an ideal device with which to drive a loudspeaker. Some writers, describing the performance of amplifiers of high damping ratio, have spoken of the "apparent" lack of lows and have gone on to imply that the ear of the listener needs retraining; that the lows are there all right, just less conspicuous because of lack of resonance. Actually the lows *are* attenuated more and more as damping is applied. With the circuits of *Figs.* 6 and 7 the loss of lows will be very pronounced with some

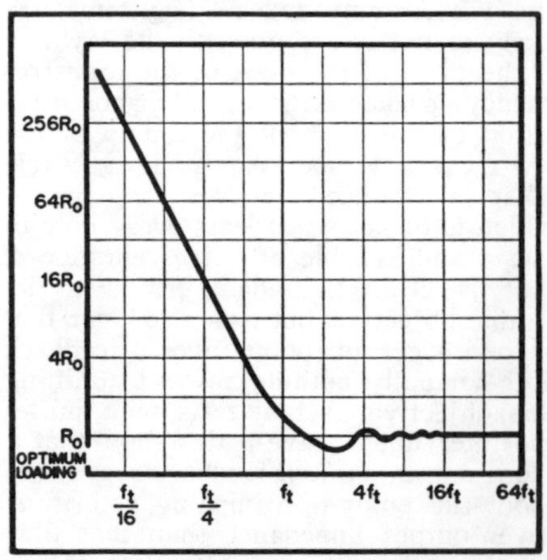

Fig. 9. Variation with frequency of resistive component of air loading on diaphragm, using the force-current analogy.

speaker arrangements, though negligible with others, as will be explained.

It has to do with variation, with frequency, of the air loading on the speaker cone. In (B) of *Fig.* 2, the resistance and reactance contributed to the equivalent electrical circuit of the motional impedance by the mass and mounting of the voice-coil and cone have been depicted as fixed. On the other hand, resistance and reactances arising from air loading have been depicted as variable. In general, the circuit values due to definable *mechanical* parameters are relatively fixed. Those due to *acoustic* values may vary with frequency, even the equivalent inductance and capacitance. Over a wide range of frequencies where the speaker (if considered as a perfect piston) gets a big enough "bite" of air, the loading due to air is almost purely resistive and is constant with frequency. For the equivalent circuit within this range C_{ma} and L_{ma} could be left completely out, and R_{ma} could be shown as fixed. However, if the applied frequency is lowered far enough, the "bite" will cease to be big enough at some frequency that depends on the size of the cone and the resistive loading will start to change. Whether it may be said to go "up" or "down" depends upon the acoustical-electrical analogy one uses (e.g. force-current or force-voltage). At any rate, the net effect is that the resistance "seen" by the electric circuit, R_m in *Figs.* 2 and 4, goes *up* with decreasing frequency.

Figure 9 shows the variation of R_m with frequency on the assumption that R_m is entirely due to the air loading on the cone. (In different speakers, the contribution of R_{mm} will flatten the actual rise of the curve to various degrees.) The "turnover" frequency, f_t, will vary from about 520 cps for a 15-in. speaker to about 1,660 cps for a 5-in. speaker. It can be seen that loudspeaker response will drop off about 6 db per octave below this frequency if an effective constant voltage is applied across Z_m (since the power output will then be inversely proportional to R_m). But the ordinary amplifier, with considerable output resistance adding its effect to that of the voice-coil, will drive a speaker in such a manner that there will not, in general, be such a low-frequency attenuation. The reason is that the reactive part of the motional impedance, as at (C) of *Fig.* 2 "looks", above resonance, like a capacitance shunting the load. Together with output resistance and voice-coil resistance it forms a power-eating tone control circuit that flattens out the response between speaker resonance frequency and f_t. Then above f_t, where the tone-control effect is no longer bucked by a changing radiation resistance, one would expect a falling-off of response. There is less than might be anticipated, however, because speaker cones are designed to quit functioning as rigid pistons at about this point.

At any rate it can be seen that some apparent internal resistance in the amplifier is essential for flat frequency response from the usual direct-radiator speaker. This should make it clear why it is that some speaker manufacturers have specified limits on the amount of feedback to be employed in amplifiers suitable for use with their respective speakers. It also explains why exponents of a high damping-ratio sometimes contradict themselves by placing a pad between amplifier and speaker.

But suppose that the benefits of perfect damping are desired without reduction of low-frequency response. The speaker system must be given a big enough bite of air so that loading will remain constant down to the lowest frequency to be reproduced. Horn loading presents one possibility. But with a horn (or horns) pains must be taken to see that the mouth area is adequate at frequencies well below taper cutoff. To fulfill this condition down to 100 cps or below requires a huge installation. Horn systems designed for home use are mostly short on mouth area and as a result the diaphragm loading fluctuates violently with frequency. This may do no harm where there is series resistance to smooth out the output, but, as has been shown, with the negative-impedance amplifier the output will fluctuate exactly as the cone loading does. Another possibility is to use a large number of direct-radiator speakers. Theoretically, to be down not more than 4 db at 60 cps would require a cone area of 5,900 sq. in. in an infinite wall. This would correspond with the staggering total of fifty-four 15-in. speakers. But it is not as bad as that. By mounting in a corner near the ceiling or floor, one-fourth as much radiating area will suffice. By making use of the effective radiating area of a bass-reflex port, the number of speakers can be cut in half again. As it figures out, six 15-in. speakers in a bass-reflex cabinet in a corner will do very well. Since the power each speaker will be required to handle will be small, bargain-basement speakers will do; the total investment can be quite small.

If space or purse does not permit a cone area large enough for constant air loading, the only recourse left is to add some bass-boost, to decrease the amount of damping, or to compromise and do both. The proper turnover frequency, f_t, for the boost and the amount of boost required will vary with the speaker and with the installation. The circuit of *Fig.* 8 will give a mild amount of bass-boost, depending on the setting of R_f.

If the reader is impressed by the lure of improved damping, he may still want to adapt for negative impedance just for its novelty value. When the indicated simple changes have been made, it will result in an amplifier whose output voltage actually goes *down* when the load is removed. This phenomenon is so fascinating that the writer has had to demonstrate it for friends over and over again.

Loudspeaker damping is an extensive subject and this article is probably not really the "last word." But it should finally lay the ghost of some old fallacies, and it does point the way to securing perfect damping by electrical means. It would be hard to ask for anything more in this direction. In the writer's opinion, it's hard to improve on perfection.

Dynamic Negative Feedback

ULRIC J. CHILDS

A further study of the effect of introducing some current feedback in an amplifier to counteract the varying impedance of the loudspeaker. The author gives a practical circuit which is adaptable to any feedback amplifier.

AUDIO ENGINEERING has gone a long way since the day when any amplifier with an extended frequency range was considered to be "high fidelity." However, a number of misconceptions still remain. One of them is the question of loudspeaker damping—reduction of overshoot and lag of the speaker cone caused by its inertia, especially at low frequencies when cone excursion is large and the period between impulses is long enough to allow noticeable wave trains to exist. Because only the initial pulse of the wave train is caused by amplifier output signal, the remainder of the movements create signals that are entirely spurious and are detectable as muddiness in the music.

Negative feedback has been widely used in the last few years to lower the internal impedance of the amplifier so that the signal generated by the loudspeaker as a motor during its overshoot is short-circuited by the amplifier, and the same dynamic braking effect is obtained as though a motor armature were shorted with the field excited. However, as Warner Clements pointed out in the foregoing three pages, there is still a considerable resistance in series with

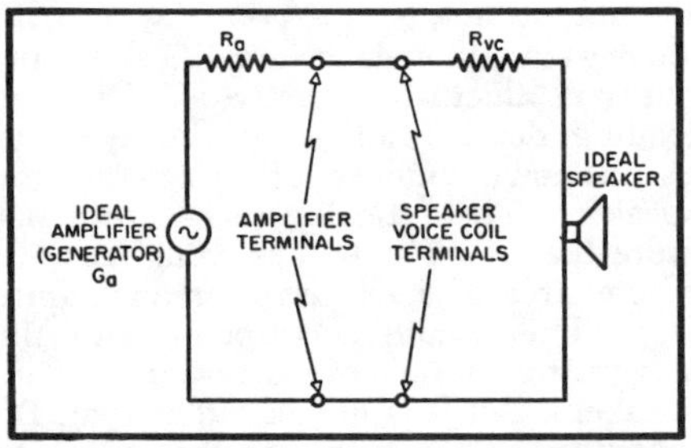

Fig. 1. Equivalent circuit of output-transformer secondary and loudspeaker in idealized system.

the "short circuit"—the comparatively large internal ohmic resistance of the voice-coil winding itself. Thus ideal dynamic braking is quite impossible with negative feedback of the usual type, since, even with a zero amplifier internal impedance, the average 16-ohm speaker voice coil still places about 10 ohms in the series circuit.

There are two possible methods of achieving perfect or nearly perfect damping in a loudspeaker. The first, obviously only a theoretical possibility, is to use a speaker whose voice-coil wire has no resistance. The second, suggested by Clements and explored in greater detail by the writer, is to add a new type of speaker-control network. The amplifier may be made so sensitive to the movements of the speaker that when the speaker tends to overshoot, the amplifier output is automatically reduced to cancel the overshoot; when the cone lags the signal the output is increased to make the speaker follow the excitation signal. This is done by controlling the inverse feedback present in the system so that it is largely dependent on the behavior of the speaker and varies from instant to instant.

Although Clements chose to call this kind of feedback positive it is speaker-controlled *negative* feedback, and because its value is dynamic rather than fixed, *dynamic negative feedback* seems to the writer to be an appropriate term.

Equivalent Circuit

Figure 1 is the equivalent circuit of an amplifier with or without ordinary negative voltage feedback. If the negative feedback is high, the value of the amplifier internal impedance R_a which appears in series with the amplifier as an ideal generator G_a, is low, and the output voltage at the amplifier terminals does not vary greatly between load and no-load conditions.

The ultimate in this type of design would be reached when $R_a = 0$, but this is impossible by ordinary means. However, even with a zero internal amplifier resistance, the internal resistance R_{vc} of the voice coil remains and is a severely limiting factor in the "short-circuit" current available for damping.

The ideal solution as a basic idea is quite simple: the voltage generated by the speaker due to overshoot, for example, must be applied independently to the amplifier in such a way that it will appear at the amplifier output terminals in the correct phase and at the correct amplitude to restrain the cone from overshooting. Stated in another way, the output of the generator G_a itself must be controlled by the speaker within wide limits of phase and amplitude so that the circuit will cancel R_a and R_{vc}. The speaker then looks back into an impedance of zero ohms and is perfectly damped.

Examination of this scheme indicates that under no condition can speaker damping be perfect. Using a philosophy applicable to any closed-loop control system, no error correction can be had until an error exists, since the error is necessary to create the correction signal. A more cogent limitation is oscillation, which must exist if damping is perfect. However, the method can be used within limitations to obtain improvement in speaker performance which is obvious to the discriminating listener.

Practical Circuit

Figure 2 shows a basic circuit used to obtain dynamic negative feedback. It is similar to the Clements circuit, but the *net* feedback is not positive. R_1 and

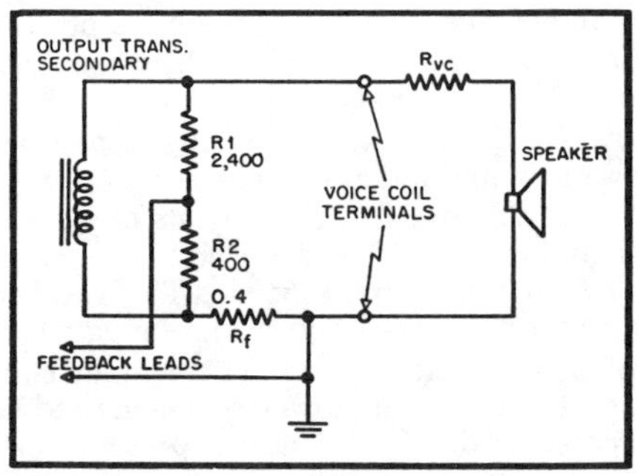

Fig. 2. Circuit which can be adapted to any feedback amplifier to provide improved damping.

R_2 are a voltage divider across the secondary of the amplifier output transformer. The feedback lead is connected to the junction of R_1 and R_2, with ground at right of the current feedback resistor R_f. The voltage across R_2 due to the output signal is, of course, in the same phase as that from the output transformer, while the voltage drop across R_f due to signal current is in opposing phase with respect to the junction of R_2 and R_f. The net voltage appearing across the feedback leads is therefore that of R_2 minus the R_f drop. Because the voltage across R_2 is much greater, the net feedback voltage is in the same phase as the amplifier output voltage. It is fed to an early stage at the correct point to make it *negative* feedback.

When the speaker overshoots it generates a voltage 180 deg. out of phase with the signal voltage. This counter-e.m.f. causes a reverse current to pass through R_f, reducing the existing signal drop across R_f. The result is an increase in voltage across the feedback leads, an increase in negative feedback to the amplifier, and a drop in amplifier output voltage.

The negative feedback is dynamically controlled at each instant by the current through R_f in such a manner as to cause the amplifier output to counteract the spurious actions of the speaker cone; the degree can easily be controlled by varying the value of R_f—the larger the value the greater the corrective effect.

Even with quite small values of R_f, the effective amplifier internal resistance may be zero or even negative. An overshooting (high-resistance) speaker causes increase in negative feedback and decreased output voltage; a lagging (low-resistance) speaker causes less negative feedback and more output. This contradicts Ohm's Law, according to which the output voltage of a generator with internal resistance falls as the load resistance decreases. This means that the internal amplifier resistance R_a is negative. It is still amenable to Ohm's Law except, that instead of causing a voltage drop, R_a causes a voltage *rise*, the amount of which depends on the value of the negative resistance.

The corrective effect may theoretically be raised as necessary to compensate exactly for all spurious movement in the speaker. In this case the effective internal resistance of the amplifier would be negative and exactly equal in magnitude to the sum of the R_{vc}, R_f, and any transformer secondary resistance. Regarding the speaker as an ideal zero-resistance generator in series with the voice-coil wire resistance, the generator would look back into a genuine zero impedance and, in effect, would be perfectly braked dynamically. Such a condition is, however, impossible of attainment, even in an amplifier with no phase shift at any frequency, because oscillation would result, as shown by the following demonstration.

Figure 3 is an equivalent circuit of the amplifier, consisting of zero-resistance generator G_a and internal resistance R_a, connected to a load R_{vc}. Assuming that R_a has a negative value, as would be necessary if it were to cancel out as much as possible of R_{vc}, the amplifier output voltage will rise when the load is connected. The load voltage e is therefore higher in amplitude than the open-circuit voltage E, and the output regulation of the amplifier may be expressed in db at 20 log e/E. Then

$$e = E \frac{R_{vc}}{R_{vc} + R_a}$$

and rearranging to find the ratio

$$\frac{e}{E} = \frac{R_{vc}}{R_{vc} + R_a} \qquad (1)$$

Eq. (*1*) may now be used to find the regulation with any desired amount of

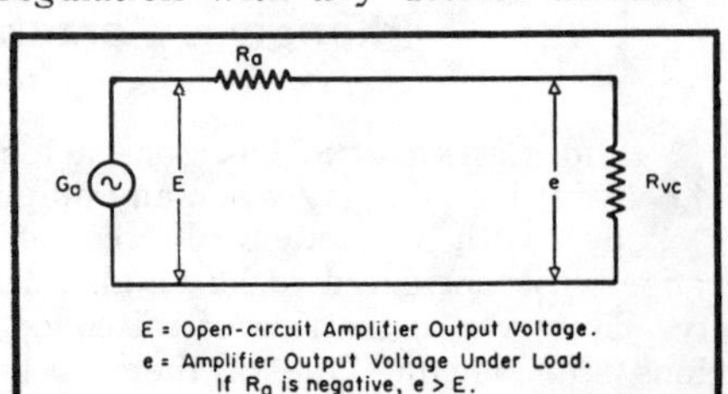

E = Open-circuit Amplifier Output Voltage.
e = Amplifier Output Voltage Under Load.
If R_a is negative, e > E.

Fig. 3. Equivalent circuit to show effect of complete cancellation of voice-coil resistance by means of feedback.

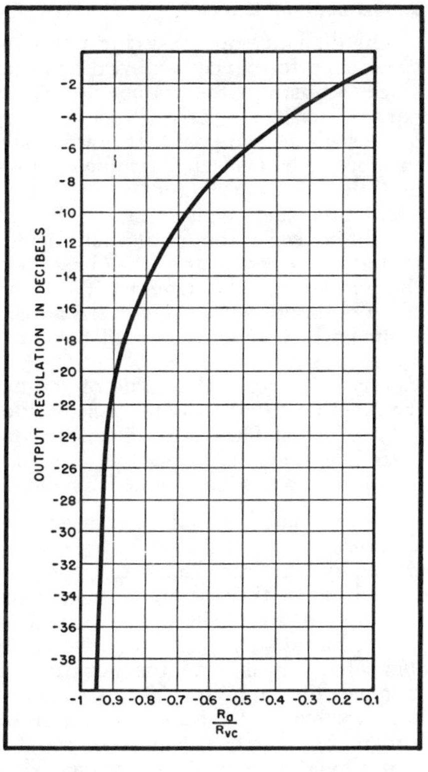

Fig. 4. Curve showing output regulation for varying ratios of amplifier output impedance to speaker voice-coil impedance.

R_{vc} being cancelled out by a negative value of R_a. The result, if expressed in db, is a negative number since the regulation is opposite in direction to ordinary positive regulation, i.e., the voltage rises instead of falling when a load is applied.

If one-half the value of R_{vc} is to be cancelled, then

$$R_a = 0.5R_{vc}$$

$$\frac{e}{E} = \frac{R_{vc}}{R_{vc} - 0.5R_{vc}} = 2 = (-)6 \text{ db}.$$

For example, if the speaker voice coil resistance is 10 ohms and the internal resistance of the amplifier is made –5 ohms to cancel half of R_{vc}, the amplifier output voltage will rise 6 db between no load and full load.

To obtain complete cancellation of R_{vc}, $R_a = -R_{vc}$. Then

$$\frac{e}{E} = \frac{R_{vc}}{R_{vc} - R_{vc}} = \frac{1}{0} = \infty \qquad (2)$$

which means that adding any load whatever, no matter of how high a resistance, causes a rise in output voltage to infinity, which is oscillation. Therefore, it is obvious that perfect damping simply cannot be achieved.

But additional figures on this same subject are of interest. How far is it practical to go? How well *can* a speaker be damped with dynamic negative feedback? The graph of *Fig.* 4 was prepared by using Eq. (*1*) and converting the resulting voltage ratio to db. It shows what the output regulation of the amplifier must be in order to cancel given portions of R_{vc}. Note that to cancel as much as 0.8 R_{vc} the regulation must be –14 db. The decibel regulation figure indicates how much of the negative feedback in the loop will be cancelled when the dynamic negative feedback circuit is used. Thus, to cancel 8/10 of R_{vc}, this loop must have sufficient negative feedback from the junction of R_1R_2 of (*Fig.* 2) to keep net feedback negative even when a large portion of the negative feedback is removed for use in correcting spurious speaker cone movement. The figure for negative decibel regulation becomes inordinately large as the damping is increased, approaching infinity toward perfect damping, as shown in Eq. (*2*).

It appears, therefore, that only a limited amount of dynamic negative feedback may be used with the very best practical amplifiers, both for the reason just mentioned and because too great a figure causes the designer to run into severe difficulties with oscillation due to phase shift and response nonlinearity at extreme frequencies, even those far above and below audibility.

The Childs amplifier, which the writer has designed (and redesigns from time to time) and constructs for use in better-quality custom installations has flat frequency response to around 100,000 cps and down to 10 cps. About 3 db of dynamic negative feedback has been incorporated in recent models with no danger of instability and with noticeable improvement in the already impressive performance. Even so, intermodulation distortion at the 15-watt level rises from 0.1 per cent to 0.2 per cent.

The graph of *Fig.* 4 and Eq. (*1*) are additionally useful to the designer for measuring the dynamic negative feedback caused by a particular value of R_f in *Fig.* 2. Each time the value is changed, the open-circuit and loaded output voltages of the amplifier may be measured and converted to decibels. Reference to the graph will then indicate what portion of the speaker's internal resistance has been cancelled. Knowledge of the speaker's resistance will, of course, indicate the exact value in ohms of the internal resistance of the amplifier.

Perfect Damping Impossible

The above proof that perfect damping is impossible may be somewhat puzzling in view of Clements' statement that it could be obtained, accompanied by apparent verification with an oscilloscope and a square wave measurement. The error made by Clements was in assuming that an oscilloscope placed across the actual voice-coil terminals could indicate motion of the speaker. The following simple demonstration shows that it cannot do so and that there is, in fact, no practicable way of depicting on an oscilloscope what the speaker cone is doing without the use of fairly elaborate acoustic measurements under ideal conditions or possibly a special type of bridge circuit.

In *Fig.* 5, (A) and (B) show the amplifier output circuit and the speaker, each equated to an equivalent circuit. The amplifier becomes a zero-resistance generator G_a with internal resistance R_a in series, and the speaker consists of another ideal generator G_s with internal voice-coil resistance R_{vc}. (C) shows the amplifier connected to the speaker with a 10-ohm potentiometer R in series with the two and an oscilloscope connected between the zero-voltage reference point (ground) and the arm of R. The purpose of the demonstration is to show that for any value of negative amplifier resistance we can cause the oscilloscope to show no voltage generated by the speaker, giving us a false impression that the speaker is generating no spurious-movement voltage. The only requirement is that we must be able to place the above-ground lead of the oscilloscope at any desired point on the internal resistance of the speaker. Since it is impossible to do this, we have added R, which is merely additional voice-coil resistance placed out in the open so we can get at it.

The circuit is again a simple voltage divider as shown in more conventional form at (D). We are not interested in amplifier signal but only in speaker-

Fig. 5. Equivalent circuits for study of 'scope patterns as a measure of the effectiveness of output impedance cancellation.

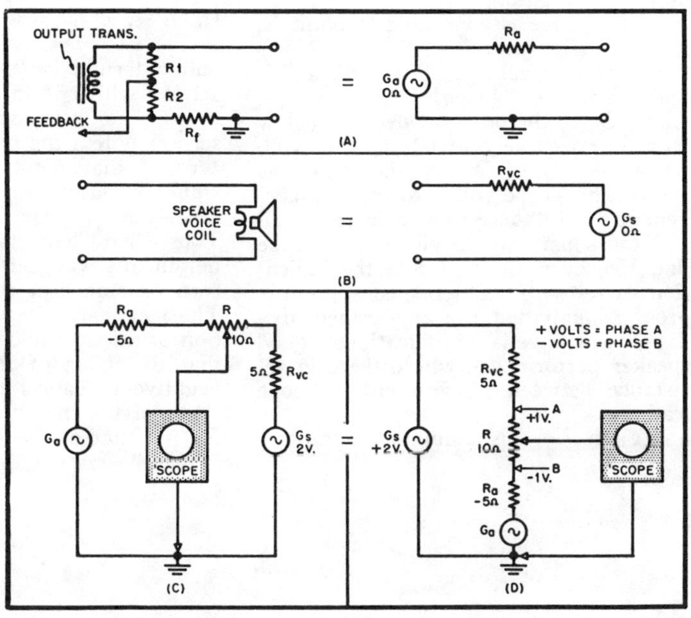

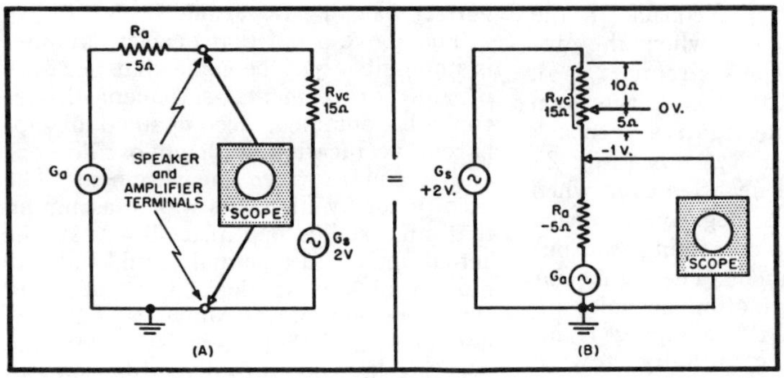

Fig. 6. Equivalents for further study of 'scope indications.

generated voltage, so G_a may be considered as simply a short circuited element for the moment.

The voltage in the circuit appears in two phase relationships 180 deg. apart—Phase A being the positive speaker-generated voltage, and Phase B being the negative voltage appearing across the negative resistance of the amplifier R_a. If the speaker is ringing due to insufficient damping on a square wave it may be seen on the 'scope if it is connected across either a positive or negative voltage. In the former case the actual voltage generated by the speaker will be seen; in the latter the counter-voltage generated by the dynamic negative feedback will be seen, with the "wriggles" atop the square wave in reversed phase. With the 'scope across zero volts, however, neither is seen, but only the square-wave signal put out by the amplifier. If the amplifier is a good one, the signal will appear to have flat tops.

The total resistance around circuit (D) of *Fig.* 5 is $5 + 10 - 5 = 10$ ohms. Since the speaker is generating 2 volts we may say that each 5 ohms drops 1 volt (and each –5 ohms drops 1 volt in the opposite phase).

If the 'scope is connected between ground and point A it will show

$$E_R + E_{Ra} = 2 - 1 = 1 \text{ volt positive.}$$

Thus the 'scope is showing the speaker-generated voltage (Phase A) less a certain amount of correction voltage (Phase B), the resultant giving no useful information.

If the 'scope is connected to point B, it shows E_{Ra}, which is –1 volt. This is the correction voltage (Phase B) and it will show the "wriggles" in reversed phase. Since the points above and below R are respectively at +1 and –1 volt, when the arm of R is at the center, as shown, the 'scope will show no speaker-generated or correction voltage at all and the square wave will appear to be flat. However, the fact that the indication varies with the 'scope connection is proof enough that the appearance of a flat-topped wave is no indication of good speaker performance where there is resistance between G_s and either 'scope lead.

(A) in *Fig.* 6 is a similar circuit, this time showing an ideal speaker G_s with an internal resistance R_{vc} of 15 ohms and the 'scope connected across the speaker terminals. This duplicates the scheme used by Clements to adjust feedback. Redrawing the circuit in conventional voltage-divider form at (B) and finding the voltage across the 'scope by the same means as before, the 'scope shows –1 volt (Phase B), i.e., 1 volt of correction signal. It could be made to indicate zero by going into the voice coil and connecting it to the tap on R_{vc} labelled 0 volt, but this is an obvious impossibility.

It should be clear by now that in order for the 'scope to indicate zero, it must be connected at a point where the speaker-generated and the correction voltages cancel and produce zero. In the case of a perfectly damped speaker this point would be the terminals of the ideal speaker G_s which are hidden from access. It should also be clear that with the 'scope set at any point it can be made to indicate zero by adjusting the amplifier output impedance R_a but such an indication *will not indicate what the speaker is doing.* A third interesting point is that the 'scope cannot indicate zero when connected directly across the amplifier

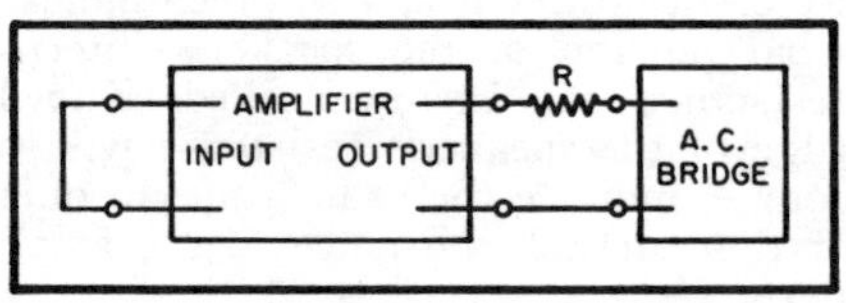

Fig. 7. Set-up for measuring output impedance.

output terminals (across G_a and R_a in series) which is in practice also across the speaker terminals (G_s and R_{vc} in series) unless the balanced voltages are zero at that point, which means that amplifier output resistance R_a must be *zero*—not negative and not positive. Thus when Clements adjusted his amplifier in this way, he adjusted it only for zero resistance at the output terminals. The fact that voltage rose with connection of a load could only have indicated that the value of R_a was just enough negative to cancel out only the resistances of the output transformer and of R_f, not that of the voice coil.

Adjustment of Circuit

The difficulty of measuring the actual speaker performance is no real handicap, since adjusting the value of R_f the current feedback resistor does not require such measurement. It is impossible to approach perfect damping anyway and the limiting factor in practice is oscillation due to phase shifts. Even if the limit were the oscillation demonstrated by *Figs.* 3 and 4 and Eq. (*1*), the adjustment still would not be made to an optimum point but toward a maximum beyond which it would be impossible to go. The obvious approach is merely to increase the value of R_f until the amplifier oscillates and then back off to a safely stable value. After this has been done the actual values attained may be measured and calculated as shown.

An alternate method of measurement is interesting and useful because of its simplicity and because it may be used to measure either positive or negative values. An a.c. bridge and a resistor are required. The resistor should be somewhat larger than the negative resistance the amplifier is expected to have and should be measured first on the bridge by itself. Then the resistor, bridge, and amplifier are connected as shown in *Fig. 7* and the bridge is balanced. The input terminals of the amplifier are shorted and the only signal source used is that within the bridge. The resistance of the amplifier is the indication of the bridge at balance minus the resistor value.

If, for example, the resistor is 10 ohms and in the circuit of *Fig. 7* the bridge balances at 5 ohms, the amplifier internal resistance is $5 - 10 = -5$ ohms. A positive internal amplifier resistance may be measured simply by connecting the bridge to the amplifier without a resistor and accepting the indication of the bridge at balance.

All of the information given in this article has been checked extensively in the laboratory by independent empirical means. The writer has tried to be careful to include no assumptions that cannot be proved without so labelling them.

While it is true, as has been pointed out by Clements, that improved speaker damping demands a price in some decreased bass, the bass heard before was at least partly due to underdamping and consisted in some part of spurious cone excursions which tended to "compensate" for poor speaker-air coupling at low frequencies. The writer has not been able to detect any bass deficiency but has noticed a decided increase in clean and crisp sound. It is felt, in any case, that allowing a speaker to overshoot to compensate for deficient acoustic coupling between cone and air is a poor way of trying to get something for nothing.

It's Positive Feedback

WARNER CLEMENTS

The author, who offers positive feedback as a means to nullify the effect of voice-coil impedance in the article starting on page 34, takes issue with the succeeding article on the same subject. He states his case clearly and concisely.

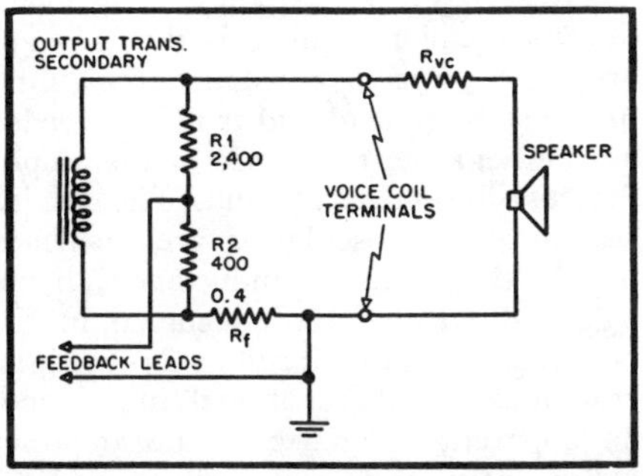

Fig. A. Convenient means for applying positive current feedback along with negative voltage feedback. (Fig. 2 from page 38).

IN THE ARTICLE commencing on page 34, the writer described his circuit for obtaining vastly improved loudspeaker damping. In the Childs' article commencing on page 37, several points of disagreement were brought out in further discussion. While he agreed on the general merits of the circuit, at several points his findings were at variance with those of the writer. Some divergence was to be expected. By comparison with more conventional feedback arrangements which have been in almost universal use for many years, this one is virtually unproved and untested. Unfortunately the issues which Childs raised are vital and fundamental ones. It is the purpose of the present article to try to reconcile the conflicting findings to everybody's satisfaction; and in so doing to cast additional light on the principles that govern the operation of the circuit. It will not be necessary to study the mathematics presented in order to follow the thread of the argument.

Figure 2 of Childs' article is here reproduced as *Fig.* A. The reader can confirm that it is identical with *Fig.* 7 of the writer's original article, except that the optional feedback inductance is not shown. First, let's consider the matter of the role of positive feedback. If R_f in *Fig.* A is made equal to zero, the circuit becomes an ordinary negative feedback circuit. R_f, then, is what makes the difference between an entirely conventional arrangement and the high damping circuit under discussion. As R_f is increased from zero, the gain, distortion, and instability all increase. It has been conventional for many years to call feedback that increases the gain "positive" feedback.

The writer's original article recommended using enough negative feedback to more than cancel out the increased distortion due to the positive feedback. In a sense, then, the article was calling for more negative than positive feedback and would seem to have been clear enough on this point. But to say that "net" feedback is always negative under these circumstances is not quite correct, as will be shown.

The clincher is that negative feedback is not really necessary at all for the operation of the generalized circuit. *Figure* B shows circuit and formulas applicable to a non-feedback amplifier. (Of course, if the amplifier were enclosed in a "little black box" no one would know from the outside whether or not some inverse feedback had been sneaked in; the formulas would still be applicable using measured gain and output resistance.)

Perhaps the most important point at issue is the possibility or impossibility of perfect loudspeaker damping. The question is a fascinating one from a theoretical standpoint. But more than that, it is of immediate practical importance to the designer or constructor who wants to know if, in increasing R_f, he can possibly go *beyond* the setting representing best damping.

"Perfect" Damping

Let it be understood that here and throughout the article when we speak of "perfect" damping we mean "theoretically perfect." In actuality, damping cannot be 100 per cent perfect for the reason that there is no such physically realizable thing as a lumped circuit constant.

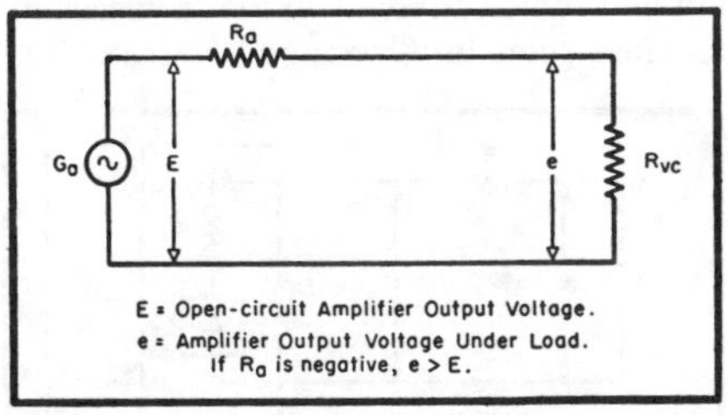

Fig. C. Equivalent circuit used by Childs to establish point of oscillation. (Fig. 3 from page 38.)

Childs bases his proof that perfect damping is impossible on his *Fig.* 3 circuit which is here given as *Fig.* C. He considers the voltage ratios in the circuit and states that under the condition that R_a equals the minus of R_{vc} the circuit will be oscillatory. Here we must observe that just as surely as two and two equals four, so does two minus two equal zero. A positive resistance plus a quantitatively equal negative resistance equals zero resistance and may be so considered in circuit computations. Then we have an ideal generator facing a zero resistance.

But just a few paragraphs earlier Childs has cited an ideal generator fac-

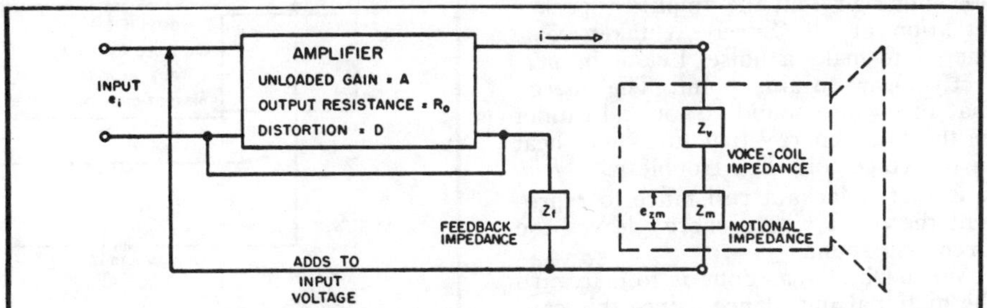

DIAPHRAGM VELOCITY IS PROPORTIONAL TO:

$$e_{zm} = iZ_m = \frac{Ae_iZ_m}{Z_m + Z_v + R_o - Z_f(A-1)} \quad (1)$$

FOR PERFECT DAMPING MAKE:

$$Z_f = \frac{R_o + Z_v}{A-1} \quad (2)$$

AT VALUE OF Z_f GIVEN BY (2),

$$e_{zm} = Ae_i \quad (3)$$

OUTPUT IMPEDANCE LOOKING TOWARD AMPLIFIER FROM SPEAKER TERMINALS:

$$R_o' = R_o - Z_f(A-1) \quad (4)$$

AT VALUE OF Z_f GIVEN BY (2),

$$R_o' = -Z_v \quad (5)$$

DISTORTION OR GAIN-INSTABILITY:

$$D' = D\frac{R_o + Z_m + Z_v + Z_f}{Z_m + Z_v + R_o - Z_f(A-1)} \quad (6)$$

AT VALUE OF Z_f GIVEN BY (2),

$$D' = D\frac{R_o + Z_m + Z_v + Z_f}{Z_m} \quad (7)$$

Fig. B. Arrangement showing negative feedback is not essential to speaker damping.

ing a zero resistance as a case of perfect "dynamic braking." Thus he has used the same criterion for oscillation that he used for perfect damping. There's no use inquiring which use is correct, for in a strict sense both are wrong. With regard to damping, it depends on the configuration of the circuit. A series-resonant circuit is perfectly damped when driven through an infinite-resistance (i.e. constant-current) source. A parallel-resonant circuit is perfectly damped when driven from a zero-impedance (constant-voltage) source.[1] The

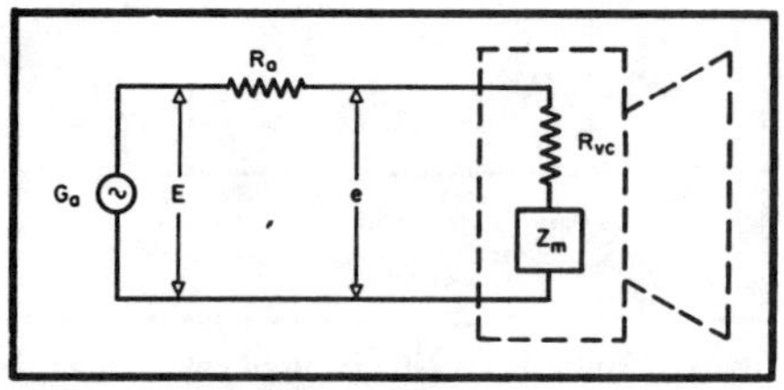

Fig. D. Correct representation of an equivalent generator driving a loudspeaker load.

motional impedance of a loudspeaker is roughly a parallel-resonant circuit on the electrical side.[2]

Oscillation, too, depends on circuit configuration. Oscillation necessarily involves the movement of energy between at least two storage elements. The energy must move back-and-forth, without diminution. But storage, in an electrical circuit means reactive elements (capacitor, inductance). The circuit of *Fig.* C, as drawn, will not oscillate at *any* value of the negative resistance. With R_a equal to the minus of R_{vc} the current is, of course, infinite. Here we have an interesting mathematical anomaly. The only energy coming into the circuit is through R_a and the only energy leaving the circuit is through R_{vc}. At this point one begins to suspect that the diagram does not represent the amplifier-speaker situation at all. Surely it takes *some* energy to make a noise. Let us be perfectly clear on this point. The energy that turns into sound is not and cannot be the same energy that turns into heat in the voice-coil. The trouble with *Fig.* C is that it lacks a resistance to represent the point where energy leaves the circuit as sound.

What Childs has done is to disregard the motional impedance. Since this cannot be justified, let us put it back in as in *Fig.* D and see what happens to his Eqs. 1 and 2:

$$\frac{e}{E} = \frac{Z_m + R_{vc}}{Z_m + R_{vc} + R_a} \qquad (1)$$

$$\frac{e}{E} = \frac{Z_m + R_{vc}}{Z_m} \qquad (2)$$

Where $R_a = -R_{vc}$

One loudspeaker, designed to drive a horn, is rated at 16 ohms and has a voice-coil resistance of only 4 ohms. This means that properly air-loaded and at the rated frequency the motional impedance should be 12 ohms. It is probably actually closer to 8 ohms, but still a hefty item to leave out of one's calculations. Any high-efficiency 16-ohm speaker will have several ohms of motional impedance. Substituting typical values in Eq. (2), one can see that at a condition of perfect damping no excessive negative output regulation is required. It follows that the graph Childs gives as his *Fig.* 4 and the instructions for using it are without validity. Aside from the considerations above one would suspect this anyway, as Childs' method treats all speakers with a given voice-coil resistance alike, even though their effectiveness as energy sinks may vary considerably.

Performance Proofs

After all the talking is through, the proof of the pudding still remains in the eating. Suppose we measure the negative output resistance of an amplifier by the method explained by Childs and illustrated in his *Fig. 7*. Suppose we set that negative resistance at, say, 10 ohms. Then, according to Childs, if we connect the amplifier to a speaker having a voice-coil resistance of exactly 10 ohms, the combination should oscillate. The writer has on hand four different amplifiers that will pass this test *without* oscillation. In fact, with a good speaker, each of them can be set from 10 to 30 per cent *beyond* the critical point (that is to a negative resistance exceeding the voice-coil resistance by that amount) before oscillation starts.

Why didn't Childs make similar findings? Paradoxically enough, it was probably because the Childs amplifier is *too good*. It apparently has excellent bandwidth. Meanwhile the feedback connection of the high-damping circuit is direct-coupled low-impedance and could not be expected to impair that performance. The probable result is that when Childs set his amplifier to a given negative impedance, it maintained that negative impedance out to frequency extremes at which the motional impedance fell to nearly zero. At these frequencies the voice-call resistance is the only appreciable positive resistance remaining in the speaker load and if it is exceeded by the negative resistance of the amplifier, oscillation may result. The needed reactances for oscillation are furnished by: 1) the voice-coil inductance (if not balanced out by an inductance in the positive feedback circuit); 2) vestigial reactances in the motional impedance; 3) apparent reactance of the amplifier output caused by phase-shift; and 4) distributed parameters and electrical lines. These reactances combine in various ways to make one or more series-resonant circuits.

These considerations are no cause for concern within the normal audio range, where the resistive part of the motional impedance remains high. All that is necessary to ensure stability at and beyond the point of perfect damping is to ensure that the performance of the amplifier as a negative impedance does not exceed the performance of the speaker; in other words to restrict the frequency range of the amplifier or positive feedback circuit to within the frequency range of the loudspeaker. This need

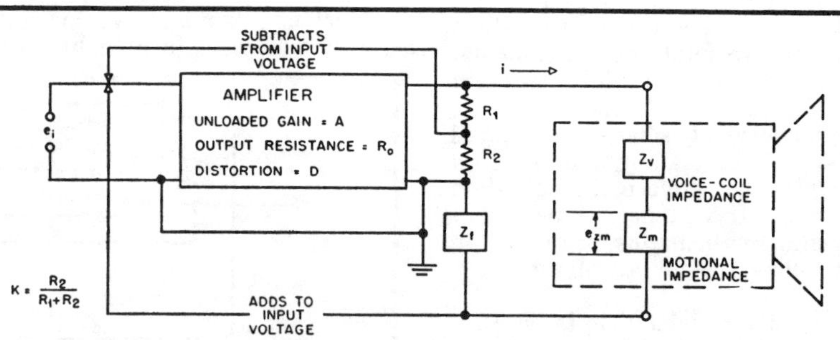

DIAPHRAGM VELOCITY IS PROPORTIONAL TO: $e_{zm} = iZ_m = \frac{Ae_iZ_m}{Z_m(1+AK)+Z_v(1+AK)+R_o-Z_f(A-1-AK)}$ (1)

FOR PERFECT DAMPING MAKE: $Z_f = \frac{Z_v(1+AK)+R_o}{A-1-AK}$ (2)

AT VALUE OF Z_f GIVEN BY (2), $e_{zm} = \frac{Ae_1}{1+AK}$ (3)

OUTPUT IMPEDANCE LOOKING TOWARD AMPLIFIER FROM SPEAKER TERMINALS: $R_o' = \frac{R_o - Z_f(A-1-AK)}{1+AK}$ (4)

AT VALUE OF Z_f GIVEN BY (2), $R_o' = -Z_v$ (5)

DISTORTION OR GAIN INSTABILITY: $D' = \frac{Z_m+Z_v+R_o+Z_f}{Z_m(1+AK)+Z_v(1+AK)+R_o-Z_f(A-1-AK)}$ (6)

AT VALUE OF Z_f GIVEN BY (2), $D' = \frac{D}{1+AK} \cdot \frac{Z_m+Z_v+R_o+Z_f}{Z_m}$ (7)

Fig. E. Combined positive and negative feedback. Formulas given here are applicable to Fig. A without change. Two other variations of the basic circuit are possible and have slightly different circuit equations.

---

[1] Vannevar Bush, *Operational Circuit Analysis*. New York: Wiley, 4th. ed., Chap. IV.

[2] Keith Henney, *Radio Engineering Handbook*. New York: McGraw-Hill, 3rd. ed., p. 906.

cause no raised eyebrows. It is analogous to what many good designers do when they retrict the bandwidth of an amplifier to within that of the output transformer to prevent the negative feedback from becoming positive feedback at frequency extremes.[3]

The equations of *Fig.* E are very instructive. Eq. (3) shows that at the setting for perfect damping the voice-coil velocity will be exactly proportional to the input voltage. This will hold regardless of non-linearities in the suspension system, provided the flux density is constant throughout the gap. As far as the voltage across the motional impedance is concerned, this is a "no-error" condition and, as Childs says, no error correction can be had unless an error exists. But what Childs has completely overlooked is the fact that while the *voltage* may have no error, the *current* may have plenty, which is registered as a drop across Z_f and fed back. This can be shown by dividing Eq. (3) by Z_m to obtain the current and by using for Z_m its complex parallel-component values.

The writer is indebted to Childs for pointing out that a square wave across the speaker terminals does not represent the setting for perfect damping. As Childs correctly states, it is impossible to deduce anything from the amplitude of the transient wiggles unless the negative impedance of the amplifier output is known. Nor, as has been shown here, can it be assumed that the edge of oscillation represents the ultimate in damping. Circuit users who have no facilities for measuring a.c. resistance will just have to adjust for best sound.

This, in view of the bass response situation, may be the best method anyway. As explained in the original article, bass loss with a direct-radiator speaker may run as high as 6 db per octave below midband,[4] which cannot be ignored. Each listener will have to compromise to his own taste between clean lows and missing lows.

In conclusion, the writer wishes to acknowledge a very special debt to Æ whose forward-looking policies have kept its readers in the forefront of the rapidly-advancing audio art.

[3] L. B. Arguımbau, *Vacuum-Tube Circuits.* New York: Wiley, Chap. VIII.

[4] H. F. Olson, *Elements of Acoustical Engineering.* New York: Van Nostrand, 2nd. ed., p. 126.

Improved Cathode Bias Circuit Affording Fixed Bias

JOHN A. MULVEY

OFTEN IN THE INTERESTS of economy, simplicity, or dependability—or all three—the designer of an audio output stage will decide to sacrifice the claimed advantages of fixed bias and use cathode bias. The conventional fixed-bias arrangement usually requires a separate transformer, or a bias-tapped power transformer, and a separate rectifier with its associated filtering network, besides the need for fusing the plate circuit of the output tube to protect it in the event of bias failure. Most fixed bias circuits take time to develop a bias voltage after a set is turned on, and thereby are apt to allow a certain period of improper operation to the output tubes. Though this interval of time may be small, it becomes significant when it recurs repeatedly as the set is turned on and off, as any amplifier usually is. With some tubes—the 6AS7G, for instance—the manufacturers specifically do *not* recommend fixed bias operation, since even a short period of high current drain from the cathodes before they are fully heated will seriously harm the tube. These considerations sway most designers to ignore fixed bias.

The accompanying diagram shows a simple inexpensive method used by the author for achieving fixed bias with what is basically a cathode bias circuit. The idea is to return the grid circuits of the output tubes through a network built around the usual cathode bias resistor. This can be made to provide a fixed bias voltage between grid return and cathode. By choosing a suitable resistance value for the cathode resistor such that the voltage developed across it due to the total cathode current will be considerably more than required for bias voltage, a circuit such as that shown can be wired in parallel to it that will give almost perfect regulation to a voltage developed between some midpoint in the branch and the cathode. A filter capacitor could be substituted for the OC3/VR105 shown. However, this would delay somewhat the bias voltage developed for the grid. For such a substitution, a 6800-ohm resistor should be used instead of the 2700-ohm resistor shown. The use of a capacitor in this manner does provide one advantage in addition to a small saving in time of assembly and in expense, and it does provide a favorable automatic bias change with changing line voltage. The potentiometer shown is for balancing plate currents in the two tubes or in the two sections of such a tube as the 6AS7G.

Modifications suggest themselves for other tubes which require less bias voltage. The OA3/VR75 will provide an economy of required plate voltage.

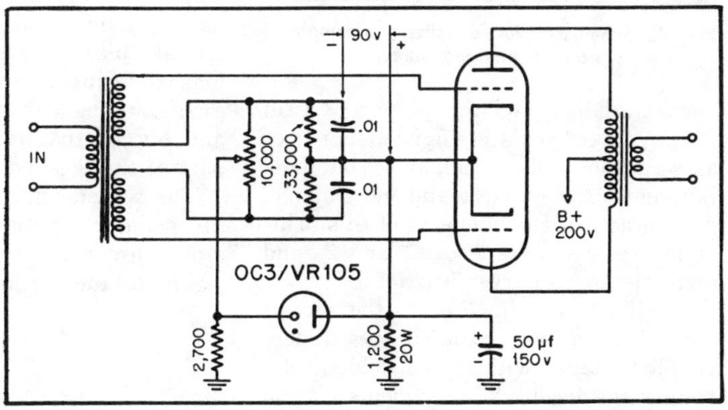

Fig. 1. Output stage arrangement to provide fixed bias from a cathode-bias circuit.

Equalized Pre-Amplifier Using Single-Stage Feedback

LAWRENCE FLEMING

Single-tube circuit which offers the advantages of simplicity and easily varied turnover frequency.

BASS EQUALIZATION for magnetic pickups is often accomplished by simple resistance-capacitance networks. The *RC* network is most commonly inserted between two triode stages in the preamplifier. Another widely-used arrangement employs the *RC* network in the feedback link between the plate of the second stage and the cathode of the first. Both arrangements require thorough filtering of the plate supply to the two stages, and the second type of circuit tends to be "hummy" because the first cathode is off ground.

A third possible location for the equalizing *RC* network is in a feedback

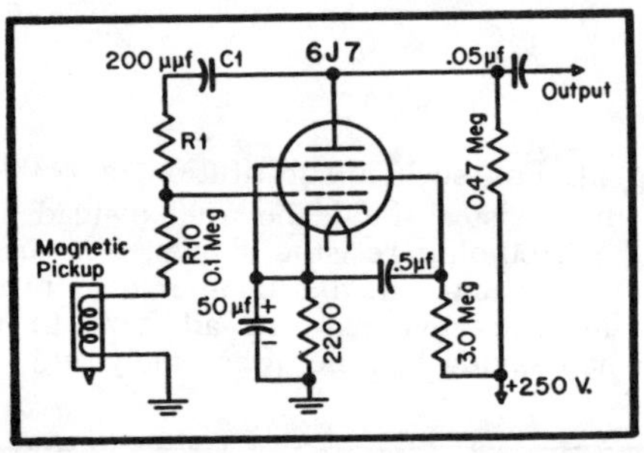

Fig. 1. Schematic for equalized preamplifier using a pentode stage.

connection between the plate and grid of one stage.[1] This configuration is not as well-known as it should be, but it is inherently the quietest and most stable. Two practical embodiments of this principle are shown in *Figs.* 1 and 2, and their measured curves in *Fig.* 3.

As illustrated in *Fig.* 1, the circuit comprises a conventional resistance-coupled stage with three added elements: the feedback components C_1 and R_1 and an isolating resistor R_{10}. Because freedom from hum is particularly sought, it is essential that the cathode be grounded or well bypassed to ground and that one side of the phonograph pickup be grounded. Hence, the feedback must be introduced in parallel with the grid circuit. This makes the input impedance low, as is the case with the phase inverter tube in the well-known "floating paraphase" circuit. The signal therefore is fed to the grid through an

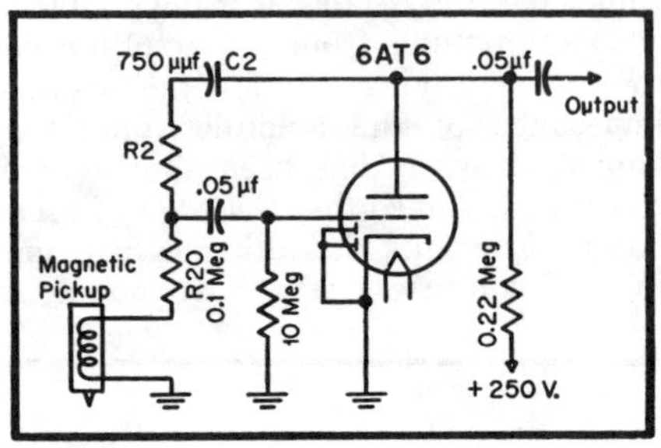

Fig. 2. High-mu triode circuit employing the same principles as that of Fig. 1.

isolating resistor R_{10}, and it is required that the impedance of the pickup be low compared to this isolating resistance at all frequencies in the desired range. For the average type of magnetic pickup having an inductance of the order of 0.2 henries, this requirement is easily met. The circuit is not practicable with crystal pickups, nor is the low-frequency equalization necessary. The circuit of *Fig.* 1 employs a 6J7 tube for minimum microphonics and hum.

High-Mu Triode Circuit

Figure 2 illustrates a practical circuit for a 6AT6 or any other triode having a mu of about 70. "Contact potential" bias is used here, but cathode bias could just as well be employed. The grid coupling capacitor has to be good and big so that it will not counteract part of the bass boost.

The screen and cathode bypassing in the pentode circuit of *Fig.* 1 has to be ample for the same reason. Referring again to the feedback components, the gain at high frequencies—where the reactance of capacitor C_1 is low—is the gain of a simple feedback amplifier having the fraction $R_{10}/(R_1 + R_{10})$ of the output fed back to the input. The crossover frequency is approximately at the point the reactance of C_1 is equal to resistance R_1. Below this frequency there is less and less feedback, and finally the gain approaches asymptotically the gain without feedback, as in other *RC* equalizing circuits.

The measured curves of *Fig.* 3 show the value of the feedback resistance for crossover frequencies of 300 and 800 cps, for both the circuits described.

It will be noted that the asymptotic low-frequency gain is made equal for each of the two crossover frequencies in the curves of *Fig.* 3. This results in the gains being unequal at the higher frequencies. It is thought that this scheme is advisable, because below the crossover frequency it allows the gain to increase continuously down to about 40 cps. If the gains are made equal at high frequencies, the curve for the high crossover frequency will start to level off sooner on the low end than the curve for a lower crossover. This effect can be reduced by sacrificing more gain.

If, however, it is desired to vary the crossover point while holding the gain constant at high frequencies, this can readily be done by leaving R_1 constant (*Fig.* 1) and varying the capacitor C_1. To change the turnover from 300 cps to, say, 600 cps, merely reduce the capacitance C_1 from 200 $\mu\mu$f to 100 $\mu\mu$f.

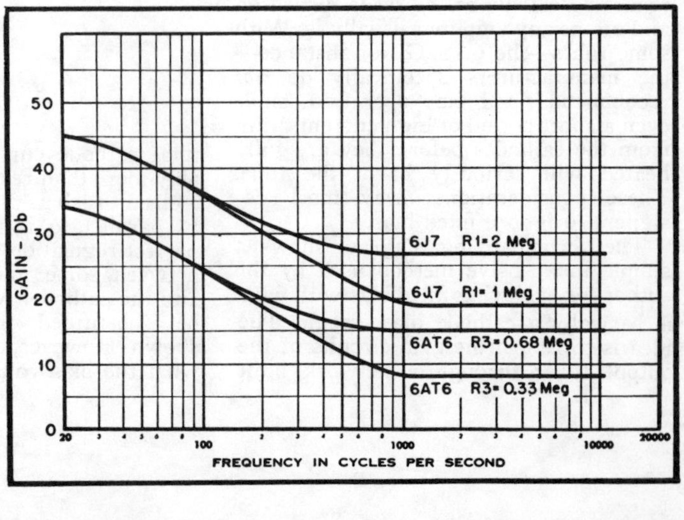

Fig. 3. Response curves for circuits of Figs. 1 and 2.

[1] J. Ellis, "Bass Compensation," *Wireless World,* Sept. 1947.

Two Preamplifiers for Magnetic Pickups

GEORGE ELLIS JONES, JR.

For those who are never satisfied with existing preamplifiers, these should be attractive. One provides variation in specific turnover points with several roll-off characteristics, while the other provides a continuously variable turnover point.

Proper reproduction of sound from phonograph records requires, in general, two types of compensation since both low-frequency attenuation and high-frequency pre-emphasis are introduced before or during the cutting of a "master." Since the recording characteristics vary among different manufacturers and even among different recordings offered under the same label, it is desirable that the playback system afford adjustable compensations. This need exists no matter what kind of pickup is used and obtains for all three recording speeds.

Tone-control systems, including those which provide both boost and attenuation independently for bass and treble, are not particularly suited to this purpose.

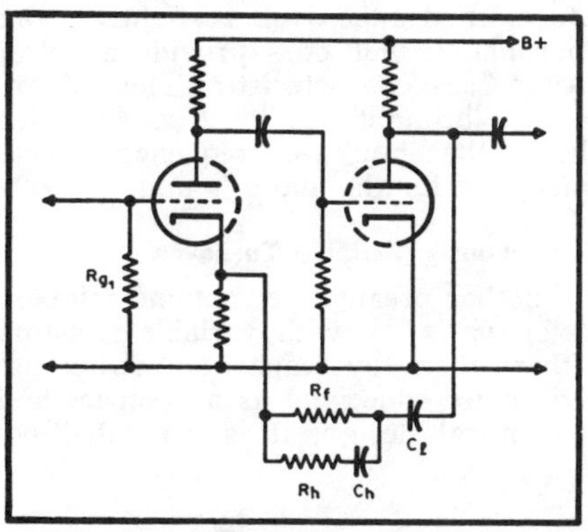

Fig. 1. Basic circuit of first type of preamplifier with adjustable turnover points and roll-off characteristics.

It may be observed that these systems affect primarily the rate of gain change with frequency and to a lesser extent the range over which the gain remains constant. The author prefers to use a preamplifier with adjustable compensation feeding through a volume control to a flat main amplifier, rather than a preamplifier with some fixed average compensation followed by a tone control system and then the main amplifier. No additional tubes are needed in an adjustable preamplifier as compared to one with only fixed compensation. On the other hand, a tone control system providing both boost and cut may be expected to include vacuum tubes so that additional noise, distortion, and varying phase shift will be introduced.

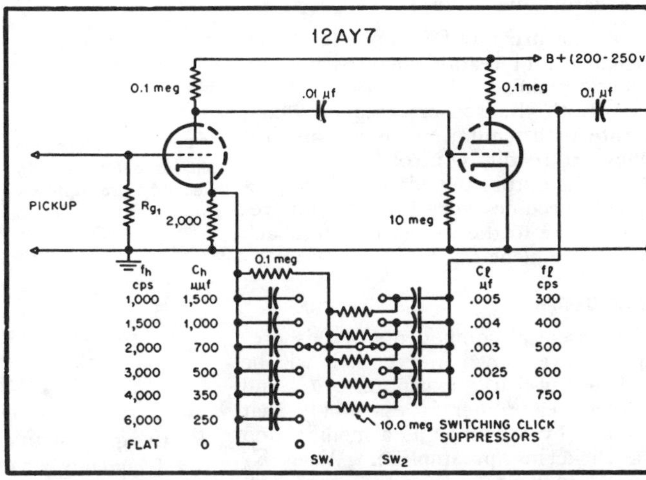

Fig. 2. Complete schematic of preamplifier shown in basic form in Fig. 1.

In the past, three methods have been presented for obtaining desired high-frequency roll off:

(a) Loading the pickup with an adjustable resistive load so that the inherent inductive reactance of the pickup will cause a roll-off of 6 db per octave. The frequency above which this roll-off obtains is influenced not only by the value of the load but also by the magnitude of the pickup inductance, which varies significantly among manufacturers.

(b) Placing a high impedance network between the pickup and the preamplifier input.[1] The operation of this net is based on the assumption that over the frequency range of interest the pickup's inductive reactance will be much

[1] St. George and Drisco, "Versatile phonograph preamplifier," Audio Engineering, March, 1949, p. 14.

Fig. 3. Internal construction of first type of preamplifier.

smaller than the network impedance. This will be only approximately true; and in addition, the relatively high impedance of the network is a possible source of hum pickup.

(c) Placing a high-frequency attenuator after a preamplifier with a flat high-frequency response.[2] With this arrangement the output voltage at high frequencies from the preamplifier may be of sufficient amplitude to be distorted. Furthermore, the output impedance of the high-frequency attenuator will be much greater than that of a well designed preamplifier.

The preamplifier shown disassembled in *Fig.* 3 obtains the desired high-frequency roll-off by adjustment of the feedback loop. This unique method for removing high-frequency pre-emphasis has several advantages:

(a) The preamplifier input is of low impedance and readily adapted for any magnetic pickup.

(b) Both the roll-off frequency and the rate of attenutation above this frequency are readily controlled.

(c) The output impedance of the preamplifier becomes very low at high frequencies due to the increase in feedback at these frequencies.

Circuit Design

The essential components of this preamplifier are shown in *Fig.* 1. It will be noted that with the exception of C_h and R_h, the preamplifier is essentially that presented by Sterling[3] as a modification of a Pickering preamplifier. When R_h is equal to zero, a high-frequency roll off of 6 db per octave occurs; the gain having fallen 3 db when the frequency has risen to that given by the relationship:

$$f_h = \frac{1}{2\pi R_f C_h}$$

In the event that R_h is one fourth of R_f the roll-off rate is about 3 db per octave over a range of frequencies from:

$$f = \frac{1}{2\pi(R_h + R_f)C_h} \text{ to: } f = \frac{1}{2\pi R_h C_h}$$

Because of the loading effect of the feedback loop on the second triode, the gain falls below unity at high frequencies before the phase shift becomes 180 deg. Oscillation does not occur, nor does square-wave excitation reveal any ringing. The high-frequency roll-off is selected by switching one of several available R_h and C_h combinations across R_1, so as to complement the recording pre-emphasis.

For low-frequency compensation, a rising characteristic of 6 db per octave is obtained by the combination of C_l and R_f, the gain having increased 3 db when the frequency has fallen to that given by the relationship:

$$f_l = \frac{1}{2\pi C_l R_f}$$

This technique for obtaining low-frequency compensation is quite common, but does have the disadvantage of reduction in feedback and increase in output impedance with decreasing frequency. Furthermore, if the turnover frequency, f_l, is increased much above 500 cps, the rising characteristic does not extend to the very lowest frequencies. Low-frequency turnover is established by switching an appropriate capacitor value for C_l into the feedback loop.

Figure 2 is a complete schematic for this preamplifier as constructed by the author. The chassis was fabricated from sheet aluminum and when assembled places the control shafts on the side opposite the tubes and cable connections. This is a convenient arrangement, since the preamplifier can thus be mounted beneath the phonograph motor board or against the rear of the control panel. The schematic does not indicate the decoupling resistor and electrolytic filter capacitor since their values will depend on the B-supply voltage available. For the tube, the author has used with equal success a 12AY7 or a selected 12AT7. The second tube is but slightly more microphonic but does provide a better rising bass characteristic. This advantage is brought out by *Fig.* 4, which shows the gain vs. frequency curves calculated for the author's unit.

Continuously Variable Turnover

Another preamplifier obtains compensation variation with variable resistors rather than by capacitor switching. Originally conceived as a compact and economical design, it is presented be-

[2] H. T. Sterling, "Simplified preamplifier design," AUDIO ENGINEERING, November, 1949, p. 16.

[3] H. T. Sterling, *Op. cit.*

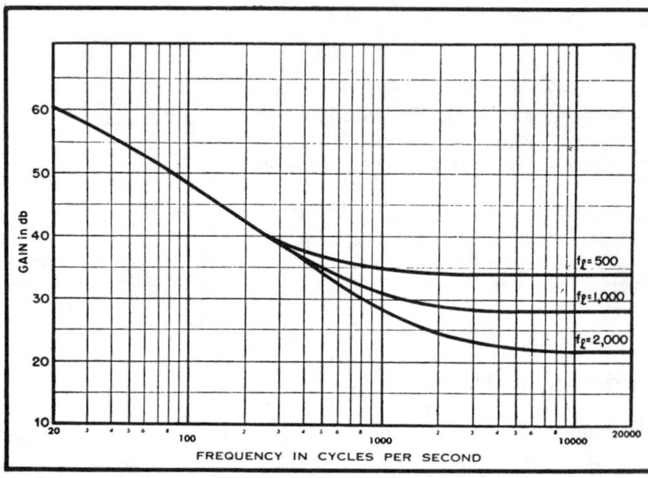

Fig. 6. Response curves obtained with second preamplifier.

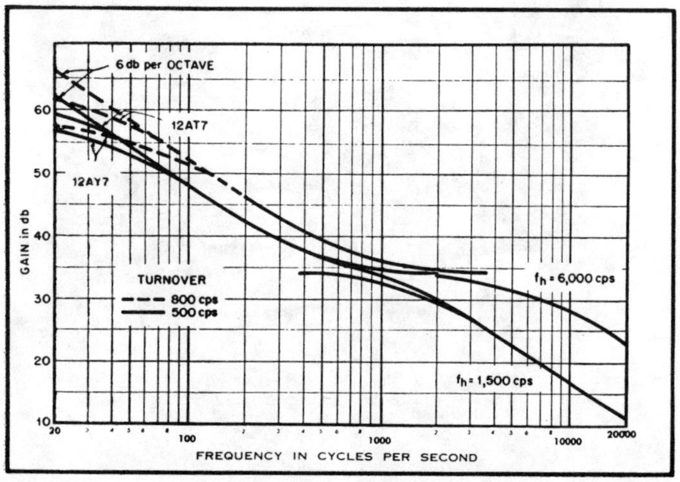

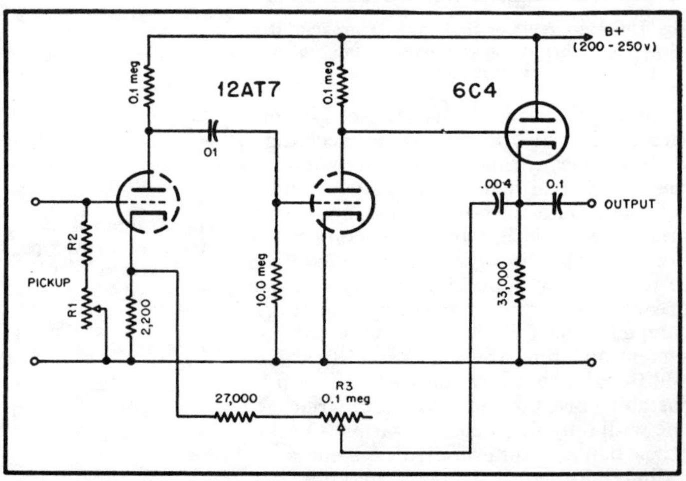

Fig. 4 (left). Typical response curves obtained with first preamplifier. Fig. 5 (right). Schematic of second preamplifier, with continuously variable turnover point.

A Simple Preamplifier and Tone-Control Unit

DAVID H. O'BRIEN

Details of a unit which can be constructed with a minimum of time and material and which was designed for use with the Musician's Amplifier.

The famous English "Williamson" amplifier circuit, in its Americanized version called the "Musician's Amplifier," has become very popular in the United States during the last two years, and justifiably so. It is capable of ten watts output; it has extremely low distortion; and it has a low generator impedance, of the order of 0.6 ohms at the 16-ohm tap. These features combine to make an excellent power amplifier for driving a high-quality loudspeaker.

However, there has been one important link missing for this amplifier. As yet no good preamplifier and tone control circuit has appeared which would complement the excellent characteristics of the power amplifier, and which was designed especially for it.

This article deals with a tone control and preamplifier unit which uses only three tubes, or even two, as will be pointed out later, and does not contribute enough measureable distortion to mar the excellent characteristics of the power amplifier. This unit has been used and tested with four different versions of the Musician's Amplifier and has been found to work equally well with all of them.

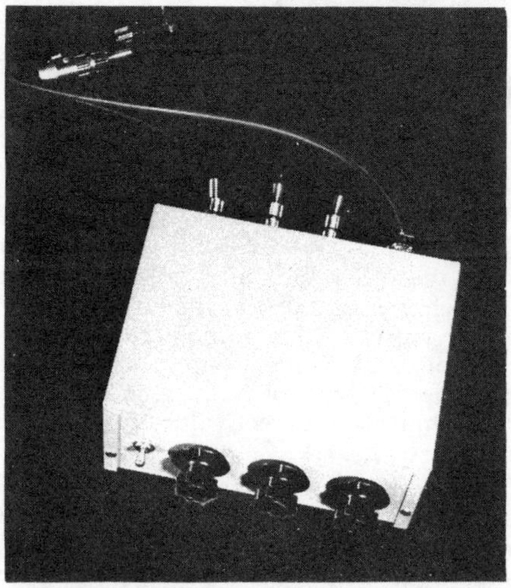

Fig. 1. External view of the author's embodiment of his tone control and preamplifier unit.

Tone Control Section

The tone control section consists basically of a two-stage cascaded triode amplifier, as shown in *Fig.* 3. the output stage consisting of a 12AX7 with both sections paralleled. These two cascaded stages have negative feedback applied over them, and by means of two frequency-selective networks in this feedback loop, it is possible to obtain bass and treble boosts of the order of 30 db at 35 and 20,000 cps without boosting the mid-range (1000 cps) over 1½ db.

The bass section consists of a selective T network with R_{17} across it to vary the effectiveness of the network. The treble section consists of a 0.25-meg. control, R_{21}, with its movable tap grounded. When the control is turned fully counterclockwise, C_{11} is then effective in a T network from the center point of the two 27,000-ohm resistors, R_{19} and R_{20}, to ground. This is actually a low-pass filter which, being in the feedback loop, prevents high frequencies from passing through and thus allows

Two Preamplifiers for Magnetic Pickups (Cont'd.)

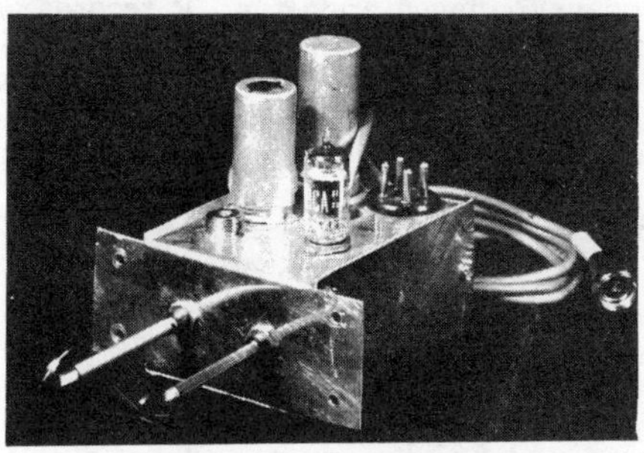

Fig. 7. External view of second preamplifier.

cause of several other advantages. First, the degrees of compensation for bass and treble correction are continuously variable. Second, bass amplification rises at a 6 db per octave rate to very low frequencies, no matter what turnover point is selected.

A high-frequency roll-off of 6 db per octave is obtained by loading the pickup with R_1 and R_2 in *Fig.* 5. The variable resistor should have a log (audio volume control) taper with maximum resistance in full clockwise position. R_1 should be the maximum resistance recommended by the pickup manufacturer for flat response, and R_2 should be one tenth of R_1. The following values will work well with three magnetic pickups tested:

Pickup	R_1	R_2
Audak	50,000	4700
G. E.	15,000	1500
Pickering	25,000	2700

The continuously variable low-frequency turnover is established by setting R_3, which is a 0.1 meg log taper control, wired for maximum resistance in full clockwise position, in which position the low-frequency turnover is about 400 cps. This method of determining the low-frequency turnover does not actually affect the gain below the turnover point. Rather, the gain above turnover, which remains constant for any given control setting, is decreased, and the turnover frequency is increased. *Figure* 6 shows this effect much better than words describe it. *Figures* 7 and 8 show this preamplifier assembled and disassembled. As with the previous unit, the chassis was built from sheet aluminum, although ready made boxes suitable for such work are available. This second preamplifier does have one distinct disadvantage in that a change in the low-frequency turnover setting should be accompanied by an adjustment of the amplifier volume control if the level above the turnover is to be maintained constant. The cathode follower was inserted to provide a low source impedance to the feedback net so that it would not unduly load the amplifier when R_3 was in a counterclockwise position.

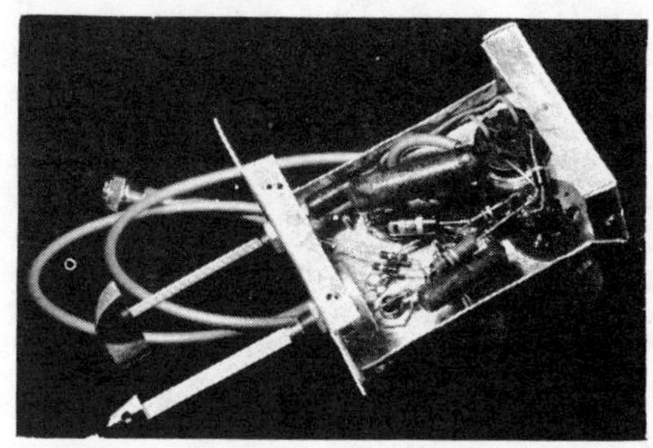

Fig. 8. Chassis wiring of second type of preamplifier.

The purpose behind both of these designs was to provide a maximum of control action with a minimum of nonlinear elements. Either of these preamplifiers will provide quiet, compact, and effective compensated gain between a magnetic phonograph pickup and an amplifier.

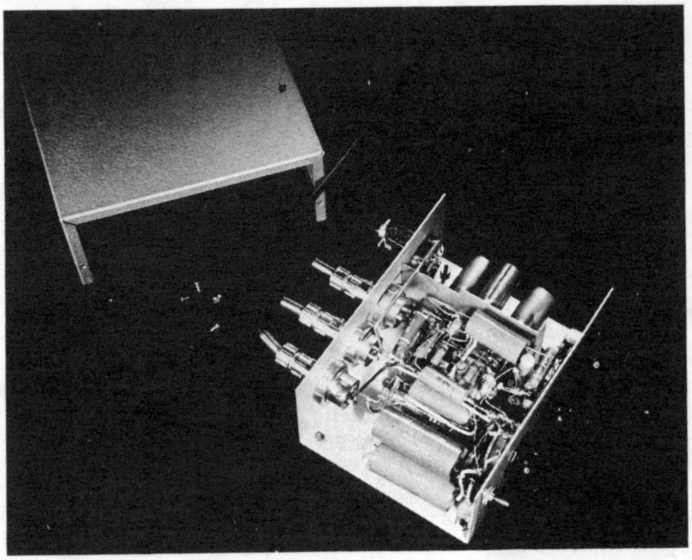

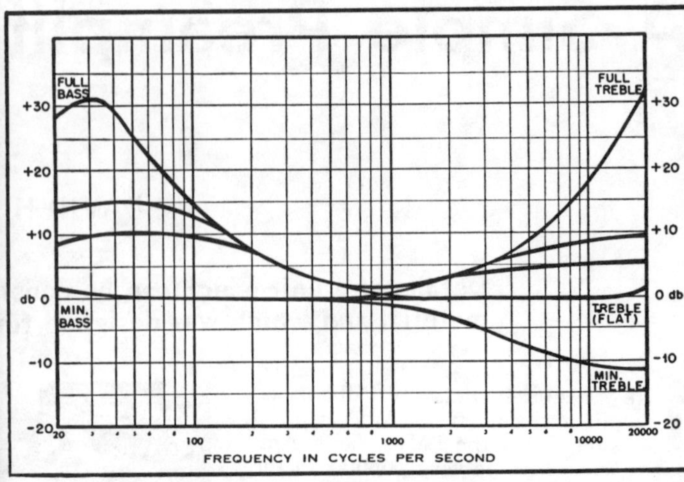

Fig. 2 (left). Internal view of the unit shown in Fig. 1. Fig. 4 (right). Response curves of the tone-control section.

a high-frequency boost. When the control moves fully clockwise, the .001 µf capacitor, C_{12}, bypasses the highs to ground.

The treble control originally used was a standard audio taper which was reversed in its connections relative to common practice in order to give smooth action on both boost and cut and also to have the mid position (50 per cent rotation) give flat response.

A reverse audio taper is preferred, since it will give the conventional clockwise boost and counterclockwise cut.

Resistor R_9 in series with the capacitor C_5 across the cathode resistor of V_2, is a high-boost network—which exactly compensates for the high-frequency droop caused by the capacitance of a 6-feet length of good rubber-insulated shielded cable which connects to the power amplifier. If the tone control unit is to be used on the same chassis as the power amplifier, and not as a remote unit, this network should be omitted.

Capacitor C_6 in series with the feedback loop prevents direct-current flow from the cathode of V_2 through the input circuit of the main amplifier to ground, and avoids noise when the bass control is turned. R_{11} is also in series with C_6, and isolates the T network from the cathode when the bass control is set for flat response.

Resistor R_{16} in series with the output of the paralleled 12AX7, serves to isolate the feedback network from the high-frequency-cut capacitor in the first stage of the power amplifier. If this is omitted, the feedback will automatically compensate for the effect of the capacitor, resulting in no high-frequency cut and an increase in distortion.

The last stage of this tone control unit uses a 12AX7 with both sections in parallel because it was found that a 50 per cent reduction in distortion was realized over a single section. On a harmonic distortion test this unit with the controls set for flat response added only 0.1 per cent distortion to the overall amplifier at 10 watts output. If the increase in distortion could be tolerated, one could omit the 6AQ6 and replace it with the other half of the 12AX7. The only changes needed in the circuit would be to increase the output section plate resistor to 0.1 meg. and increase the cathode resistor of the same stage to 2000 ohms, the correct values for a single section of a 12AX7.

The reluctance cartridge preamplifier is of the feedback type so common that it needs no description. The resistor R_1 should be selected in accordance with the recommendation of the manufacturer of the pickup to be employed. The phono-radio switch is of the conventional DPDT toggle type, with one half used to ground whichever channel is not being used, thus preventing crosstalk.

Construction

Construction practice must be left up to the individual as this unit may be built up in many physical forms. *Figures* 1 and 2 show how the author arranged the components in a small aluminum box. The power cable was separated from the output cable. Also all leads should be kept as short as possible, especially to the tone controls. The usual practice of grounding the circuit to the chassis at only one point near the high-gain input shoud be followed. One-watt resistors are recommended, as there is less chance of damaging them during soldering, and in high-gain circuits of this type, low noise is essential.

No bass-cut circuit was included in this control unit as it would require a much more complicated circuit, and it also has been the author's experience that a bass cut is seldom, if ever, used in high-quality reproduction.

One note of special interest is that when the treble control is turned to full cut, the response from a Pickering, G. E., or Audax cartridge will reproduce the Cook Series 10 Test Record LP side within one db from 100 to 10,000 cps. This gives an accurate high-frequency compensation for LP records as well as approximating the new A.E.S. playback standard. The accompanying curve shows the typical response, measured at the voice coil, of a Pickering

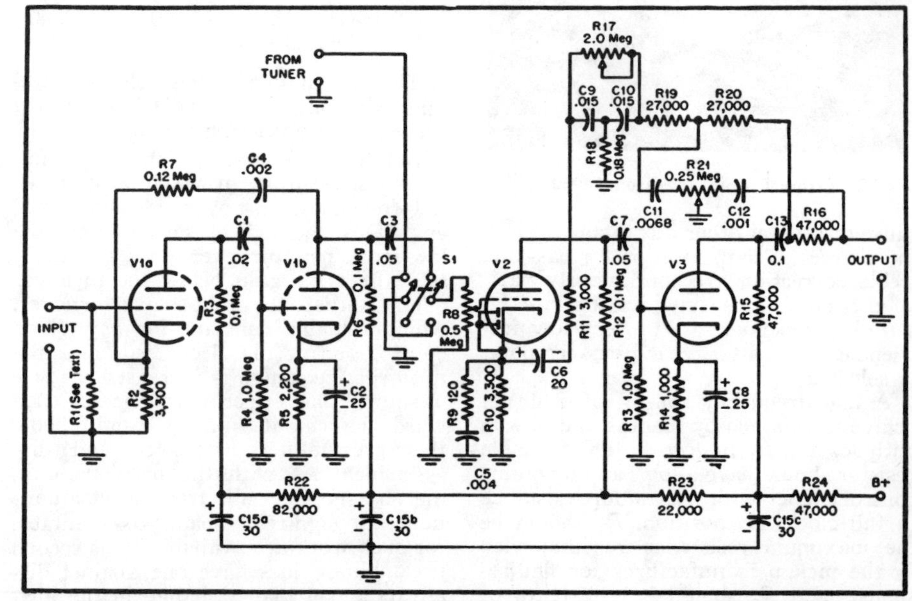

Fig. 3. Schematic of the preamplifier and tone-control which can be used with any suitable power amplifier.

For The Discriminating Listener: An Audio Input System

WAYNE B. DENNY

Preamplifier with flexible control provides facilities for practically any type of input, and is designed to feed a wide variety of equipment.

MANY EXCELLENT AUDIO AMPLIFIERS are currently available, together with improved accessories for the high-quality reproduction of recorded and radio programs. Much of this equipment is ideally suited to custom installations, but there are many situations where reasonably priced commercially built apparatus does not provide sufficient flexibility to permit proper matching of units and sufficient control of both the frequency and dynamic ranges of the signal. The system to be described was engineered in response to requests from serious listeners who desired something better than the usual commercial reproducer but who could not afford so-called "professional" equipment. Nor would these listeners, many of them musicians, tolerate equipment whose operation requires considerable technical skill. As finally evolved, the audio control system to be described meets this need by providing equalized high gain for variable reluctance pickups; controllable low-frequency attenuation or boost; controllable high-frequency attenuation or boost; volume expansion; extremely low noise level; and an output impedance sufficiently low to permit almost any number of power amplifiers to be bridged across the output without loss of gain or frequency response. This is achieved with only six manual controls, making the operation simple enough for the most inexperienced personnel. Moreover, the cost is moderate.

The diagrams of *Figs.* 1 and 2 show the circuits of the audio unit and power supplies. One of the power supplies provides filtered d.c. heater power for the first six tubes in the low-level section of the audio unit. Using one of the bias supply transformers now available on the surplus market, this unit provides about 72 volts at 150 ma for the heaters connected in series. In order to keep heater-cathode potentials at a minimum, the heater string is grounded at its electrical center, i.e., between V_2 and V_1. This places essentially zero heater-cathode bias on the two tubes operating at the lowest signal levels. There is no reason why all tubes could not be tied into the heater string, provided the 6.3 volt types are replaced by suitable 150 ma equivalents and sufficient additional voltage is available. The other power supply—entirely conventional—provides 250 volts for the plate circuits and 6.3 volts a.c. for the remaining four heaters. Both power supplies occupy the same chassis.

Turning next to the signal circuits it will be seen that the phonograph preamplifier is conventional in every re-

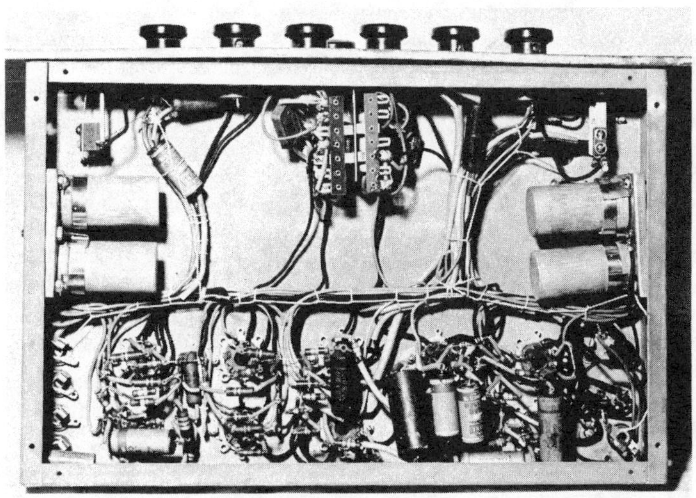

Controls, from left to right, are: input selector, input level, h-f equalizer, l-f equalizer, volume, and expander. Signal circuits are wired point-to-point to reduce stray capacitance; power circuits are cabled.

Simple Preamplifier and Tone Control Unit (Cont'd.)

D-140S cartridge reproducing the 78-r.p.m. side of the Cook Series 10 test record. The curves of the action of the bass and treble controls are also shown.

It must be pointed out that when tone correction on the order of 30 db is available, it should be used carefully since it is easy to exceed power amplifier capabilities at the frequency extremes. However, when this unit is used with care it gives sufficient flexibility.

A 100-ohm hum balancing potentiometer was added to the original Musician's Amplifier circuit across the 6.3 volt winding in order to balance out a small amount of residual hum which originated as heater cathode leakage from the first stage of the magnetic cartridge preamplifier.

PARTS LIST

C_1	.02 µf, 600 v. paper
C_2, C_6, C_8	25 µf, 25 v. electrolytic
C_3, C_7	.05 µf, 600 v. paper
C_4	.002 µf, 600 v. paper
C_5	.004 µf, 600 v. paper
C_9, C_{10}	.015 µf. 600 v. paper
C_{11}	.0068 µf, 600 v. paper
C_{12},	.001 µf, 600 v. paper
C_{13}	0.1 µf, 600 v. paper
C_{15} a, b, c	30-30-30 µf, 450 v. elect.
R_1	See text
R_2, R_{10}, R_{11}	3300 ohms, ½ watt
R_3, R_6, R_{12}	0.1 meg, 1 watt
R_4, R_{13}	1.0 meg, ½ watt
R_5	2200 ohms, ½ watt
R_7	0.12 meg, ½ watt
R_8	0.5 meg, audio taper pot
R_9	120 ohms, ½ watt
R_{14}	1000 ohms, ½ watt
R_{15}	47,000 ohms, 1 watt
R_{16}	47,000 ohms, ½ watt
R_{17}	2.0 meg, audio taper pot
R_{18}	0.18 meg, ½ watt
R_{19}, R_{20}	27,000 ohms, ½ watt
R_{21}	0.25 meg, reverse taper pot
R_{22}	82,000 ohms, 1 watt
R_{23}	22,000 ohms, 1 watt
R_{24}	47.000 ohms, 2 watt
V_1	12AX7
V_2	6AQ6
V_3	12AX7 (both sections in parallel)

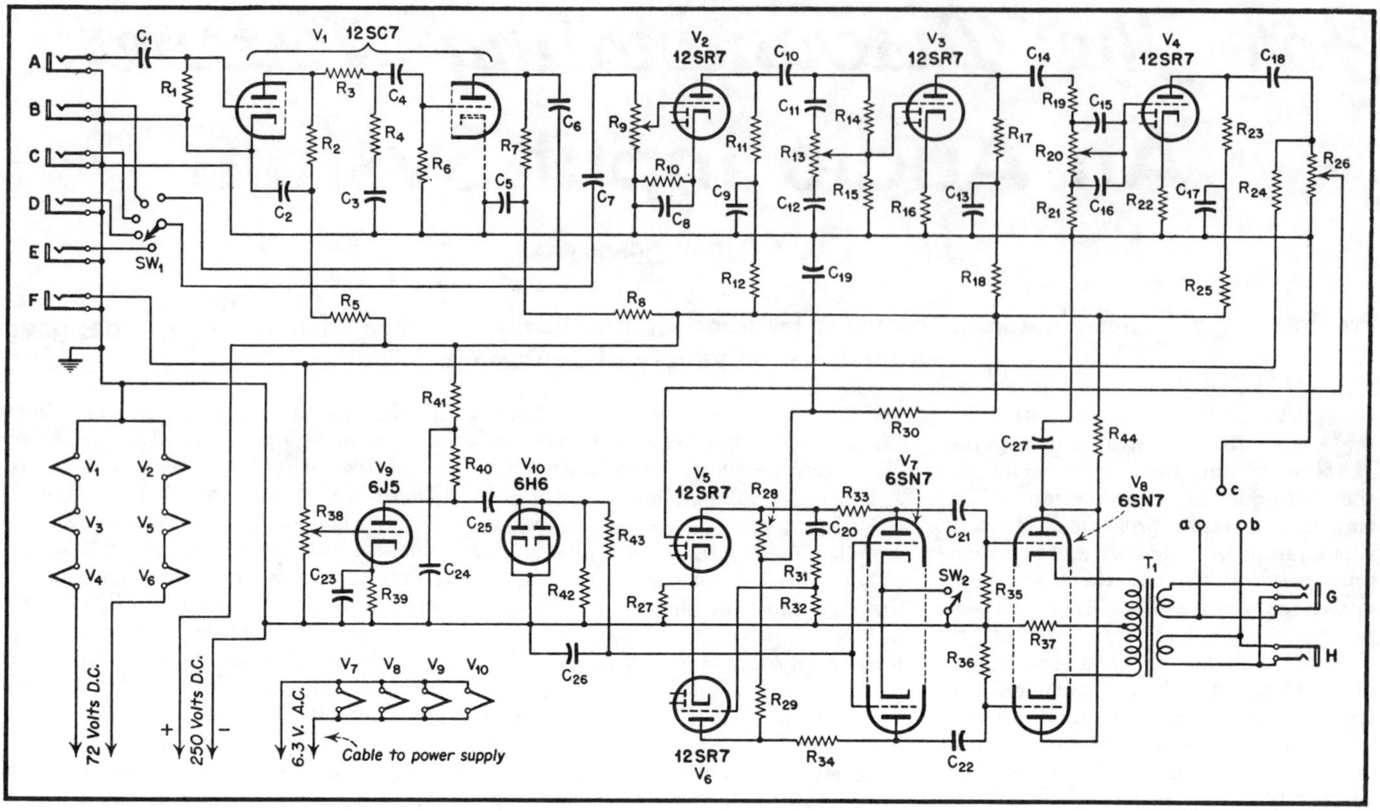

Fig. 1. Schematic of preamplifier unit. Note flexibility of control and provision for volume expansion if desired.

spect, providing fixed bass boost of 3 to 4 decibels per octave. By shorting out C_3 this boost can be eliminated, making the pre-amplifier suitable for use with microphones.

The input selector switch, Sw_1, permits selection of any of five inputs. The additional complication of a mixer system appears to be unnecessary, considering the purpose for which this device was designed. R_9 is used primarily to provide the correct input level for the grid of V_2. The plate of V_2 feeds the high-frequency control circuit consisting of C_{11}, C_{12}, R_{13}, R_{14}, and R_{15}. V_3 compensates for equalizer loss and isolates the high-frequency and low-frequency control circuits. The latter comprises R_{19}, R_{20}, R_{21}, C_{15}, and C_{16}. Both equalizing circuits are adapted from an equalizer developed by Logemann.[1] Each frequency control circuit provides approximately 15 db of boost or attenuation, the frequencies of maximum effect being 50 and 10,000 cps.

The main volume control is R_{26}. This is used in preference to R_9 because manipulation of R_9 disturbs the operation of the volume expansion circuit. V_5 and V_6 comprise a voltage amplifier and phase inverter. Expansion occurs in the plate circuits of these tubes. Push-pull operation of the expander is essential if extraneous control voltages are to be eliminated from the output signal. The expander circuit is a push-pull version of one originally developed fifteen years ago by the late McMurdo Silver. The operation of the expander can be understood as follows: In the plate circuit of V_5, the resistor R_{33} and the plate resistance of the top section of V_7 comprise a voltage divider. The variable arm of this divider, the plate resistance of V_7, is controlled by the potential at its grid. Similarly, R_{34} and the plate resistance of the other section of V_7 comprise the other half of the push-pull circuit. The side amplifier and rectifier for developing the variable control voltage for the expander are entirely conventional, using a 6J5 and a 6H6. The expansion control, R_{38}, has a self-contained switch, Sw_2, which permits V_7 to be effectively removed from the circuit when desired. With SW_2 open, the gain is the same as with maximum expansion during the loudest signals.

Time Constants

The attack time of the expander is determined by the values of R_{43} and C_{26}. Their product (megohms times microfarads) gives the time constant

[1] Logemann, H.,"Simple RC Equalizing Circuits." *The Review of Scientific Instruments*, March 1948.

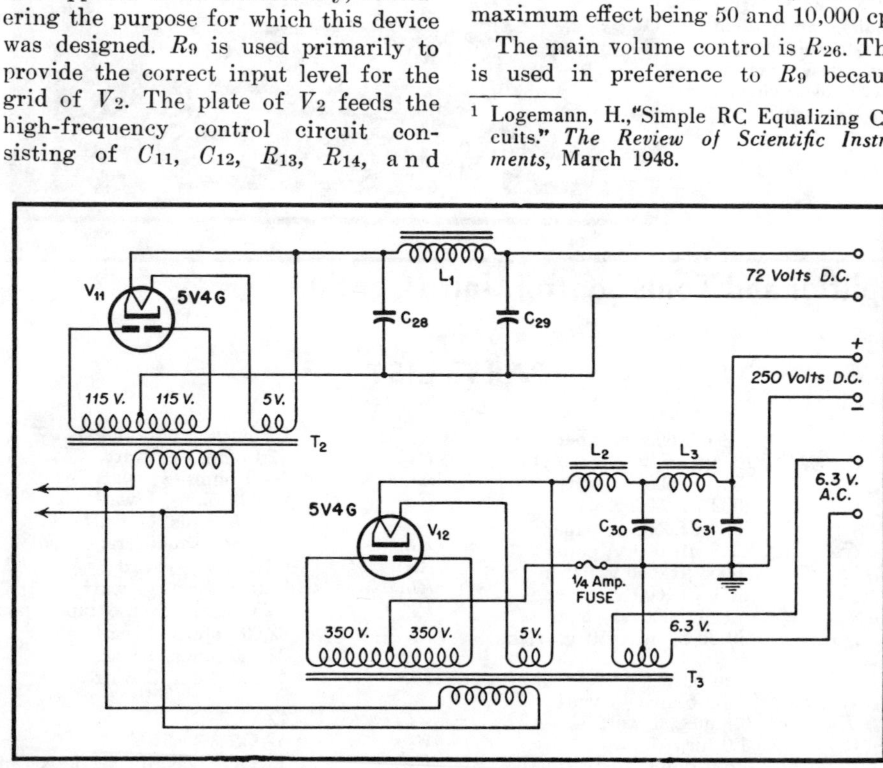

Fig. 2. Schematic of power supply used with preamplifier of Fig. 1.

Complete input system installed in a typical operating cabinet.

in seconds. The release time is determined by the product of C_{26} times the sum of R_{42} plus R_{43}. This permits the release time to be longer than the attack time. Since the signal circuits of the expander are entirely push-pull, the attack time can be made quite low, even though the release time is half a second or longer. Proper attack and release times are a matter of opinion, but in general depend somewhat on the acoustics of the listening room and the damping constants of the loudspeakers. After careful listening, R_{43} was made 0.15 meg, C_{26} was made 0.5 μf, and R_{42} was made 1.0 meg. These values provide a fast attack, together with a decay of slightly over one-half second. However, other values may prove more suitable in other situations or these elements might be made variable.

When properly used, expanders add much to the enjoyment of program material which has previously been compressed. It should not be supposed, however, that all such material requires expansion. Improperly used, expanders can detract considerably from the quality of the signal. The tendency seems to be to use entirely too much expansion. Although the circuit just described is capable of providing about 20 db of expansion, it will be found that a lower value is preferable. From 3 to 10 db is normally sufficient.

Output Circuits

After passing through the expander, the signal is impressed on the grids of V_8, acting as a push-pull cathode follower. This circuit is unusual. First, the cathode follower serves to isolate the varying impedances of the expander from the output transformer. Second, since the gain of the cathode follower is less than unity, it is practically impossible for extraneous control voltages to drive the grids beyond the linear portions of their characteristics. Thus, the changing control potentials at the plates of V_7 cannot cause distortion. Third, the output of the cathode follower presents a low impedance to its load, and this is reflected to the secondary of T_1. Fourth, the inclusion of T_1 causes any control voltages to cancel in the primary without being passed on to the secondary. Fifth, T_1 isolates the output connections from chassis ground, thus eliminating the ground loop and the accompanying hum which so often occur when various amplifiers are connected in cascade without isolating transformers. The outout circuit permits either a balanced or an unbalanced connection. For balanced output, terminals *a* and *b* are strapped together. Both terminals, together with the ground terminal, *c*, are located near the output jacks. Connection is made to jack *G* for balanced output. An undistorted signal of approximately 10 volts rms is available from each half of the transformer secondary. Jack *H* provides unbalanced output.

Mechanical Construction

The entire amplifier unit was built on a steel chassis 11 x 17 x 3 in. with bottom plate. The ten tubes are mounted near the rear edge, permitting additional equipment to be stacked above deck with controls accessible from the front. All low-level signal circuits are completely shielded. An a.c. switch was *not* included on either audio or power chassis; an external master switch is more convenient besides isolating the audio circuits from a possible source of hum. One of the input jacks is located on the front panel, permitting temporary connection of input devices without removal of the amplifier from its cabinet. Similar considerations resulted in placing the low-level output jack *F* on the panel. Experience has shown that locating one input jack and one output jack on the panel is most useful when making comparison tests between tuners, pickups, and power amplifiers. This feature is highly recommended.

The two power supplies are mounted together on another steel chassis 8 x 17 x 2 in. with bottom plate. Aside from considerations of space, separation of signal circuits from the power supply is essential for quiet operation. The amplifier has a connecting cable terminating in an octal plug, the corresponding socket being placed on the power supply chassis. The cable may be any convenient length if sufficiently heavy wire is used, permitting the power supply to be placed out of sight at a remote point.

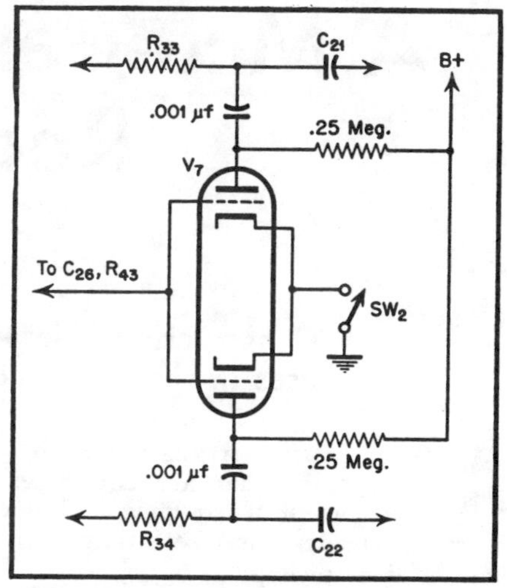

Fig. 3. Modification of V_7 circuit to make the expander function as a noise suppressor.

Wiring follows customary practice for audio units. In no case is any part of the chassis used as part of the grid return circuits, grounding occurring only at one point. This precaution results in low noise and inaudible hum.

Operation

After several months' trial, the input system described has proved itself in every respect. While twelve tubes may appear excessive, each tube has a definite function. Heavy filtering eliminates hum and instability. The hum level of the unit (including the preamplifier) was measured at 65 db below one volt with all gain controls wide open and bass boost at maximum. In use, the signal-to-noise level is considerably better, since the maximum amplification is far greater than nec-

External view of amplifier, showing part of chassis space occupied by a tuner. Design was based on the provision for such additional equipment.

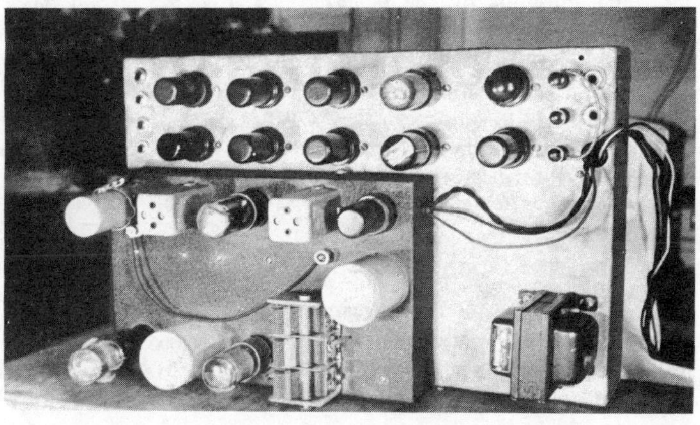

A Mixer and a Preamplifier for the Recording Enthusiast

G. H. FLOYD

Constructional details of two simple but useful items which will find many uses in the experimental work of anyone who works with audio in any form.

A TWO-CHANNEL MIXER and a high-gain preamplifier are two necessary pieces of equipment if one does semi-professional recording work with a "live" pickup.

Two microphones are often needed if a large pickup area is involved. In order to use two microphones properly, a mixer is required. Further, when two pickup points are involved it is more than likely that at least one of the microphones will have to be placed at a point remote from the recorder. This is no great disadvantage if the microphone so placed has a low-impedance output, as the pickup cable can be run for several hundred feet without serious frequency discrimination. On the other hand, if the microphone has a high-impedance output the connecting cable must be kept short unless a preamplifier is used. Such a preamplifier can be designed so that a long connecting cable may be used between it and the recorder, the preamplifier being placed near the microphone.

A mixer is also invaluable for re-recording work. That it, it may be used to mix two different sources, feeding the output to a single recorder. With this setup it is quite simple to make an edited copy of similar recorded material. In addition, oral comments may be mixed with previously recorded material to form a sort of running commentary.

The purpose of this article is to describe how to build a one-tube two-channel mixer and a one-tube high-gain preamplifier. Since the design of these units is such that no audio transformers are required the use of expensive input and output transformers is avoided. The resultant performance is better than that obtained with all but the most expensive audio transformers.

One-Tube Two-Channel Mixer

Figure 2 shows the circuit diagram for the mixer to be described. Potentiometers R_1 and R_2 serve as individual gain controls for the two inputs. Each input feeds a control grid in the dual-triode 12AT7 miniature tube. Resistors R_5

Audio Input System (Cont'd.)

essary. The unit has been used with a variety of power amplifiers and loudspeakers. A power amplifier was designed especially for use with the unit, consisting of two 6J5's in push-pull, transformer coupled to triode-connected 6L6's. The latter, operated as cathode followers, develop over 4 watts at the voice coil with a plate potential of 290 volts and a plate-to-plate load of 3800 ohms. This power is more than sufficient to drive a highly efficient corner-type loudspeaker using a single eight-inch cone coupled to two horns. The first horn is driven from the back of the cone and has a cut-off frequency of 40 cps. The second horn operates from the front of the cone and has a cut-off frequency of 300 cps. The horns increase the loading, raise the efficiency, and reduce intermodulation by restricting the excursions of the cone.

Noise Suppression

In the course of designing this unit, it was found that a simple modification of the expander circuit permits its use as an effective automatic scratch suppressor. As shown in *Fig.* 3, the plates of V_7 were connected to $B+$ through 0.25-meg resistors. Small capacitors were then inserted between the plates of V_7 and resistors R_{33} and R_{34} as shown. The expander then became frequency selective, operating only at the higher frequencies, the lower registers receiving maximum amplification at all times. If desired, a switching arrangement could be incorporated to permit either expansion or noise suppression. It is suggested that a high-pass filter be placed ahead of R_{38} so that the "gate" operates only at high frequencies. This additional complication, plus the fact that with the new low noise recordings the scratch problem is not so important, caused the writer to decide against incorporating the noise suppressor in the final version.

An audio control unit of the type described is not something which should be built by the novice. However, those who are often called upon to provide audio facilities for special purposes, particularly where several input devices are involved, may find this unit the answer to many of their problems. To date, several versions of the unit have been built, and all have operated extremely well. This, and the fact that no "trick" circuits are involved, makes it possible to recommend it to the serious audio enthusiast.

Parts List

C_1, C_4, C_6, C_{10}, C_{14}, C_{18}, C_{20}, C_{21}, C_{22}, C_{25},—0.1 μf., 400-volt paper
C_2, C_5, C_9, C_{13}, C_{17}, C_{19}, C_{24}, C_{27}, C_{30}, C_{31}—30 μf., 450-volt electrolytic
C_3, C_7 —.01 μf., 400-volt paper
C_8, C_{25}—20 μf., 20-volt electrolytic
C_{11} 50 μμf. mica
C_{12} 500 μμf. mica or paper
C_{15} .002 μf., 200 volt paper
C_{16} .02 μf., 200 volt paper
C_{26} See text
C_{28} 4 μf., 200 volt electrolytic
C_{29} 200 μf., 200 volt electrolytic
R_1, R_6—3 meg
R_2, R_5, R_7, R_8—0.1 meg
R_3, R_{33}, R_{34}—.25 meg
R_4—25,000 ohms
R_9, R_{26}—1-meg potentiometer
R_{10}, R_{16}, R_{22}, R_{39}—2000 ohms
R_{11}, R_{17}, R_{23}, R_{28}, R_{29}, R_{32}, R_{40}—50,000 ohms
R_{12}, R_{18}, R_{21}, R_{25}, R_{41},—20,000 ohms
R_{13}, R_{20}—2-meg potentiometer
R_{14}—2 meg
R_{24}, R_{31}, R_{35}, R_{36}—0.5 meg
R_{27}, R_{37}—1000 ohms
R_{30} 10,000 ohms
R_{38} 1-meg potentiometer with switch (Sw_2)
R_{42}, R_{43} See text
(All fixed resistors—1watt)
L_1 8H, 175 ma
L_2, L_3 8H, 100 ma
T_1 PP plates to PP grids, 1:1 ratio, with split secondary
T_2 115-0-115 volts at 175 ma, 5 volts at 3 amperes
T_3 350-0-350 volts at 75 ma, 5 volts at 3 amperes
V_1 12SC7
V_2 V_3, V_4, V_5, V_6 12SR7
(Note: 12J5GT's can be substituted if correct socket changes are made.)
V_7, V_8 6SN7
V_9 6J5
V_{10} 6H6
V_{11}, V_{12} 5V4G

and R_6, together with capacitors C_4 and C_5, form a frequency-compensated attenuator which will be discussed in more detail later. Two outputs are available, one being 20 db down from full output.[1]

The power supply uses selenium rectifiers in a full-wave voltage-doubling circuit. Transformer T_1 has a 120–volt secondary as well as a filament winding. The output voltage under load is approximately 250–300 volts. An a.c. outlet is provided in the mixer as a convenience, because recording work usually involves a number of components requiring a.c. power.

Operating convenience is of great importance, so that the mechanical design of the mixer should follow personal preferences of the user. Two features are desirable in any case, however. The cabinet should be large enough so that it cannot be moved or jarred easily. The knobs on the gain controls should be large and easy to handle, especially if they are to be used for long periods of time. The skirted knobs shown have raised indicator points so that the settings of the controls can be determined by touch.

Fig. 1. The simplicity of the front panel of the mixer is made possible by mounting the connectors and switches on the chassis so they are accessible by opening the cabinet top.

Fig. 3. Detail of the mixer chassis. The input connectors marked "L" and "R" correspond with the left and right gain controls.

The rest of the mechanical design as shown reflects the personal preference of the author. The mixer shown in *Fig.* 1, for example, has no controls on the front panel other than the gain controls. The on-off switch and the input and output connectors are mounted on the chassis and accessible through the hinged top.

Rubber mounting feet are placed on the bottom and back of the cabinet so that the mixer can be used in one of two positions. *Figure* 3 shows how the chassis is mounted midway on the cabinet front panel, so that the gain controls are not so close to the bottom of the cabinet that they are difficult to handle.

Figures 3 and 4 indicate the placement of parts quite clearly. These parts are mounted on a conventional 5×7×2 chassis. The cabinet is 8 in. wide, 7½ in. high, and 8 in. deep.

Component layout is not critical but two precautions are in order. The first is an obvious one. Keep the input circuits separated from the output circuit. Secondly, use separate ground points for the two input circuits and output circuit. That is, the ground connection for R_1, R_3, C_1 and the Input No. 1 connector should be made at only one point on the chassis. Also R_2, R_4, C_2, and Input No. 2 connector should tie together at a second point on the chassis. The same is true for R_6, C_5, C_6 and the two output connectors. If the circuit is wired in this way hum introduced by ground currents is avoided.

The potentiometers shown in the under-chassis view are actually dual units. These were used only because single units with a logarithmic taper were not immediately available at the time the mixer was built.

Shielded wire is used for one connection only, between the potentiometer at the right of *Fig.* 4 and its associated input connector. The use of shielded wire at this point may be unnecessary but it is a worthwhile precaution.

Output Attenuator

In any audio-frequency amplifying system which is made up of a number of units, flexibility of application is provided by fixed attenuation steps. The mixer incorporates a single 20-db output attenuator. In order to maintain constant attenuation over a wide range of frequencies, a frequency-compensated attenuator is valuable. This consists of R_5, R_6, C_4 and C_5 (see *Fig.* 2).

Although this type of compensation is not new, a brief statement of how it works might be in order. Assume that capacitors C_4 and C_5 were not in the circuit. Further assume that output cable with a total capacitance of 1000 μμf is connected to the 20-db output connector. There is now a large amount of capacitance in shunt with R_6, and only a very small amount of stray capacitance is shunt with R_5.

If a low frequency is impressed on this voltage divider the division of voltage is determined almost entirely by the ratio of $R_6/(R_5+R_6)$. However, if a high frequency (such as 15,000 cps) is considered, the voltage division will no longer be proportional to the resistance ratio because of the capacitance added by the output cable.

Fig. 4. Under-chassis view of the mixer.

Uniform attenuation can be accomplished by the addition of shunt capacitance such as C_4 and C_5. In order that these capacitors divide the voltage in the same ratio as the resistors it is only necessary to make the products R_5C_4 and R_6C_5 equal.

An additional benefit accrues from the use of shunt capacitors in that moderate additional capacitance can be tolerated across the output connectors. This means that the capacitance of the output cable need not be of too great concern. For example, 500 μμf can be

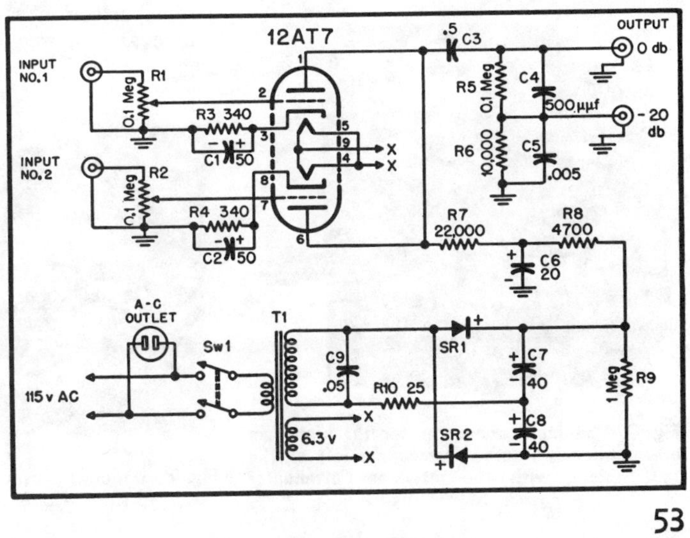

Fig. 2. Schematic of two-channel mixer and power supply.

[1] In the schematics, photographs, and text, reference is made to voltage ratios, expressed in decibels. The decibel quantities stated are intended only as voltage ratios defined by the equation $db = 20 \log_{10} (E_1/E_2)$, and bear no particular relationship to the commonly understood zero reference level of one milliwatt in a 600-ohm load.

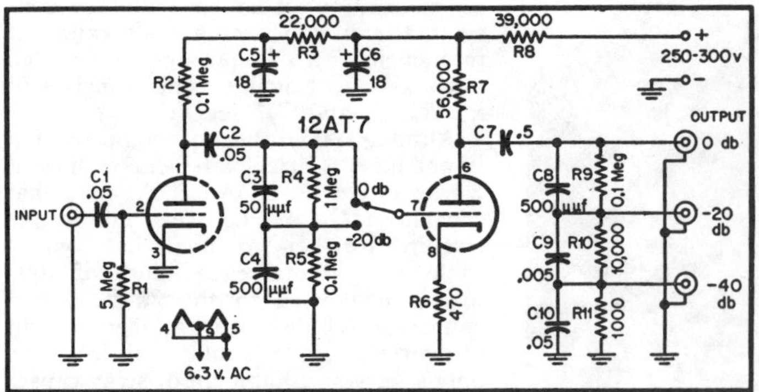

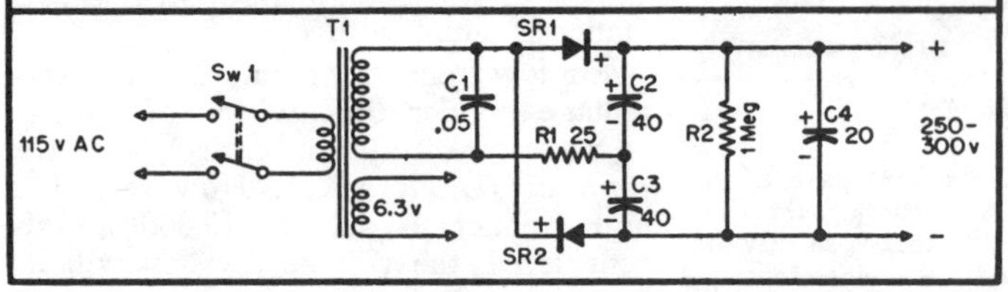

Fig. 5 (above). Schematic of the high-gain preamplifier, and Fig. 6 (below), its power supply.

tolerated across the – 20 db output connector and 1500 μμf across the 0 db output connector without seriously disturbing the frequency response. (Low-capacitance microphone cable has between 20 and 30 μμf per foot. Some microphone cable may have a capacitance as high as 150 μμf per foot.)

It is also possible to *replace* the fixed capacitance in the circuit with an identical amount of capacitance in the output cable, which means that 5,000 μμf of cable capacitance could be added across the – 20 db output if C_5 is removed. This is desirable only where a given cable of known capacitance is used. Of course, the capacitance represented by C_5 can be made up of any amount of external capacitance, up to 5,000 μμf, and the remainder made up of a fixed capacitor in parallel with R_6. Under any circumstances the important thing is to keep the product $R_6 \times (C_5 + \text{cable capacitance})$ equal to the product R_5C_4 when the – 20 db output connector is used.

Performance

Working into a load of 100,000 ohms or more, the maximum voltage gain for either channel is 10. Under the same conditions the voltage gain measured at the – 20 db output point is unity.

When the mixer feeds a circuit of 0.1 megohm resistance shunted by no more than 1500 μμf, the frequency response is uniform within ± 1.5 db from 20 to 20,000 cps at the 0-db output terminal. When using the – 20-db output terminal with total capacitance (C_5 plus cable) as specified, the frequency response is uniform within ± 0.5 db from 20 to 20,000 cps.

HIGH-GAIN PREAMPLIFIER

Figure 5 is the schematic for the preamplifier proper, and *Fig.* 6 the schematic for the preamplifier power supply. These are shown separately because each is built as a separate piece of equipment.

The circuit of the preamplifier is quite usual in all respects, except for the interstage frequency-compensated attenuator. A single-pole double-throw switch is used to cut in 20 db of attenuation. This attenuation is desirable only if the input signal, after amplification by the first section of the 12AT7, is great enough to cause overload of the second section. The frequency compensation network is the same, in principle, as that described previously.

The output circuit also uses a frequency-compensated attenuator, capable of 0 db, – 20 db and – 40 db attenuation. The associated shunt capacitors are, respectively, 500, 5000 and 50,000 μμf. Proper selection of output cable capacitance and subsequent readjustment of the associated shunt capacitor will permit the use of an output cable of almost any length.

Construction

The preamplifier has been made as small as practicable, since it is likely to be used in places where it might be considered unsightly. The smaller it is, the less obvious it is.

The entire unit is built on the removable top of a 4×4×2 utility box. Input and output connectors, electrolytic capacitor, tube socket, interstage switch, and power cable all mount on the 4×4 top panel. *Figures* 7, 8 and 9 show the parts placement. The under-chassis photographs have been taken from two points in order to show the wiring details more completely.

The same wiring precautions mentioned for the mixer should be observed in the wiring of the preamplifier.

The two 8-μf electrolytic capacitors shown taped together form a part of C_5 and C_6 (*Fig.* 5). Capacitor C_5 consists of one of the 10-μf sections in the mounted capacitor and one of the 8-μf capacitors. C_6 is similarly made up of two such capacitors. These 8-μf units were added after the preamplifier had been placed in service, because additional filter capacitance seemed desirable. A neater job could be obtained if a dual 20-μf mounted capacitor replaced the combination just described. Rubber mounting feet are used on the bottom of the preamplifier case to reduce the effects of excessive shock or vibration.

The schematic of the power supply used with the preamplifier is shown in *Fig.* 6. The circuit is similar to that of the power supply used with the mixer. The preamplifier power supply is not shown in the photographs. The author's unit is made on a 4 by 4 by 2 inch chassis. The center of the filament circuit should be connected to ground.

Performance

Working into a load of 100,000 ohms or more, the voltage gain at the 0-db output point is 600 (interstage switch at 0 db). The respective voltage gains at the – 20 db and the – 40 db output points are 60 and 6.

The frequency response of the preamplifier is ± 1.5 db from 20 to 20,000 cps when the output circuit is as described previously for the mixer.

Fig. 7. The input connector for the high-gain preamplifier is located just to the left of the tube shield, with the interstage attenuator switch to the right.

Fig. 8. Under-chassis view of the preamplifier to show details of the input side.

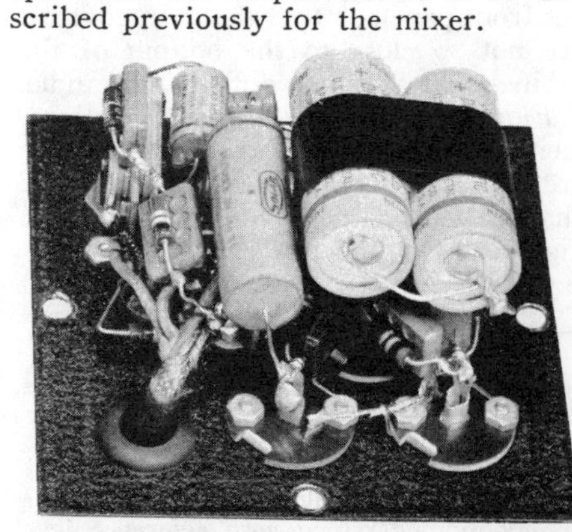

Fig. 9. Under-chassis view of preamplifier showing details of the output side.

D.C. Heater Supply for Low-Level Amplifiers

L. B. HEDGE

Hum problems in preamplifier and other low-level stages may be reduced or completely eliminated by the use of one of the d.c.-supply circuits described by the author.

THE USE OF DIRECT CURRENT to supply the heaters of pre-amplifier and other low-level, high-gain amplifier tubes is standard practice in professional and high-quality equipment design. Direct current in this application is effective in reducing hum and it has the added advantage of making critical stages less sensitive to individually selected tubes when replacement is required. This is no small advantage, as anyone who has tried to locate a quiet replacement for a defunct a.c.-heated pre-amplifier tube will appreciate. A d.c. supply for the low-level end of a good audio system is sound design practice, and it need be neither difficult nor expensive.

Basically there are two schemes for direct current heating in an alternating-current-supplied amplifier — a separate rectifier and filter may be used to supply the heaters, or they may be supplied from the regular high-voltage unit which furnishes the plate power for the amplifier. The second plan is the one most commonly used—the low-level tubes are connected in series at the low-voltage end of the plate supply circuit—but it suffers from several disadvantages:

1. It requires a heavy-duty power supply unit.
2. It reduces the available plate-voltage by the amount of the drop across the heater circuit.
3. It warms the d.c. heated cathodes with the plate voltage applied.
4. It requires a constant plate-current load.
5. In event of tube failure it may induce excessive further damage.

The heavy-duty power supply may be no disadvantage—if the plate load is in excess of 150 ma the excess will have to be bled across the heater circuit. The second and third items listed are obvious, and obviously not good. The constant-plate-load requirement may not be a serious drawback, but in some involved devices (such as magnetic recording-playback units) it may complicate switching requirements. The fifth item is often neglected—a tube which breaks down in such a way as to draw excessive plate current may easily cause a heater burn-out in one or more of the series-connected low level tubes—and a burn out in one of these heaters, however caused, will leave the no-load plate supply voltage between the heater and cathode of each tube remaining on the supply side of the burn-out.

The other alternative—a separate rectifier to supply the low-level heaters—may be provided in several ways, each of which in large measure avoids the first objection listed for the series plate-supply scheme, and completely avoids the others. With presently available surplus transformers, selenium rectifiers and chokes, and with standard components not much more expensive, a separate d.c. heater supply is both practical and economical.

Typical Circuits

Basic circuits for d.c. heater supply are shown in *Fig.* 1. The plate-supply connection is diagrammed at (A). With a string of n 12-volt heaters (150 ma) the voltage drop from X to $B-$ will be $12.6n$, and the resistors R_1 and R_2 are selected to provide this condition. If the normal plate current drain is less than 150 ma. $R_2 = 0$, and R_1 is adjusted to give the 150-ma current through the heater string. If the normal plate current is over 150 ma, R_1 is adjusted to draw a few ma (as a safety bleeder across the plate supply filter) and R_2 is adjusted to by-pass the excess current. Adjustment in either case may be made with a milliameter in the heater loop or with a voltmeter between X and $B-$. The available plate-supply voltage between $B-$ and $B+$ will be the supply voltage across $X\,Y$ less the drop between X and $B-$ in the heater loop.

Figure 1 (*B*) is the wiring diagram of a separate heater-supply unit shown in *Fig.* 2, built by the author as a stand-by and bench unit. A 110/110-volt isolation transformer supplies a half-wave selenium rectifier followed by a capacitor-input filter, a load-adjusting resistor, and a safety bleeder. The unit will supply one to six 12-volt, 150-ma heaters—the load-adjusting resistor being set to the

Parts List for the Mixer and Preamplifier

Two-Channel Mixer

C_1, C_2	50 µf, 25 v. electrolytic
C_3	0.5 µf, 400 v.
C_4	500 µµf, 400 v.
C_5	5000 µµf, 400 v.
C_6	20 µf, 450 v. electrolytic
C_7, C_8	40 µf, 150 v. electrolytic
C_9	.05 µf, 400 v.
R_1, R_2	0.1 meg potentiometer, log taper
R_3, R_4	340 ohms, ½ watt
R_5	0.1 meg, ½ watt
R_6	10,000 ohms, ½ watt
R_7	22,000 ohms, 1 watt
R_8	4700 ohms, 1 watt
R_9	1.0 meg, ½ watt
R_{10}	25 ohms, 1 watt
Sw_1	DPST toggle switch
SR_1, SR_2	100-ma selenium rectifier
T_1	Power transformer, 120 v. sec. at 75 ma, 6.3 v. at 1.5 amps.

High-Gain Preamplifier

C_1	.05 µf, 200 v.
C_2, C_{10}	.05 µf, 400 v.
C_3	50 µµf, 400 v.
C_4, C_8	500 µµf, 400 v.
C_5, C_6	18 µf, 450 v. electrolytic (see text)
C_7	0.5 µf, 400 v.
C_9	.005 µf, 400 v.
R_1	5.0 meg, ½ watt
R_2	0.1 meg, ½ watt
R_3	22,000 ohms, ½ watt
R_4	1.0 meg, ½ watt
R_5, R_9	0.1 meg, ½ watt
R_6	470 ohms, ½ watt
R_7	56,000 ohms, ½ watt
R_8	39,000 ohms, 1 watt
R_{10}	10,000 ohms, ½ watt
R_{11}	1000 ohms, ½ watt
Sw_2	SPDT toggle switch

Power Supply

C_1	.05 µf, 400 v.
C_2, C_3	40 µf, 150 v. electrolytic
C_4	20 µf, 450 v. electrolytic
R_1	25 ohms, 1 watt
R_2	1 meg, ½ watt
Sw_1	DPST toggle switch
SR_1, SR_2	100-ma selenium rectifier
T_1	Power transformer, 120-v. sec. at 75 ma, 6.3 v. at 1.5 amps.

(The author originally specified T_1 as G. E. Type K68J661. However, this transformer is difficult to obtain, and similar results can be obtained from any of the following:

Thordarson T22R12
Merit P-3045 (Heater winding 2.0 a.)
Stancor PS8415 (Heater winding 0.6 a.)
Triad R-2C (Heater winding 0.9 a.)

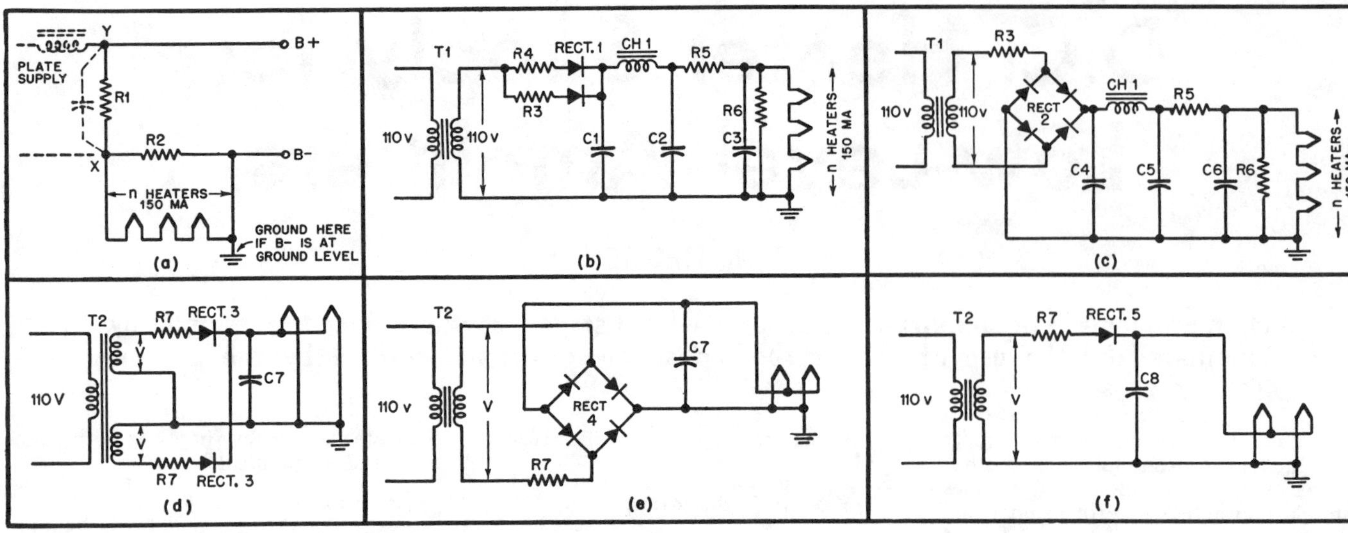

Fig. 1. Typical circuits for supply of direct current for heaters of low-level amplifier stages.

proper value. The rectifier might be one 150-ma (or more) unit or, as shown, two parallel 75-ma units. A resistor of 50 to 100 ohms should be connected in the rectifier circuit to limit the peak current in the rectifier. The value of the resistor is not critical, but if parallel units are used separate and equal resistors (as shown at B in *Fig.* 1, R_3 and R_4) should be used since small differences in the resistances of the rectifier units may otherwise result in large unbalance in the currents flowing in them, and overload of one of the units.

Figure 1 (C) shows a full-wave bridge rectifier in a supply unit similar to that of (B). The rectifier should have a rating of 150 ma or more as a bridge unit—it can be made up from four 75-ma half-wave rectifiers.

The filters shown in *Fig.* 1 (B) and (C) will give high-voltage output and a well filtered supply. Omission of the input capacitor (C_1 or C_4) will lower the output voltage and reduce the size of the resistor R_5. For fixed installation, R_5 should be selected using, as with the plate series scheme, either a milliameter in the heater circuit or a voltmeter across the string. For an adjustable laboratory unit R_5 may well be a 500-ohm 20-watt power rheostat. The output capacitor (C_3 or C_6) effectively puts R_5 in the filter section, and R_6 is included to discharge the capacitors if the unit should be turned off with the heater string open.

The low-voltage end of the heater string should be the most sensitive (lowest level) tube supplied, and the heater circuit should be grounded at the socket of this tube to discourage hum and/or noise from "riding in" on the heater supply.

The small selenium rectifier and the high-capacitance low-voltage electrolytic also make practicable the provision of d.c. heater current from the low-voltage windings of a filament or combination power transformer. The basic schemes for this type of supply are shown in *Fig.* 1 (D), (E), and (F).

If two 12.6-volt windings or a center-tapped 25-volt winding are available on the transformer the full-wave rectifier shown at (D) may be used to supply several 12-volt heaters—two 6.3-volt windings may be similarly connected to supply a few 6-volt tubes. Each rectifier will carry only half of the d.c. output current in this arrangement, but the permissable voltage drop precludes the use of an effective network filter and the capacitor C_7 should be at least 200 µf for each 150-ma load on such a 12-volt supply. C_7 in a 6-volt supply wired this way should be not less than 1000 µf for each 300 ma drawn.

The bridge connection shown at (E) may be used to supply 12-volt tubes from a 12.6-volt winding or from two 6.3-volt windings connected in series or 6-volt tubes from a single 6.3-volt winding. Filter capacitor requirements are the same as for the full-wave connection of (D), and each leg of the bridge rectifier will carry only one-half of the d.c. load.

Simplest Form

Simplest of the low-voltage d.c. supplies is that shown at (F)—a single half-wave rectifier with a single capacitor as the filter. In this circuit C_8 should have a minimum capacitance of at least 1000 µf for each 150 ma of d.c. load at 12 volts, or 4000 µf for each 300 ma at 6 volts, and the ripple present in the heater circuit will, at best, be fairly heavy.

It should be noted that filament windings used in any of the low voltage circuits (D), (E), and (F) need not be reserved for the d.c. supply alone—they may be feeding a.c. to other tubes as well, provided only that the added rectifier drain does not overload the windings. Grounding the negative heater lead to the lowest-level tube at the socket of that tube is still to be recommended with these low-voltage supplies—but if one of these circuits is added to an existing piece of equipment care must be taken to insure against short-circuiting the supply by grounding the d.c. output when the heater circuit is grounded elsewhere in the equipment.

Voltage to supply a bridge rectifier of the type shown at (E) may sometimes be contrived in a set where it is not obviously available. A transformer with a 6.3-volt filament winding and 5-volt rectifier winding, for example, may be made to serve by substituting a selenium or cathode type 6.3-volt rectifier for the 5-volt tube, and connecting the two windings in series to provide 11.3 volts to the bridge. Operating a low-level amplifier tube at slightly under its rated heater voltage will usually provide better rather than poorer quality in that stage, and it will tend to reduce electrical and microphonic noises in the stage and to further reduce tube replacement variations.

Fig. 2. Separate heater supply unit built by the author.

PARTS LIST

C_1, C_2, C_3 150v, 80-µf electrolytic
C_4, C_5, C_6 150v, 40-µf electrolytic
Ch_1 8 henry, 150 ma choke
R_1, R_2, R_5, R_6 See text
R_3, R_4 50-ohm, 2-watt resistors
R_7 10-ohm, 2-watt resistor
$Rect_1$ Selenium rectifier (1–150 ma or, as shown, 2–75 ma)
$Rect_2$ " " (150 ma bridge, 75 ma each leg)
$Rect_3$ " " (rated ½ of d.c. load)
$Rect_4$ " " (bridge rated for full d.c. load, each leg rated for ½ of d.c. load)
$Rect_5$ " " (rated for full d.c. load)
C_7 25v elect. capacitor. For 6.3v, 1000 µf per 300 ma load
For 12.6v, 200 µf per 150 ma load
C_8 25v elect. capacitor. For 6.3v, 4000 µf per 300 ma load
For 12.6v, 1000 µf per 150 ma load

A Continuously Variable Loudness Control

E. E. JOHNSON

A new approach to the problem of adjusting frequency response simultaneously with changes in level in order to compensate for varying sensitivity of the ear.

It is well known in audio circles that the human ear is very sensitive to both low and high frequencies at reduced volume levels. The accepted standard used in compensating for this hearing deficiency in audio systems is a set of curves at different levels known as the Fletcher-Munson curves. These curves show the amount of low-frequency and high-frequency boost that is required above some mid-range level to make the sound output of an amplifier appear balanced at all volume control settings.

Many attempts have been made to obtain the required compensation by use of single or multiple tapped volume controls, stepped loudness controls, and various types of bass and treble boost circuits. None of these has given the performance of a truly continuously variable loudness control. The tapped volume control affords compensation only when its contactor is at the tap but does not provide proper compensation when located away from the tap. To obtain wider spread of compensation, two or three taps are used, but such controls are more difficult to manufacture and, therefore, are more expensive. The stepped type loudness control does not provide full flexibility of adjustment and is relatively expensive. The bass and treble boost circuits require multiple adjustments with change of volume for ideal compensation.

The control described in this article is a continuously variable loudness control that may be assembled easily from standard parts available widely from radio parts distributors. It may be wired into most audio systems as easily as an ordinary volume control.

Description

This new loudness control consists of three variable resistance units—R_1, R_2, and R_3—operated from one common shaft and in combination with the proper resistors and capacitors, as shown in *Fig.* 1.

The panel section R_1 functions as a standard volume control supplying a variable voltage to the other sections which form the frequency-compensating networks. The center section R_2 forms one variable leg of a potentiometer circuit and the rear section R_3 forms the other leg. Fixed resistor R_4 acts as a limiting resistor to keep the input impedance as constant as possible when the control is set near maximum output. The center control in combination with capacitor C_1 forms the arm of the variable voltage divider network which decreases in impedance as the frequency increases, causing the output voltage to rise at frequencies above 1000 cps. The response curves for this control at various settings are shown in *Fig. 2*. The monitor level figures represent the actual listening level at which the compensation most closely follows the Fletcher-Munson curves, one of which is shown dotted for a level of +60 db. (Normal listening level in the average living room will range from +65 to +75 db.)

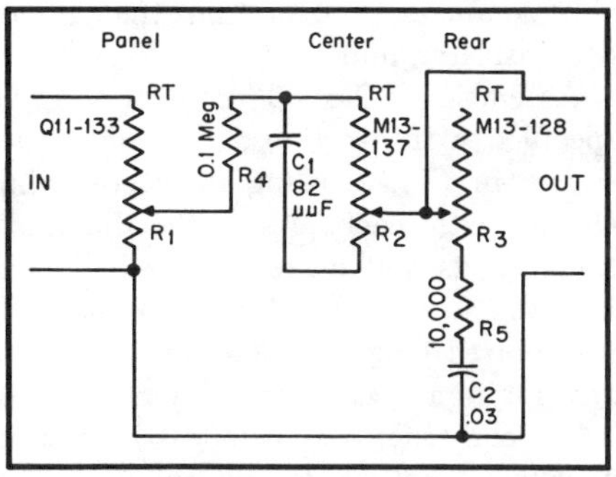

Fig. 1. Schematic of new loudness control which may be assembled from standard parts.

The rear section R_3, fixed resistor R_5, and capacitor C_2 form the arm of the variable voltage divider network that increases in impedance as the frequency is decreased, causing the output voltage to rise at frequencies below 1000 cps, as shown in *Fig. 2*.

This arrangement offers a truly continuously variable loudness control that can be used to improve the sound quality of many radio, FM and TV receivers, as well as many sound systems. It must be pointed out, however, that there is an insertion loss of 6 db which must be compensated for in low-gain systems if full output is required. Also, the unit is not satisfactory for operation in the plate circuits of high-impedance tubes since its input impedance is not constant at or near full volume settings. Impedance is fairly constant if the control is used up to approximately 75 per cent of its rotation.

The three-section control required is readily assembled with a standard IRC type Q Volume Control and two IRC Multisections. The Multisections are rear control sections so designed that they may be added to Type Q Controls or to other Multisections in the same manner as switches are attached. Simple assembly instructions are included with each Multisection. A pictorial schematic

Fig. 2. Response curves for control at various level settings.

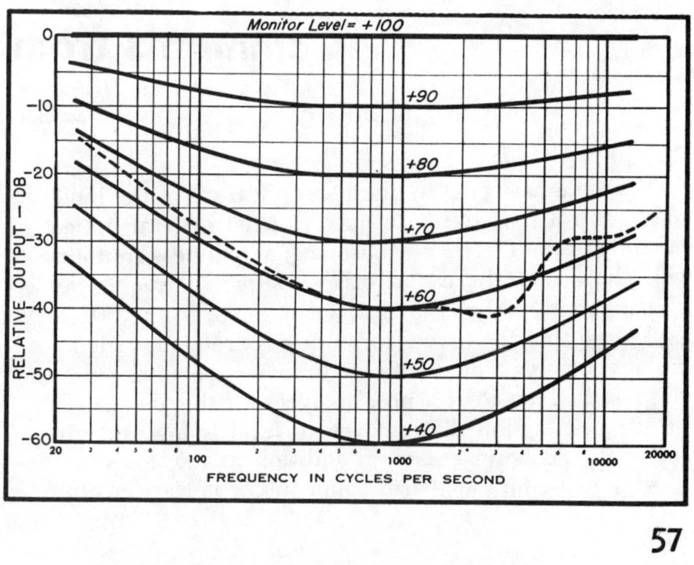

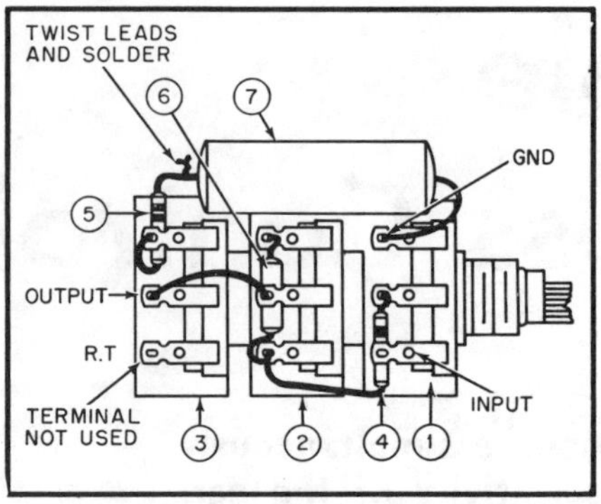

Fig. 3. Pictorial schematic showing exact wiring of components.

of the assembled unit is shown in *Fig.* 3, while its over-all appearance is shown in *Fig.* 4.

Demonstration Unit

Those who wish to demonstrate the remarkable difference this new control can make in producing a pleasing, well-balanced sound output at low volume levels over the results of an ordinary, uncompensated volume control may assemble a very effective demonstrator unit with the use of the new IRC Concentrikit, an arrangement which permits the quick assembly of a wide variety of concentric controls. When used in conjunction with Multisections, it will provide triple-single concentric control, shown in schematic form in *Fig.* 5.

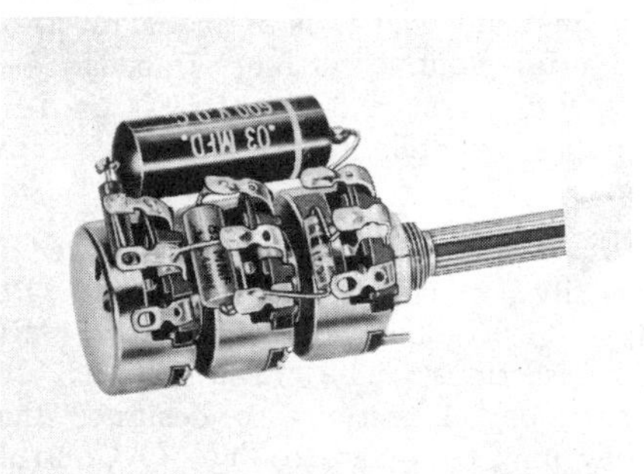

Fig. 4. External appearance of completed control.

The outer shaft of this demonstrator unit varies the panel section R_1 which is an ordinary uncompensated volume control. The inner shaft varies the three rear sections, R_2, R_3, and R_4, which comprise the continuously variable loudness control. By means of a d.p.d.t. slide switch the output is adjusted for the same volume level through each control at 1000 cps and direct comparison may be quickly made at low level. Appearance of the completed demonstrator unit is shown in *Fig.* 6.

Following are simple assembly instructions for both the continuously variable loudness control and the unique demonstrator unit.

Assemble to the "Q" control the two specified Multisections, in the order shown in *Fig.* 3, using instructions included with each. Assemble the additional parts and make all required connections as shown, solder, and cut shaft to required length. Install and wire into any high-gain audio amplifier.

To construct the demonstrator unit, assemble Concentrikit by following instructions included using B13-133 (R_1) as panel unit and B11-133 (R_2) as rear unit. Omit cover on rear. Assemble M13-137 (R_3) and M13-128 (R_4) per instructions included with each Multisection. Attach this assembly in place of rear cover on the above Concentrikit being sure inner shaft rotates both sections R_3 and R_4. Assemble the additional parts and make all required connections to the last three controls R_2, R_3, and R_4 in exactly the same manner as described previously for the loudness control. An additional connection is required between the most counter-clockwise terminal of the panel section and that of the second section to form a common ground.

Assemble d.p.d.t. switch as shown by photograph of completed control, *Fig.* 6,

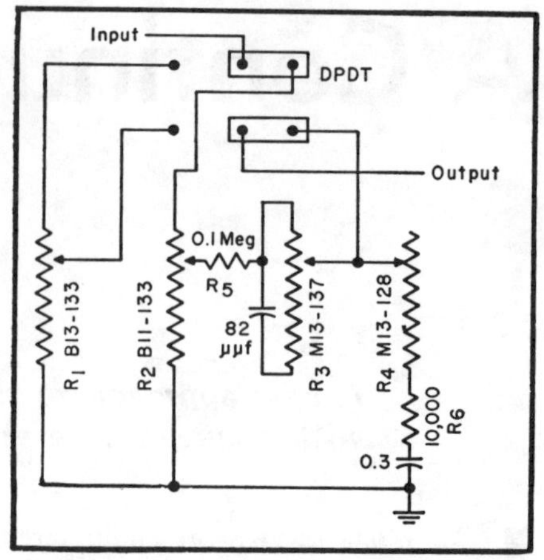

Fig. 5. Schematic of demonstrator unit used to show advantages of the new loudness control.

and make necessary connections as shown by schematic, *Fig.* 5. It is advisable to ground the metal case of switch, if that type is used, to reduce possibility of hum pickup during operation of switch.

(All wiring to and from control should be as short as possible and should be shielded to reduce hum pickup. Use low-capacitance wire to avoid loss of highs due to shielding. The complete assembly can be mounted in a small steel box to form a well shielded unit.)

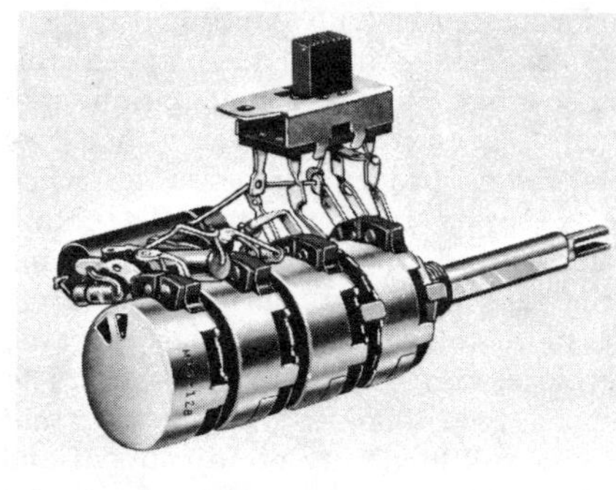

Fig. 6. External appearance of demonstrator unit.

Comments from Editor's Report

December, 1952

The time has come to discuss one feature of the loudness control that seems to have escaped general notice. A study of the Fletcher-Munson equal loudness contours will show that the ear is less sensitive to frequencies above about 3000 cps than over the range from 1000 to 3000, and that the curves for all levels are almost identical above this point. Some loudness controls have been proposed which correct progressively for this deficiency, on the assumption that as over-all level is lowered, the highs should be increased in addition to the lows.

May we point out that the human ear is less sensitive to these high frequencies at *all* levels, and that it is with these same ears that we listen to live music. Thus if the music is reproduced with a flat system (above 3000 cps) it should be presented to the ear just as it is from a live source, since the F-M curves are almost exactly the same in the high-frequency range. We submit, therefore, that no correction should be applied to the loudness control except that for the low frequencies.

This observation is addressed to those who go for the idea of the loudness control, and may be overlooked by those who do not. However, we still prefer them.

A Two-Tap Bass and Treble Compensated Volume Control

WILLIAM O. BROOKS

A simplified analysis of the method of determining the constants for a loudness control which will simulate the Fletcher-Munson curves to a reasonable degree of accuracy.

ANYONE who has adjusted the uncompensated volume control of an amplifier when music was being played has noted that as the control was rotated from its full volume position to its minimum, less and less bass and treble were heard as compared to the amount heard in the maximum or full-volume position. This is borne out by the Fletcher-Munson sound pressure curves appearing in engineering texts. As the curves show, it is necessary to boost the bass—and to a lesser degree, the treble—more and more as the control is lowered to make up for the hearing curve of the ear.

By adding a tap on the volume control and running the sliding arm down to this tap we have an "L" type tone compensating network as shown in *Fig.* 1. The network can be figured so as to give correct compensation for hearing losses at this one point. Now it becomes apparent that in order to have a 100 per cent perfect control, we would have to have an infinite number of taps on the control with an equal number of networks. This is impractical from a mechanical and economical standpoint.

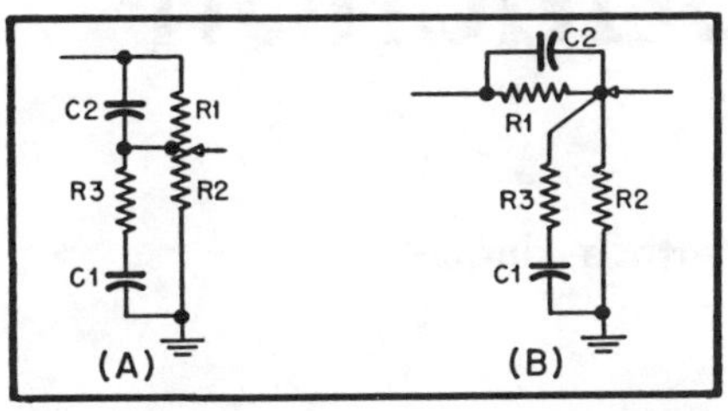

Fig. 1. Basic arrangement of components for L-pad compensation of frequency response.

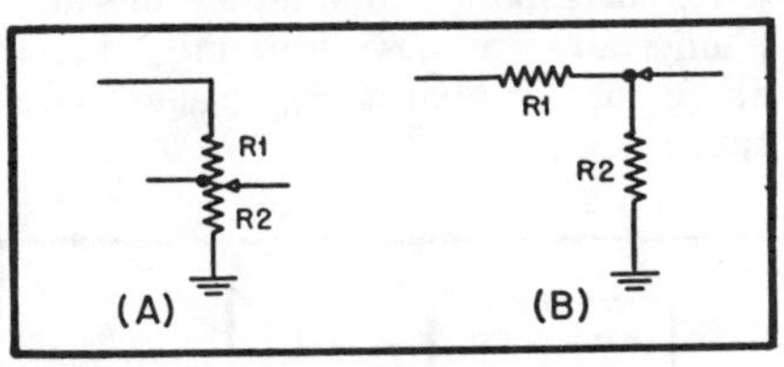

Fig. 2. Simplified equivalent of uncompensated volume control.

The next choice would be three controls back to back on the same shaft—one acting as a volume control, one as a continuously variable bass-compensation network, and the other a continuously variable treble-compensation network as described by E. E. Johnson.[1] For those in the radio manufacturing industry where cost is a great factor, a good but cheaper method might be desired.

It is possible to make two well located taps on a single control approach the desired boost requirement to a very acceptable degree. This has been done in commercial radio equipment for many years in the form of one or two tap volume controls; however, only bass compensation was obtained. On recent designs treble compensation has shown up on a single-tapped control. The development of a two-tap control with both bass and treble compensation, as proposed by the author, is to be discussed in this paper. Let us start with a single tapped volume control as shown at (A) in *Fig.* 2. If the slider is set at the tap, we have an L-pad divider made up of R_1 and R_2 as shown at (B). The voltage ratio will be $R_2/(R_1+R_2)$. If $R_1 = 0.25$ meg and $R_2 = 0.25$ meg, then

$$\frac{0.25}{0.25+0.25} = 0.5,$$ or 6 db attenuation of voltage. If C_1 is now connected from tap to ground as at (A) in *Fig.* 3, we would have a definite attenuation of voltage at some desired center frequency. Above this frequency, the reactance of C_1 decreases at a constant rate, attenuating the highs. Below this frequency, the reactance of C_1 increases at a constant rate, thus effectively boosting the bass (by attenuating the high frequencies). Since this bass-boost circuit should not function at the expense of losing treble, it is necessary to insert R_3 in series with C_1. At the frequency when the reactance of $C_1 = R_3$ the curve will begin to flatten out. This frequency is known as the bass turnover frequency. It is subject to change with each design engineer from a minimum of 250 to a maximum of 1000 cps. Since 800 cps has been chosen as the center frequency of the audio spectrum, a bass turnover frequency of 1000 cps will tend to raise the midrange too much, while a 250-cps turnover will not allow the bass to rise to a sufficiently high amplitude. It is thus desirable to stay within 250 and 1000 cps. It

[1] See his article on page 57.

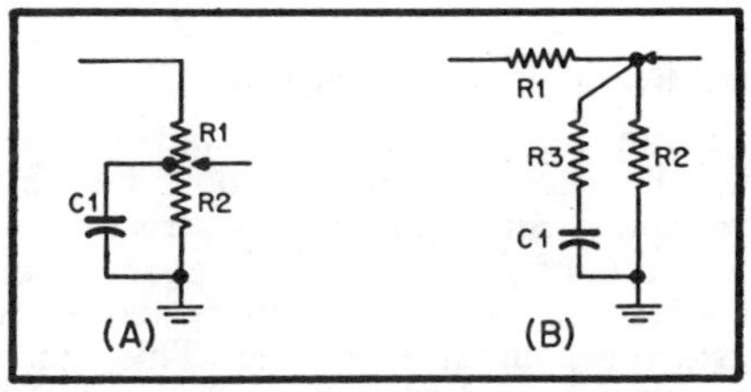

Fig. 3. Volume control with single tap, allowing for one bass turnover frequency.

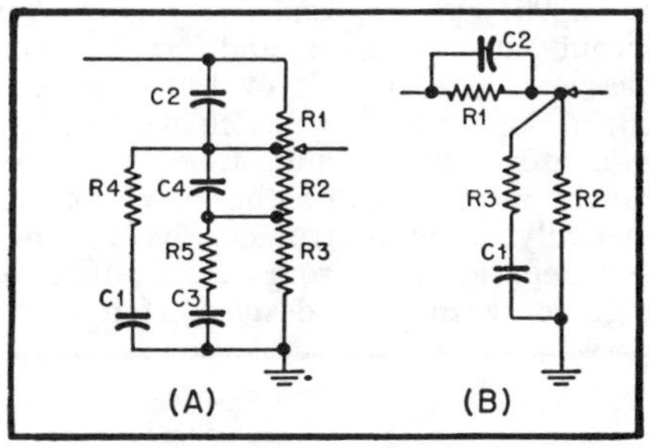

Fig. 5. Development of dual-tap control.

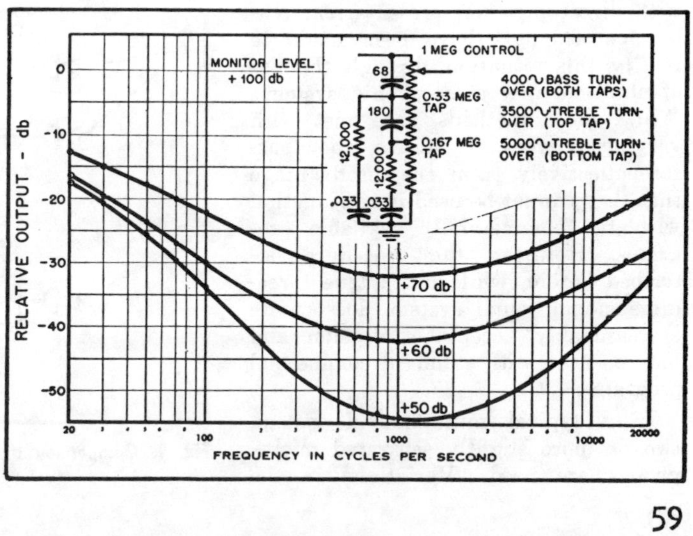

Fig. 4. Response curves for dual-tap control shown in the boxed-in schematic.

has been suggested that all manufacturers use 400 cps as the bass turnover frequency. The curves in *Fig.* 4 are plotted around this point. The next step is to add treble compensation (*Fig.* 1).

There are two reasons for adding this compensation:

1. To make up for the loss in highs due to adding the bass compensation. (The turnover of C_1 and R_3 causes a flattening out of the curve, but not 100 per cent. There is always some attenuation of highs above the turnover point.)
2. To make up for the hearing loss in highs as the volume is lowered, in accordance with the Fletcher-Munson curve.

Selection of Boost Frequencies

From the hearing curves we can pick out the necessary treble boost for this tap point. Since the tap is already set, the value of R_1 is fixed. If capacitor C_2 is added across R_1, the highs above the turnover point are boosted. This upper turnover frequency is the point at which $X_{C_2} = R_1$. (Refer to *Fig.* 1.) A suggested treble turnover frequency would be 2500 cps to make the turnover frequencies of bass and treble symmetrical on each side of 1000 cps. The curve can now be calculated or an oscillator and vacuum-tube voltmeter can be used to obtain the curve experimentally. This control can be designed as a separate unit from the amplifier, or together with it as desired. If it is designed together with the amplifier, the tone controls on the amplifier should be set to the "flat" position. Then with the volume control in maximum volume position, a flat reference curve is obtained. Let this flat curve be equal to the 100-db Fletcher-Munson curve which is flat and add the boosts accordingly as the midrange level is reduced at the different taps. If more bass compensation is needed, lower R_3 and increase C_1, thus keeping the same turnover frequency. If more treble boost is desired and R_1 is fixed by the tap, it is attained by increasing C_2, which turns over at a lower frequency and allows the curve to start rising sooner. If less treble boost is desired, decrease C_2, making the turnover frequency higher and the treble rise will start later.

In adding the networks for the second tap, the same procedure should be followed. Keep the same bass turnover frequency as before for the network C_3 and R_5 as in *Fig.* 5. The treble boost network is C_4 and R_2. Referring to the sound-pressure curves again, it can be seen that the treble increases as the bass increases with each successively lower tap but to a lesser degree. To do this, the treble turnover frequency at this second tap must be increased, thus allowing the treble to start rising later in the curve. C_4 does not return to the top of the control, but only to the tap above it. It only continues the treble rise started by the first tap.

One thing might be pointed out. Do not run the treble boost capacitor from the top of the control to the arm for continuously variable treble boost, because even though this works well as the arm is lowered from the top down to the tap, treble boosting continues below the tap while the bass boost stops at the tap. This over-boosts treble response.

The taps can be located as desired, but it is found that a nice result is obtained by setting the top tap at one-third the total resistance and the bottom tap at one-sixth total resistance. On a 1-meg control, these would be 0.33 and 0.167 meg, respectively. The curves of *Fig.* 4 show how closely the respective bass and treble boosts follow the Fletcher-Munson curves at each tap point.

A check of the hearing curves will indicate that the 100-db curve is approximately flat. For a reference level, it must be assumed that the maximum volume position as at this flat position. The nearer to this 100-db level the maximum volume of the amplifier is, the closer to reality the music will sound.

It is hoped that this article will indicate a simple but complete way of designing this type of control. The component values will change with different turnover frequencies, and are not to be taken as fixed values that must be used. It is realized that when used with certain amplifier curves, it may be desirable to under-compensate or over-compensate the volume control.

Stereophonic Reproduction

TENNY LODE

Simple method for simulating stereophonic effect with a single-channel radio or phonograph system.

A SOMEWHAT SERIOUS LIMITATION of most audio reproduction systems lies in the fact that the sound rather obviously emanates from a restricted source area. This is in direct contrast with the widely dispersed multiple-source effect usually associated with direct program presentation. Numerous attempts have been made to rectify this situation through the use of multichannel stereophonic systems. While these methods have met with varying degrees of success, they suffer collectively from the disadvantage that they cannot be used in conjunction with the commercially available programs. However, the procedure described herein, while not a true three-dimensional sound system, does create a reasonably acceptable illusion and may be used with standard commercial programs.

In a typical multichannel system, two or more slightly separated microphones are used. Variations in the position of the sound source will then cause corresponding variations in the relative output amplitudes and phases of the individual microphones. An arbitrary distinction is sometimes made as to whether the primary dependence of the system is upon phase or amplitude shifts between the various channels. This corresponds to whether the microphones are in close proximity to each other or comparatively separated in space.

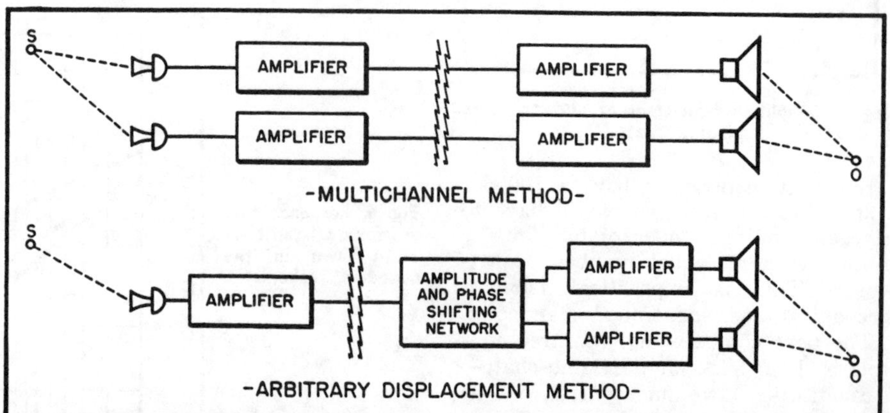

Fig. 1. Comparison between conventional and suggested methods of obtaining greater spatial realism.

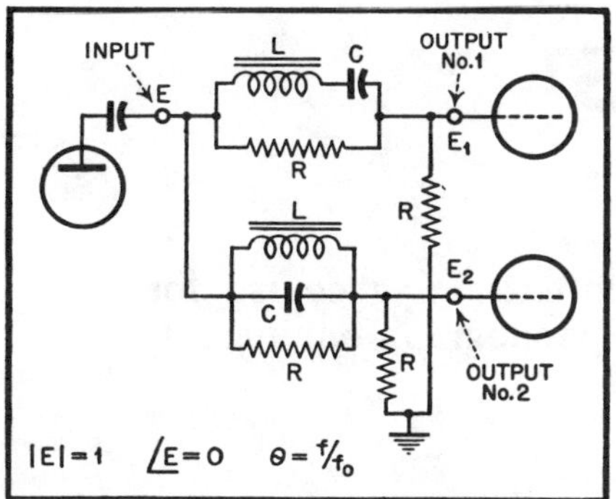

DESIGN EQUATIONS	TYPICAL VALUES
$L/C = R^2$ $f_o = \frac{1}{2\pi\sqrt{LC}}$ R in ohms L in henries C in μf f_o in cps	R = 50,000 Ω L = 12.5 H C = .005 μf f = 640 cps

$$|E_1| = \frac{\sqrt{1+(\theta-1/\theta)^2}}{\sqrt{1+4(\theta-1/\theta)^2}} \qquad |E_2| = \frac{\sqrt{1+(\theta-1/\theta)^2}}{\sqrt{4+(\theta-1/\theta)^2}}$$

$$\angle E_1 = \tan^{-1}\frac{1/\theta-\theta}{1+2(\theta-1/\theta)^2} \qquad \angle E_2 = \tan^{-1}\frac{\theta-1/\theta}{2+(\theta-1/\theta)^2}$$

$$\frac{|E_1|}{|E_2|} = \frac{\sqrt{4+(\theta-1/\theta)^2}}{\sqrt{1+4(\theta-1/\theta)^2}} \qquad \phi = \tan^{-1}\frac{3}{2}\,\frac{\theta-1/\theta}{1+(\theta-1/\theta)^2}$$

(ϕ = Phase lag of E_1 with respect to E_2)

Fig. 2. Phase- and frequency-selective network employed within an amplifier to create spatial illusion.

If the microphone outputs are then amplified separately and fed to individual properly positioned speakers, the initial directional qualities of the original tones will, to a certain extent, be preserved. In this type of system, the information pertaining to the directional distribution of the various tones is contained in the relative phase and amplitude relationships existing between the multiple transmission channels. Since, however, in a standard commercial program, no information concerning the spatial distribution of the sources is presented, any method of assigning directional characteristics to the individual tones must therefore follow a completely arbitrary pattern.

A network may be constructed which will cause relative phase and amplitude shifts to arise between its various outputs according to a predetermined pattern. If the outputs of this network are then fed through the necessary amplifiers to two or more loudspeakers, a stereophonic effect will be created. This method is contrasted with the more conventional in *Fig.* 1. Admittedly, any relationship between the apparent spatial distribution of the reproduced program and that of the original will be entirely coincidental. However, as the listener seldom has an accurate knowledge of the conditions under which the program is originated, this defect is of comparatively minor importance.

The most satisfactory method of constructing the network is to make the phase and amplitude shifts produced (and thus the apparent direction of a tone) a pure function of the frequency. This not only produces the most realistic effects, but is also by far the easiest to accomplish in practice.

The basic circuit which was used for the entire series of experiments is shown in *Fig.* 2. The values given are appropriate for operation with a triode driving stage, as the input impedance of the network is then of the order of 30,000 ohms. The outputs as shown should be fed through separate amplifiers to the individual speakers. The optimum placement of the speakers, both as to their location and orientation, is best determined by trial and error. It should be noted, however, that the most pleasing effects do not necessarily arise as a result of having the speakers pointed directly at the listener. Generally, a satisfactory effect may be attained by placing the speakers in or near two adjacent corners of a rectangular room and rotating them to determine the proper orientation.

The characteristics of the network are illustrated graphically in *Figs.* 3 and 4.

As may be seen, it is quite effective in breaking up the relative phase and amplitude relationships between the two channels without seriously affecting the composite output amplitude. The approximate distribution produced by this network is one in which the low, middle, and high tones appear to emanate from a central region, the medium low tones from one side, and the medium high tones from the opposite side. The apparent spatial "spread" thus obtained adds a great deal to the realism of the reproduced program. The effect is indeed sufficiently pronounced to render a program carried over a conventional system flat and lifeless by contrast.

The system described above, while capable of admirable results under the proper conditions, is still a decidedly inconvenient one for incorporating into existing equipment. For this reason, the following modification is offered as a compromise. The circuit, as shown in *Fig.* 5, is essentially the same. It does not, however, lend itself as readily to a detailed mathematical analysis because of the reactive nature of the load. The basic difference, as may be seen from the drawing, is that the two loudspeakers themselves now serve as the terminating impedances for the phase and amplitude shifting network, which is in turn driven directly by the output of the power amplifier. The general design equations are still valid and, as such, they are restated together with a few typical values for different impedances. In a low-impedance circuit, it is especially important to use reactances with low effective series resistances, and it will probably be necessary to wind the inductances, using heavy wire.

If the proper attention is paid to the location and orientation of the speakers, either of the systems described above is capable of adding a distinctly noticeable degree of realism to a program.

While, as mentioned earlier, there will be essentially no relationship between the *apparent* directional qualities of the reproduced program and the *original* spatial distribution at the source, this will not be greatly detrimental to the operation of the system, as seldom, if ever, will there be a chance for a comparison of the two.

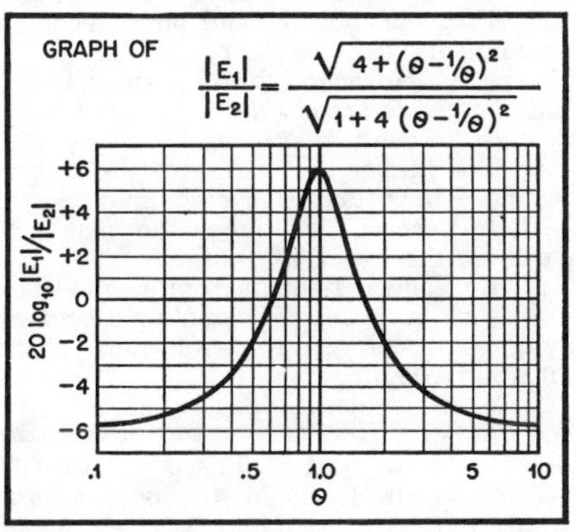

Fig. 3. Curve showing relative voltage at output #1 of Fig. 2. This curve also applies to output #2 if signs are interchanged.

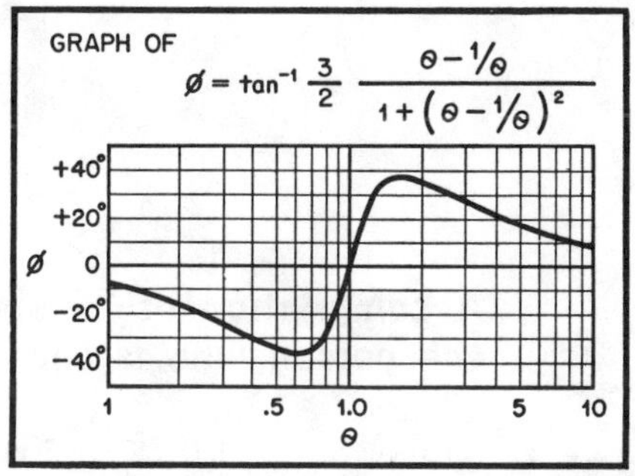

Fig. 4. Phase-shift curve for circuit of Fig. 2.

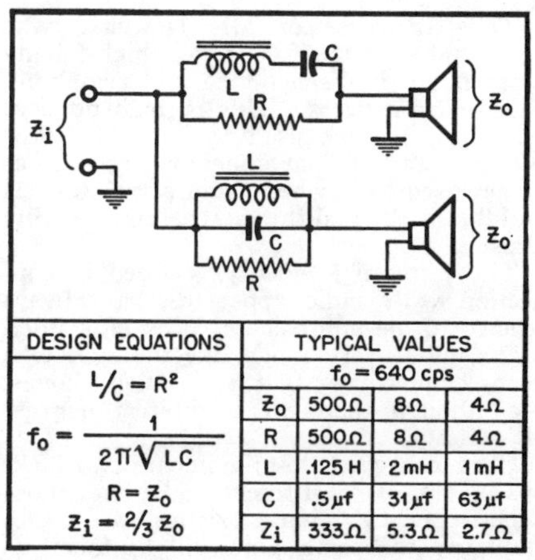

DESIGN EQUATIONS	TYPICAL VALUES			
		f_o = 640 cps		
$L/C = R^2$	Z_o	500Ω	8Ω	4Ω
$f_o = \frac{1}{2\pi\sqrt{LC}}$	R	500Ω	8Ω	4Ω
	L	.125 H	2 mH	1 mH
$R = Z_o$	C	.5 μf	31 μf	63 μf
$Z_i = 2/3\, Z_o$	Z_i	333Ω	5.3Ω	2.7Ω

Fig. 5. Schematic of suggested circuit to be used between the output of amplifier and two separate speakers.

Phono Facts

MAXIMILIAN WEIL

A compilation of 74 simple pointers which serve as a complete training course for the person who is confronted with the problem of selecting phonograph equipment.

Of the hundreds of letters received from day to day by the writer, it is astonishing to find that such a large percentage are from readers of this magazine. It is astonishing because many of these readers are in no way connected with the field of electronics, but they have gained a good understanding of audio equipment and are more critical in selection of apparatus.

Of the many questions in the field of sound reproduction that such letters contain, the most frequent ones are answered here.

It was back about 1935 that "high fidelity" became the talk of the industry. At that time, the R.M.A. issued a definition of high fidelity as—"equipment capable of reproducing frequencies to 7,500 cps or over." This was most unfortunate, as it made *wide range* and *high fidelity* synonymous. Since then, audio equipment has been built to kilocycles and more kilocycles.

On the one hand, the purchaser buys a high-fidelity audio system in the belief that the performance will be, as claimed, one of *fidelity*. That, to the purchaser, means *faithful* reproduction. To the technician, on the other hand, it means wide frequency range in most cases.

According to the above definition, the technician is not misrepresenting, if the audio system *is* capable of reproducing a wide frequency range. At the same time, however, the purchaser is in firm belief that he is buying *faithful* reproduction. Accordingly, the deal is a sort of legal fraud, with both parties innocent of it.

Let us cite a few actual cases:

Mr. Smith, who had just bought a high-fidelity audio system on sales talk such as "crisp," "sharp as a razor," "you can hear the resin on the bow," etc., is much disappointed and disillusioned.

Says Mr. Smith, "It sounds very unpleasant. I decided to check up and went to a concert. At the actual concert, the music I heard was not 'crisp,' not 'sharp as a razor,' I didn't hear the resin on the bow and, what's more, I didn't want to." Smith then puts the treble-control to work, cutting off a lot of kilocycles for which he paid that extra money.

Take the case of Mr. Newman, who just paid over $650.00 for a high-fidelity system. He is disappointed and complains. The seller answers, "This is 'high fidelity', you must become used to it . . . ." Mr. Newman says, "That's nonsense, no one has to get used to it when at an actual concert and that is the real thing, Mr. High Fidelity himself."

The term high fidelity, as used in connection with audio apparatus, has always seemed to me unfortunate. Why *high* fidelity? Sounds very much like the man who keeps emphasizing that he is *highly* honest —as though there were different degrees of honesty.

Here is a quick test: Compare two pickups A and B of different make but whose wide-frequency characteristics are substantially the same. One will be found to perform with *fidelity*, with pleasing musical quality while the other will be harsh, strident, shrill. Yet, both have the same range.

It is now generally conceded that mere extension of frequency range does not, by itself, result in faithful reproduction.

The interchangeable use of *high fidelity* and *wide frequency range* is probably rooted deeply by now. However, for the good of the audio industry, high fidelity should be redefined in order to stop continued misunderstanding.

Pickup Pointers

1. Of two singers covering the same range —each capable of reaching "High C"—one may be pleasing, the other just the reverse. With reproducers of different design, *both wide-range*, one will be pleasing, the other harsh.

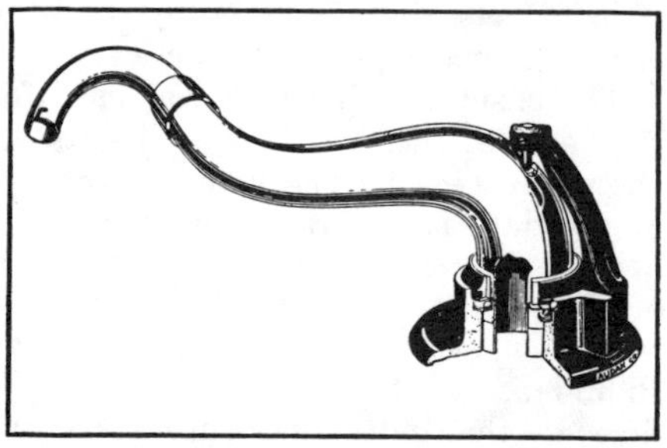

Fig. 1. One of the latest developments in acoustic tone-arms—that used on the Orthophonic instruments of the mid-twenties.

2. Most critical factor in a reproducer is vibratory momentum.

3. The larger the vibrating mass, the greater the vibratory momentum.

4. At the comparatively low frequency of 1,000 cps, the vibrating mass has to make 2,000 reversals per second—a terrific rate. But, before it can reverse in direction, it has to come to a stop, and the record groove does the stopping.

5. Obviously, the tinier the vibrating mass, the easier and cleaner the reversal and the less the "hangover."

6. Hangover causes the worst type of distortion.

7. Needle radiation is caused by stylus itself acting as a diaphragm. However, such radiation is very small and can be heard only when very near to it. Hold a steel needle, loosely, between thumb and forefinger, play it on a 78 r.p.m. record. The sound heard will be due to needle radiation.

8. Needle talk or needle noise is due to three factors:
   a. Heavy vibrating mass.
   b. Lack of sufficient lateral compliance.
   c. Lack of sufficient vertical compliance.

9. Needle noise is produced by stylus stiffness or heavy vibrating mass, or both, forcing the record to act as a diaphragm.

10. Needle noise is annoying when the record player is in the same room with the listener. However, the seriousness of needle noise is in the damage inflicted on the record groove by the factors which cause the noise.

11. The constant, inexorable pounding of the groove walls will soon show itself in increased distortion and gradual destruction of the disc.

Fig. 2. Early commercial magnetic pickup, showing similarity in construction to acoustic model.

12. A practical way to judge reproducer compliance is by listening for needle-noise from across the room, when playing a 78-r.p.m. record.

13. The greater the needle-noise, the greater the *lack* of compliance or the larger the vibrating mass—or both.

14. High vertical compliance is as important as is high lateral compliance.

15. Lack of sufficient vertical compliance greatly accelerates groove erosion.

16. Twice during each cycle the stylus hits a constricted part of the groove (pinch effect) with a terrific "bang." The effect of this is similar to what happens when the dangling chain on the rear of truck bounces along the pavement.

17. The reproducer must be designed so as not to permit frontal oscillations—to and fro movement of the stylus.

18. When a reproducer possesses the desired vertical compliance, it is important that it be so designed as not to permit vertical generation. Such generation introduces second harmonic and vertical-noise distortion.

19. For the compliance to be effective, the lateral compliance of the tone-arm should be at least equal to the lateral compliance of the reproducer.

20. Science knows of no way to determine the musical quality of a pickup. As with musical instruments, it must be put to the only test that really matters . . . listening.

21. Frequency curves and other technical impedimenta are necessary guides in the laboratory. However, the important fact is that *the ear is the absolute and final judge of musical quality.*

22. The higher the quality of the pickup, the more important is correct matching.

23. In general, the higher the quality of a pickup, the lower the output.

24. Reproducing sound from disc is based on *solid* mechanical coupling between stylus-point and groove—not mere touch, but solid coupling.

25. Most popular is the belief that the lighter the needle force, the lower the record-wear. In general, this is so down to a certain point. Reducing the needle force below that critical point will wear the record faster.

26. Even with infinite compliance, there is a definite downward limit to which the

needle force can be reduced. This downward limit is determined by the point at which "groove skating" begins.

27. Reducing the needle force to the point of groove skating will result in distortion, and will damage the record rather than preserve it.

28. Tests show that as a factor in record-wear, vibratory-momentum ranks as number one. It is even more destructive than a too low needle force.

29. In general, the pickup with the lowest vibratory-momentum will have the lowest distortion and the least record-wear, other things being equal.

Fig. 3. Modern high-quality magnetic pickup which provides for playing of two different groove widths with a needle force of 6 to 8 grams.

STYLUS

Much has been written on the subject of styli the past few years. Twelve years ago, there was an organized effort put behind the sale of "permanent" points (jewels) for use with pickups then in the homes. At that time the writer stated . . . .

> "It is safe to say that more than 90 per cent of the pickups now in use will not work properly with a jewel-point. Any attempt to use a jewel-point with such pickups will result in serious injury to the record."

This was an understatement, if anything. Yet, thousands of jewel-points were sold for use with such pickups (needle force 1/8 to 1/4 pound), ruining hundreds of thousands of discs. These styli were sold for as high as $50.00 each, net.

About four years ago, in an attempt to post the unsuspecting buyer, the writer decided on a lone crusade against the unscrupulous sale of "permanent" points. For several weeks, advertisements were run in 22 of the leading newspapers throughout the country. The cost of this campaign added up to a substantial figure. Apparently someone was being hurt by this campaign, for in October, 1949, the Federal Trade Commission called on the writer for details and explanations of the facts on which this campaign was based.

Some months later, the Commission issued injunctions ordering the vendors of jewel points to desist from making any further claims as to the number of plays, etc.

Stylus Pointers

30. *How many records can be played with a given jewel point?* This is one of the most frequently asked questions. Figures ranging from 2,500, 5,000, and even 10,000 plays have appeared in sales literature.

This is an important question and cannot be answered directly. It is like asking "How many miles of wear can be obtained out of a pair of shoes?" It will depend on the kind of shoes they are, of course, the weight of the person wearing them, the nature of the walk and—most important—the nature of the ground walked upon, such as gravel, concrete, grass, etc.

The number of plays obtainable from any jewel point depends on the kind of jewel and on the needle force. It further depends a great deal on the reproducer with which the stylus is used, the arm structure, and the nature of the material of which the record is made—this latter being especially important.

31. A jewel point that is good for 184 hours play in one case may last only half that time or even less, when used with a different reproducer, different arm, different discs, etc.

32. It is important for the stylus point to have correct radius for the particular use intended. For LP discs, a radius of .001 in. will give the best results; for 78 r.p.m. records, a radius of .0028 to .003 will give best results.

33. It is highly important for the stylus point to have the greatest degree of polish the particular jewel is capable of acquiring.

34. The included angle should be between 45 and 50 degrees.

35. With the advent of pickups with a needle force of 12 grams or less, the diamond has come to the fore as a desirable stylus.

36. The use of sapphire or diamond must be decided for each particular case. Each has some advantage over the other.

37. The natural sapphire is capable of acquiring extremely high polish.

38. The National Bureau of Standards rates the diamond as 4½ times as hard as the natural sapphire.

39. A diamond point will last longer than a sapphire point. However, *no* jewel point is permanent, be it diamond or sapphire. Therefore, periodic replacement is necessary for good reproduction and if the discs themselves are to be preserved.

40. A diamond is good for many more plays than a sapphire. However, there is no difference in the actual musical results, everything else being equal.

41. The musical performance with a compromise stylus-point will be—just as the name implies—a *compromise,* at best.

42. In the case of stylus-assemblies, the mounting of the jewel is but a small part of the procedure. The complete stylus assembly must then be *laboratory calibrated* for proper frequency response, etc.

43. *How does the user know when to change the stylus?* This is a question that keeps coming up over and over again. Here is a quick and simple method that will give a positive indication of the stylus condition:

> Obtain a lacquer disc with unmodulated grooves cut on both sides of it—one side for a 3-mil stylus and on the other for a 1-mil stylus. Both types of grooves may be on one side of the disc. Whenever in doubt, play two or three grooves on this disc. If the stylus leaves the grooves unchanged in luster and smoothness, you may be sure it is in good condition. If, however, the groove walls show score marks or any other difference when compared with the unplayed grooves, you will know that the stylus needs replacement. An ordinary magnifying glass will be of assistance.

This method is simplicity itself. It does not require dismounting of the pickup or removal of the stylus. It actually demonstrates itself.

TONE-ARM

In the old acoustic phonograph, the tone-arm had several highly important functions to perform. It was the connecting link between the reproducer ("sound-box") and the horn. This linkage had to provide lateral and vertical movements for the arm. At the same time these linkages had to be fairly air-tight. The tone-arm had to transmit the total acoustic energy generated by the reproducer diaphagm. It had to act as an acoustic preamplifier and it had to maintain a pre-determined relation between stylus and record groove.

Arm resonance—the great destroyer of record grooves—is observed only when sufficient initial exciting energy at the critical frequency is transmitted to the arm.

The stylus in the acoustic reproducer possessed great stiffness—very, very low compliance—and transmitted to the arm a large percentage of the stylus-generated vibratory energy. To avoid devastating resonance, damping was introduced between the reproducer and the arm to minimize exciter-energy from reaching the arm.

Tone Arm Pointers

44. In electronic sound reproduction, the arm has only one function—to maintain the stylus in a pre-determined relation to the record grooves.

45. The arm structure and assembly should be such as to introduce no restraint to the free travel of the stylus across the disc.

46. The greater the length of the arm, the better.

47. The greater the distance between stylus-point and the vertical pivots, the smaller the *change* in the angle of incidence between stylus and record groove.

48. The greater the distance between stylus and vertical pivots, the less the pitch-distortion due to turntable wobble, record warpage, etc.

49. Up and down movements due to pinch effect and small record surface irregularities, as well as minor eccentricity, should be absorbed by the vertical compliance of the stylus itself. A stylus capable of absorbing such movements frees the record from having to lift the total mass of the reproducer, practically eliminating record wear caused by this action. This is highly important because, no matter how light the reproducer, its total mass is considerable when compared to the tiny stylus point.

50. Since the tone arm guides the reproducer across the disc, it is important that the lateral arm compliance be at least equal to the compliance of the reproducer.

51. Where spring counterbalance is used, it should be checked periodically, preferably seasonally. This is an important factor with a highly-sensitized reproducer working with a needle force of 10 grams or less. A tiny variation of 4 or 5 grams means a variation of 50 per cent, resulting in groove skating and distortion.

52. For obvious reasons, the simpler the arm and the fewer the component parts, the better.

53. The greater the number of loosely coupled parts (not integral with the arm), the greater the possible number of resonance points.

54. The transmission of stylus-vibratory energy to the arm can be minimized, for all practical purposes, with a reproducer having high lateral and vertical compliances. Such condition prevents development of arm resonance.

55. The old practice of loading the arm with extra "dead" weight to push the resonance-point below a given frequency or to attain resonance at a certain frequency should be avoided.

56. When used with a highly compliant reproducer, arm loading introduces distortion. Furthermore, since the extra weight has to be pushed across the disc by the record grooves, serious erosion results.

57. The "McProud test" is still the best and simplest for checking the combined tone-arm and reproducer compliance (tracking). For details, see AUDIO ENGINEERING, August 1950.

58. For years past, many attempts have been made at devising a tone-arm that would maintain tangency automatically. Patent Office records show prolific activity along that line for the past 20 years. The parallelogram idea was the basis in most of these attempts. Any such mechanism would mean an increased number of component parts and joints. Thus far, such devices have all proved themselves inferior and total unsatisfactory.

These attempts overlook the highly important fact that the groove wall would have to do all the work to keep such mechanism in tangency. This would mean serious record wear. Furthermore, since the groove wall itself has to push such mechanism into tangency, obviously such tangency would then be too late to do any good. This condition is analogous to a trolley car moving slowly around a curve. The rail has to do all the work in constantly pushing the wheels to conform to the rail-radius. Everyone is acquainted with the fact that the outside rail on a curve wears out much faster

TURNTABLE

59. The first essential in a turntable is that it maintain a constant speed, if there is to be no pitch distortion due to the speed factor.

60. A simple and most practical method of testing the speed constancy of a turntable is by playing either a piano or a violin recording with long sustained notes. Where available, playing a disc with a constant frequency note—anything from 1,000 to 3,000 cps—is an excellent method to use. The human ear is very sensitive to changes in pitch (wow).

61. Ability of a turntable to maintain constant pitch is very important. It should be checked at all the speeds that are being used in actual playing.

62. The slower the speed, the more difficult it is to maintain constant pitch. Therefore, the slow speed, 33-1/3 r.p.m., should be checked carefully.

63. Where quality performance is sought, a quality magnetic pickup is assumed.

64. Up to a few years ago, pickups were made with heavy vibrating mass for high output. High output was necessary to make up for the lack of gain in amplifiers. It was also needed to overcome the stray fields due to poorly designed electric motors.

65. It is now generally accepted that today's amplifier must be built with sufficient gain to permit the use of a modern high-quality magnetic pickup.

66. Because of this, turntable motors are now engineered and built so that they will not interfere with the performance of a quality magnetic pickup.

67. Turntables designed to satisfy the factors outlined above are now available. Satisfactory multispeed turntables may now be had at a very modest cost.

68. A turntable should always be mounted so that the motor will be as far as possible from the reproducer head.

69. The turntable proper should be free of wobble.

70. When mounted, the turntable should be as level as possible.

71. A steel turntable should be solid throughout the area traversed by the stylus. That is, there should be no perforations of any kind throughout that area. As a rule, such perforations register in the loud speaker.

72. A good turntable should be free of rumble, both lateral and vertical.

73. A rumble not only registers in the speaker, but also increases record-wear.

74. Felt or rubber covering placed loosely on the turntable has a tendency to slip. This should be prevented if pitch variation is to be avoided.

Additional Phono Facts—1952

ONE OF THE IMPORTANT FACTORS in the public acceptance of LP discs is the fact that they can be played with a total absence of surface noise, commonly referred to as "scratch," that has plagued music lovers for over 50 years. It is of vital importance, therefore, to prevent these discs from becoming scratched or marred in any way.

Extensive investigation shows that the owner himself unwittingly inflicts mechanical injury—scratches—on his records.

Anything rubbed over the record will scratch the groove walls by rubbing the dust and grit into the delicate glossy surface. In the case of a very dusty record, it is better to fluff the dust off with a loosely waving handkerchief. When playing, a stylus attracts the dust, lint, and other matter lodged in the grooves. Therefore, brush the dirt off the *stylus* after each play—it will help keep your records clean.

Oddly enough, no protective envelope that would entirely protect a disc from dust and grit has yet been introduced.

Stylus Alignment

It is highly important for the stylus to be exactly vertical on the record, when viewed from the front, and that it does not lean against the groove wall.

An investigation on this condition extended over a period of eight months and included pickups and arms of all makes. Out of 74 installations, 39 were found to be badly out of alignment; that is, the stylus was tracking the groove at an angle—either right or left. An excellent method of checking stylus alignment is offered for consideration:

Place a small unframed mirror (such as may be found in a lady's handbag)—on the turntable. Now, lower the stylus point to contact the mirror in the same gentle manner you would use to lower it onto a record surface. Standing squarely in line with the front end of the arm, so that you do not see the sides of the arm, you will be able to see the stylus alignment easily, because the mirror causes the angle of misalignment to appear twice as great.

As there must be some play in the mounting of a pickup in order for the parts to fit together readily, frequent misalignment is to be expected. A slight shift of the reproducer—clockwise or counterclockwise—will cause the stylus to lean against one of the walls of the groove.

The arm also may be out of alignment. It should be adjusted so that it has no tendency to swing to the right or to the left when the pickup is supported in such a way that it does not touch the record surface. Arm misalignment can be corrected easily by shimming up one side of the base with a piece of cardboard.

De-emphasis

Last, but not least, is the question of circuitry—principally that of the correct de-emphasis network. If LP records are played with a "flat" high end, there is likely to be some noise. Most users are of the belief that the record material alone is responsible for the normally lower noise level of LP's, but such is not entirely the case. While the record surfaces are naturally quieter, a large part of the noise reduction is due to the use of pre-emphasis in recording, and the corresponding de-emphasis in playback.

In practice, pre-emphasis boosts the recorded level of the high frequencies by a predetermined amount—16 db at 10,000 cps according to the standard LP curve, or 12 db at 10,000 cps with the AES curve. In playback, therefore, a similar amount of high-frequency attenuation—de-emphasis—must be employed to restore the frequency response to normal. However, since scratch is composed largely of the higher frequencies, the ultimate effect is that the scratch is reduced. Combined with the quieter record materials, the de-emphasis serves to reduce scratch to an almost inaudible level.

Properly played—and carefully taken care of—LP records offer the best quality and a minimum of noise, but since the surface is somewhat softer than shellac's, the element of care takes on a greater importance.

RECORD CARE

1. Never use a cloth—moist or dry —to clean your records.
2. Never **rub** the record surface with **anything.**
3. When in doubt about using anything on a record—other than the stylus itself—just ask yourself if you would use it on a valued soft lacquer disc.
4. Correct any stylus-record misalignment.
5. Check stylus alignment every time the head is installed on the arm after being removed for repair, stylus replacement, or for any other reason.
6. Never play a record with a stylus out of alignment.
7. If your amplifier has no provision for proper de-emphasis, the maker of your pickup will supply the proper circuits for optimum results.

AES Standard Playback Curve

A presentation of the background behind the choice of the reproducing curve which is rapidly becoming accepted throughout the recording industry.

BASED ON THE PREMISE that the proper approach to the problem of equalizing disc recordings and transcriptions is to standardize on a *playback* curve and to let the recording engineers make their records however they see fit, knowing that they must sound properly balanced when played on this standard reproducing characteristic, the Audio Engineering Society adopted such a curve in 1950, follow- the action of the Board of Governors in approving the report of the Society's Standards Committee which consisted of the following: Gordon Edwards, chairman; S. E. Sorensen, vice chairman; James Bayless, Harry Bryant, and Russell Hanson, members of the Western Division; and Theodore Lindenberg, N. C. Pickering, A. A. Pulley and Ralph Schlegel, members of the Eastern Division. Robert Liesenberg served as alternate to Mr. Sorensen.

The standard curve, shown in *Fig.* 1, is represented by the values in Table 1.

The decision to specify a standard playback response characteristic instead of a recording characteristic was deliberate on the part of the Standards Committee. This course was chosen because of the impossible task of achieving a universal recorded characteristic compatible with all individual recording conditions and systems.

Reference to the tabulation will indicate that all points on the curve are related to 1000 cps. This reference point has been used as a standard for many years, making it evident that the maintenance and calibration of equipment would be expedited by retention of this frequency as a reference point. Furthermore, the slope of the curve at this point is sufficiently flat so that an error of 10 per cent in frequency will produce a deficiency of not more than 0.5 db.

TABLE 1

Frequency	db	Frequency	db
30	+22.5	1500	− 1.5
40	+20	2000	− 2.2
50	+18	2500	− 3
70	+15	3000	− 4
100	+12	4000	− 5.5
150	+ 8.5	5000	− 6.7
200	+ 6.5	6000	− 8
300	+ 4.5	7000	− 9
400	+ 3	8000	−10
500	+ 2	9000	−11
800	+ 0.5	10000	−12
		12000	−13.5
1000 (ref.)	±0	15000	−15.5

Permissible tolerance ±2 db

The majority of engineers active in the recording field have felt for some time that the degree of high-frequency emphasis prescribed by the NAB transcription characteristic is excessive. The trend in modern microphones and amplifiers to a wider frequency range, approaching 15,000 cps, and the use of acoustically brighter studios have made this problem much more difficult. With this extended range, the acceleration of the reproducing stylus becomes a limiting factor. Consequently, it was deemed necessary to restrict the degree of high-frequency rise used in recording. This was accomplished by making the reproducing characteristic roll off only 12 db at 10,000 cps—approximately 3 db below the NAB specification—and continuing the response out to 15,000 cps. By doing this, the high-frequency situation has been alleviated somewhat. Since microphone and studio characteristics must be considered by the recording engineer, it is required that the sum of the electrical rise in the recording equipment and the acoustical rise in the microphone must not exceed the values shown by the reciprocal of the reproducing characteristic, unless it is intended to make the high end over-brilliant.

The low-frequency characteristic was chosen to fall somewhere in the middle of the numerous low-frequency curves now in use. It is felt that the turnover frequency is low enough to keep rumble down to reasonable levels, and high enough to avoid excessive amplitude and intermodulation at low frequencies. It will be noted that no "shelving" of the characteristic at low frequencies is recommended. Again, if the recording engineer desires for some reason to have a "bassy" sound, he can easily accomplish this by making his recording characteristic tip up at the low end; conversely, he can "thin out" the sound by the opposite procedure.

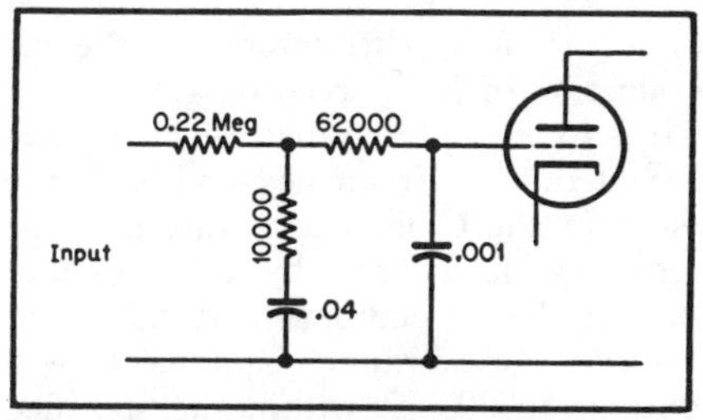

Fig. 2. High-impedance network to provide standard AES playback curve in grid circuit of amplifier stage.

The shaping of this curve can be duplicated on a flat playback system with two sections of RC equalization, as shown in *Fig.* 2 which is one possible arrangement for use in an amplifier circuit. Both of the straight portions of the curve are slopes of 6 db per octave. The intersections of these slopes with the reference axis occur at 400 cps and at 2500 cps. At these points the response is 3 db away from the reference level. Within a tolerance of ±2 db it will be seen that all turnovers between 325 and 500 cps will fall in the area covered.

The adopted response curve (within its tolerances) is sufficiently parallel to the NAB response curve so that no

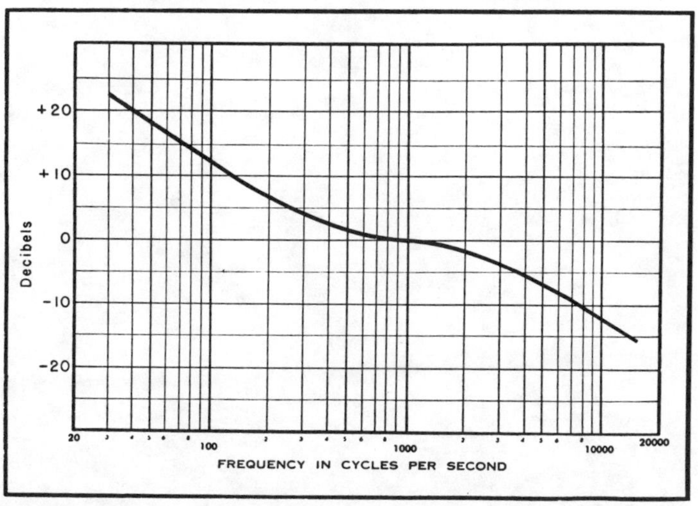

Fig. 1. Response curve of reproducing equipment for "AES" equalization.

PLAYBACK DE-EMPHASIS NETWORK

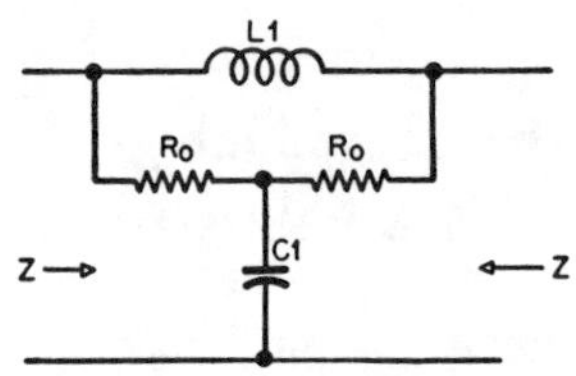

Z (ohms)	L1 (mh)	C1 (µf)	R_0 (ohms)
150	9.55	0.408	150
250	15.9	0.255	250
500	31.8	0.127	500
600	38.2	0.102	600

RECORDING PRE-EMPHASIS NETWORK

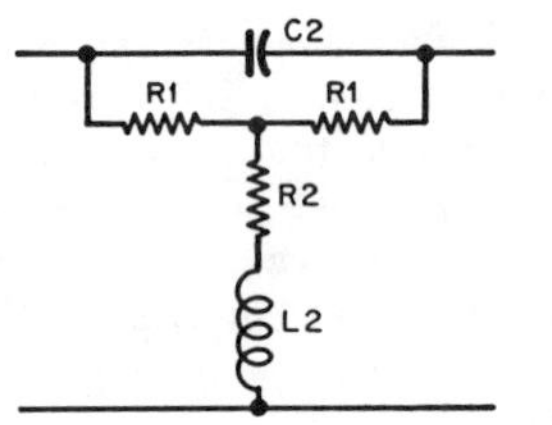

Z (ohms)	C2 (µf)	L2 (mh)	R1 (ohms)	R2 (ohms)
150	.0472	1.045	123	30
250	.0284	1.74	204	50
500	.0142	3.49	408	101
600	.0118	4.19	491	121

PLAYBACK L-F BOOST EQUALIZER

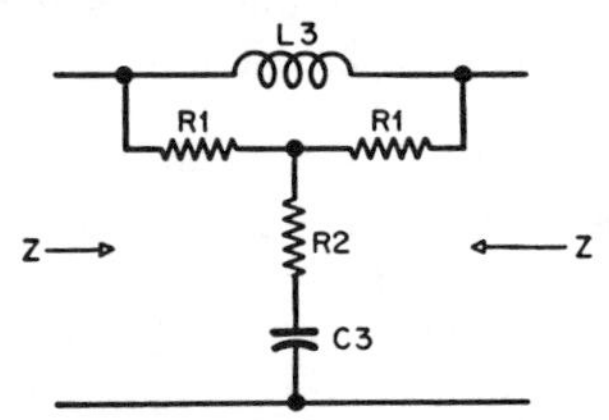

Z (ohms)	L3 (Hy)	C3 (µf)	R1 (ohms)	R2 (ohms)
150	0.545	24.08	123	30
250	0.910	14.44	204	50
500	1.81	7.22	408	101
600	2.18	6.02	491	121

Fig. 3. Constant-impedance networks suitable for line impedances indicated.

problem will be encountered in the reproduction of NAB recording.

It is to be expected that the characteristic at the low-frequency end will stop rising at the 6 db/octave rate at some frequency determined by the range of the reproducing equipment. It is felt that first-class wide-range equipment will continue to 30 cps within the specified tolerance and then flatten off as rapidly as possible. Where equipment has a higher low-frequency cutoff, it is recommended that the reproducing characteristic follow the curve to its lower limit and then drop off as rapidly as possible.

On the high-frequency end, it is recommended that the reproducing characteristic be followed to the desired upper frequency cutoff, above which point the response should drop off smoothly and rapidly. In wide-range equipment it is expected that the playback characteristic will follow the curve to 15,000 cps within the tolerance specified, and then drop off rapidly above this point.

Typical Equalizing Networks

The equalizers of *Fig.* 3 are shown in order to facilitate the construction of these networks for use in professional installations. The Playback De-Emphasis Network is designed to give the proper roll-off characteristic in circuits of the impedances shown. If used with existing equalizers in playback circuits, the high-frequency response should be set on "flat" to obtain the proper curve.

The Recording Pre-Emphasis Network is designed for insertion in circuits of the indicated impedances ahead of the main recording amplifier. It is presumed that modifications will be made in the cutter network to obtain the desired low-frequency response. For information on the methods of adjusting these circuits, it is suggested that the engineer make inquiry from the cutter manufacturer.

While most installations will already have some form of low-frequency equalizer for reproduction of existing types of records and transcriptions, it is possible that an entirely separate network will be required. The Playback Low-Frequency Boost Equalizer is designed to give a turnover frequency of 400 cps, with a total insertion loss (at 1000 cps) of 20 db. The half-loss point is 125 cps, and this equalizer will result in a slight decrease in response over the projected curve below about 70 cps. However, it falls within the limits down to 45 cps, and the decrease in response below that frequency may be an aid in reducing rumble.

All of these networks are designed to have constant impedance characteristics, and since they are symmetrical they may be used without regard to input or output connections. All networks shown are unbalanced, and usual transposition methods can be used to convert them to balanced networks if such are required in any particular installation.

Conclusion

The new standard playback curve, if accepted by the Recording Industry, can achieve at long last a common platform for the reproduction of all recordings regardless of speed, groove dimensions, or manufacturer.

Recording Characteristics

C. G. McPROUD

A discussion of the various characteristics employed in making phonograph records, together with the reasons for their use.

That the ultimate in disc recording is to make the reproduced signal as near as possible to the original seems to go without further elaboration. This applies, of course, to any recording, but this discussion is solely about disc recording systems. To achieve faithful reproduction, therefore, it would seem that if all the components were made as perfect as it is possible to make them, it would suffice to record the signal and play it back with completely *flat* equipment. However, due to a number of limitations, this is not feasible, and the curve employed in recording is definitely not flat; consequently the curves employed in reproduction must deviate from uniformity. It is the purpose of this discussion to elaborate on the reasons for deviation in both the recording and the reproducing operations.

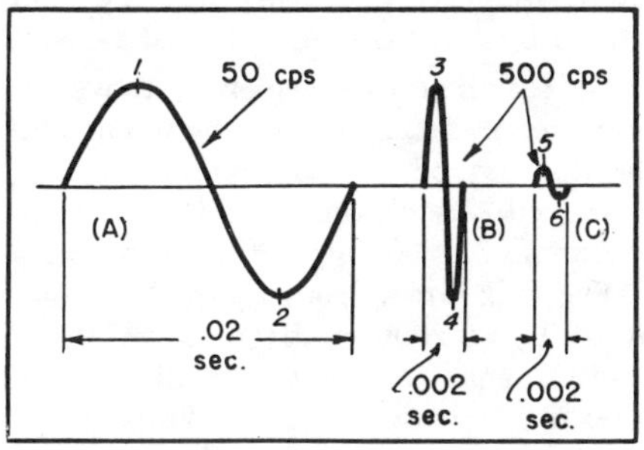

Fig. 1. Comparison between sine-wave signal plotted for 50 and 500 cps: at same amplitude, A and B; and at same velocity, A and C.

The principal variations from flatness in the recording process are due to mechanical limitations. In order to provide as high a signal-to-noise ratio as possible, it is desirable that the recorded level on the disc shall approach the maximum permissible value without overloading. Any further discussion of this problem necessitates defining certain recording terms, which will be done shortly.

It may be said, however, that low-frequency equalization is employed to reduce the possibility of overloading and that the high-frequency equalization is resorted to in order to reduce surface noise. The resulting recording curve is one in which the low frequencies are reduced in level at a constant rate, beginning at some predetermined point called the "turnover" frequency, and that the high frequencies are increased in level at some predetermined rate. The high-frequency equalization is commonly called "pre-emphasis" and must be compensated for in reproduction just as the low-frequency equalization is.

Low-Frequency Equalization

There are two basic types of recording. One of these is known as "constant amplitude" and the other as "constant velocity." Both terms apply to the tip of the recording or reproducing stylus as it traces the groove. In constant-amplitude recording, the stylus tip at a given signal level moves a fixed distance each side of its center or rest position for any frequency. Thus the amplitude of the swing of the stylus tip is constant.

In constant-velocity recording, the maximum velocity of the stylus tip at a given signal level remains constant for any frequency. Considering a sine wave as being applied to the recording head, the maximum velocity occuring during each cycle is as the stylus tip is crossing its center or rest position. *Figure* 1 is used to clarify this point. (A) shows a sine wave at a frequency of 50 cps, requiring a period of time .02 sec. in length. (B) shows a sine wave at a frequency of 500 cps, requiring a period of time of .002 sec. Both of these are shown at the same amplitude, or displacement, of the stylus from its rest position. From these diagrams, it is seen that the stylus requires .01 sec. to move from *1* to *2* in (A), and .001 sec. to move from *3* to *4* in (B), and that both of these distances are the same. Therefore, the velocity of the stylus point must be 10 times as great for the higher frequency since it must move over the same distance in 1/10 the time. If the velocity of the stylus were held constant, then the displacement or amplitude of the swing would be reduced to one tenth of its previous amount, as at (C).

Thus, in a constant-amplitude recording the velocity increases with frequency, although the displacement of the stylus point remains constant. In a constant-velocity recording, the amplitude decreases with frequency, but the peak velocity of the stylus point remains constant. In all of this discussion, it is assumed that the signal level is held at a given fixed point.

Practically all phonograph records and transcriptions are made with a combination of these two characteristics, with the change from one to the other occuring at the turnover point. In order

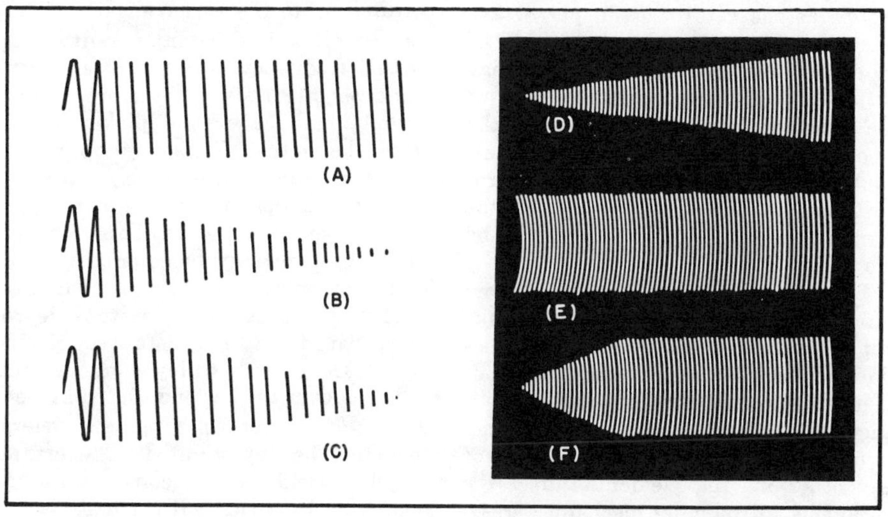

Fig. 2. Amplitude of stylus swing for swept-frequency signal at: (A) constant amplitude; (B) constant velocity; (C) commercial constant velocity. Corresponding light patterns are shown at (D), (E), and (F).

to limit the swing of the stylus at the low frequencies, recordings are normally made at constant amplitude from the lowest frequency up to the turnover point, and at constant velocity above the turnover point. *Figure* 2 shows a typical groove for a swept-frequency signal from 50 to 10,000 cps at constant amplitude at (A), at constant velocity at (B), and at "commercial constant velocity" at (C). Commercial constant velocity is the term given to a curve which is at constant amplitude up to the turnover, and at constant velocity above. If the actual record were viewed with a distant light illuminating the grooves, the pattern due to (A) would appear as at (D); that for (B) would appear as at (E); and that for (C) is shown at (F). This latter is the familiar "Christmas tree" pattern which is almost universally used to evaluate performance of recording apparatus. It is characteristic of this method of illumination that a constant-*velocity* recording will produce a light band of fixed width, and thus the width of the band at any point may be used to compare the actual "velocity" of the groove throughout the frequency spectrum.

Pickup Response

Different types of pickups respond to the recording methods in different ways. A velocity-actuated device—such as a magnetic pickup—will produce a constant-voltage output from a constant-velocity recording. This is due to the fact that the voltage is generated by the movement of a conductor through a magnetic field, or by the variation of a magnetic field which passes through a coil. Since the voltage generated in this manner is proportional to the velocity with which the lines of force and the conductor move with respect to each other, the voltage output from a magnetic pickup is flat over the constant-velocity portion of a recording, and droops at the rate of 6 db per octave over the constant-amplitude portion of the recording. Conversely, the voltage output from a crystal pickup is flat over the constant-amplitude portion of the recording, because the voltage generated by a crystal pickup is directly proportional to the *displacement* of the crystal, which is in turn actuated by the stylus.

There are a number of other factors which enter into the actual voltage output from a pickup, but these are functions of the mechanical characteristics of the device. The masses and compliances of the stylus and its supporting structure, of the pickup arm, and of any other moving elements, affect the output by introducing mechanical resonances which show up as peaks and dips in the frequency response of the pickup.

To clarify one point, however, it must be said that if a recording were made with a crystal cutter without any equalization, and played back with a crystal pickup, also without equalization, the output would be exactly like the input, provided both cutter and pickup were perfect. Similarly, if a recording were made with a magnetic cutter without equalization, and played back with a magnetic pickup—also without equalization—the output would again be exactly like the input, assuming that both pickup and cutter were perfect. However, pickups and cutters are not perfect, and in addition, the equalization previously referred to is normally applied, so some equalization must also be applied in the reproduction.

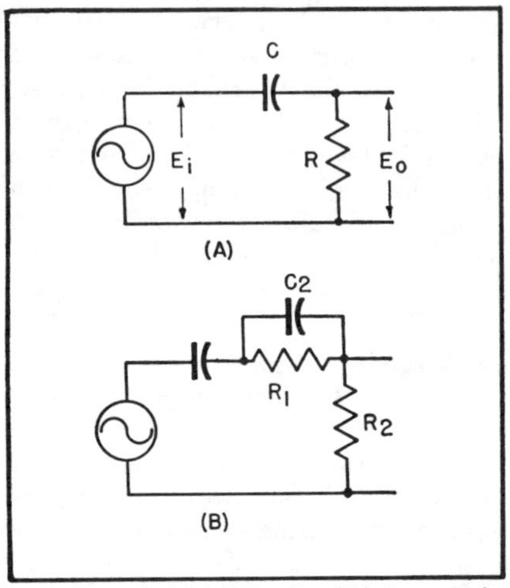

Fig. 3. Crystal pickup circuits. (A) is equivalent circuit when crystal pickup having capacitance C works into the load resistance R. (B) shows usual method of compensating for high-frequency droop.

Equalization Required

Practically all commercial recordings are made with magnetic cutters, and consequently there is a certain amount of equalization introduced to limit the excursion of the stylus, during the recording, up to the turnover point. Thus the recording is constant amplitude up to this frequency, and essentially constant velocity above. Therefore, the reproduction by a crystal pickup is flat from the lowest frequency up to the turnover point without any equalization, and droops at the rate of 6 db per octave above that point. This demands that crystal pickups be equalized on the high end only, since it is a characteristic of these devices that they reproduce flat from a constant-amplitude signal. In order that they should actually be flat over the low end, the equivalent circuit of a crystal pickup as a generator must be investigated. The crystal may be considered as a constant-voltage generator in series with a capacitance equal to that of the pickup itself. This capacitance is of the order of .0015 μf, and therefore the load resistance must be chosen so that there is adequate transmission of the low frequencies. The circuit may be likened to that of the coupling capacitor between two amplifier stages, and it is remembered that the size of the capacitor affects the low-frequency response.

The capacitor and the grid leak or load resistor may be considered as a voltage divider, with the reactance of the capacitor acting as the top section and the load resistance as the lower, as shown in *Fig.* 3. At the frequency where the reactance of the capacitor equals the resistance of the load resistance, the response is down 3 db from the maximum. This accounts for the requirement that the load resistor for a crystal pickup shall be relatively high in value, usually at least 0.5 meg, and in many instances more than this. It also accounts for the statement that the low-frequency response may be controlled by the value of the load resistor.

To equalize the response above the turnover point, some arrangement similar to that of *Fig.* 3 (B) is generally employed. Assuming that the turnover frequency is 500 cps, the response at 1000 cps is down from the 500-cps level by 6 db; at 2000 cps it is down 12 db; at 4000 cps it is down 18 db; and at 8000 cps it is down 24 db. Thus if it is desired to equalize completely up to 8000 cps, it is necessary to introduce 24 db of loss to the low frequencies by means of a voltage divider R_1 and R_2. This requires that $R_2/(R_1 + R_2)$ must equal 1/16, since this is the voltage ratio corresponding to a 24-db loss. Then, the low frequencies are reduced in level to the 8000-cps level, but this would also introduce a similar loss at 8000 cps, so this effect is counteracted by shunting a capacitor across R_1 with its reactance at the 3-db point being equivalent to the resistance R_1.

Considering actual and typical values, let us assume that R_2 is taken as 0.1 meg; therefore, R_1 must be 0.1 meg. x (16—1), or 1.5 megs. For the case under discussion, the curve is flat at 500 cps, and down 6 db at 1000 cps. By drawing this curve on a sheet of graph paper, it will be observed that the curve is down 3 db at approximately 700 cps. Thus it is determined that the reactance of the capacitor C_2 must equal 1.5 megs at a frequency of 700 cps. The actual value of the capacitance may then be calculated from the relation $C = 1/2\pi f X_c$, or it may be obtained from an inspection of a reactance chart, which is commonly used. In this case, the required capacitance is approximately 130 $\mu\mu$f, and this value will give complete equalization for a crystal pickup at high frequencies up to 8000 cps.

There is just one thing wrong with this equalization, however. Records are not normally cut with a flat frequency response all the way up from the turnover point. The NAB curve, for example, as well as the standard LP curve, are both cut on the basis of a pre-emphasis of 100 μsec, of which more later. With this curve, then, the recorded signal is already boosted by approximately 14 db at 8000 cps, and the required equalization is thus reduced to only 10 db at this frequency, and the calculations must be revised.

This discussion should give a rough idea of how equalization is arranged for one type of pickup. Let it be said that if the response of any given pickup is known for a flat frequency record, it is fairly simple to determine exactly the equalization required to obtain flat response. It then becomes necessary to know the exact recording characteristic in order to make a good match. These characteristics differ appreciably, although there are a number of more or less definite curves in common use today.

Turning for a moment to the magnetic pickup, it will be remembered that the response for constant amplitude recording droops at the rate of 6 db per octave below the turnover point. This requires a boost of the low frequencies, in direct contrast to the crystal pickup, but for a flat recording above turnover, no high-frequency equalization is required. Actually this does not obtain in practice because of the pre-emphasis employed, and some high-frequency droop must be introduced intentionally.

Low-frequency equalization for magnetic pickups may be obtained in a variety of ways, all of them about equally effective, but differing appreciably in circuit design. The output signal from these pickups is usually quite low—ranging from 10 to 100 millivolts—and some amplification is required to boost the signal up to the level of radio tuner outputs, in order to facilitate switching.

In recording for 78-rpm records, it is common practice to use a groove approximately 6 mils in width, and with 88 to 104 lines per inch. At 96 lines per inch, the centers of the grooves would therefore be 10.4 mils apart, and the lands—the spaces between the grooves—would be 4.4 mils in width. Referring to *Fig.* 4, the allowable swing for the stylus would be 2.2 mils each way, or slightly less, to provide for some solid material between peaks of the signal, as at point *X*. If the groove were exactly tangent at such points, the walls would be likely to break through on repeated playing. Naturally,

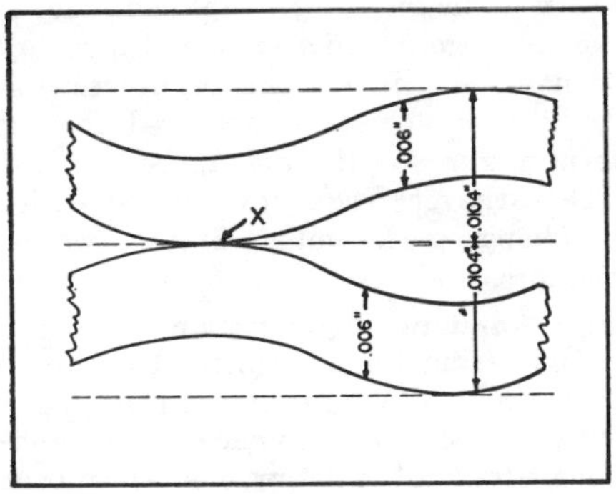

Fig. 4. Typical dimensions for a recording groove modulated slightly more than 100% at a low frequency.

the peaks of two adjacent grooves might not coincide as shown in the drawing, but it must be assumed that such a coincidence is possible, and allowance made for that possibility. Thus the allowable swing at the lower frequencies is of the order of ±2 mils.

If all frequencies were recorded with the amplitude of swing at this maximum, it would be a "constant amplitude" type of recording. At high frequencies, however, the stylus would have to make very sharp reversals, as shown at (A) in *Fig.* 5, with a resultant high velocity. As a matter of fact, it is doubtful if any stylus could follow such a groove at 10,000 cps. At constant velocity of stylus-tip movement, the groove would be more like (B) of *Fig.* 5. Practically, therefore, the groove is most usually cut with constant amplitude up to some intermediate frequency—generally somewhere between 250 and 800 cps—and at decreasing amplitude but at constant velocity above this frequency, which is known either as the transition frequency or the turnover frequency.

Equalization Methods

Methods of equalization for crystal pickups were discussed last month. For magnetic pickups, another type of circuit is required because of the difference in the output signal characteristics. Since magnetic pickups have a flat response from a constant-velocity recording, their output decreases at a fixed rate for a constant-amplitude recording, this rate being 6 db/octave. Therefore, the output for a magnetic pickup drops below turnover at this same rate, and must be compensated for.

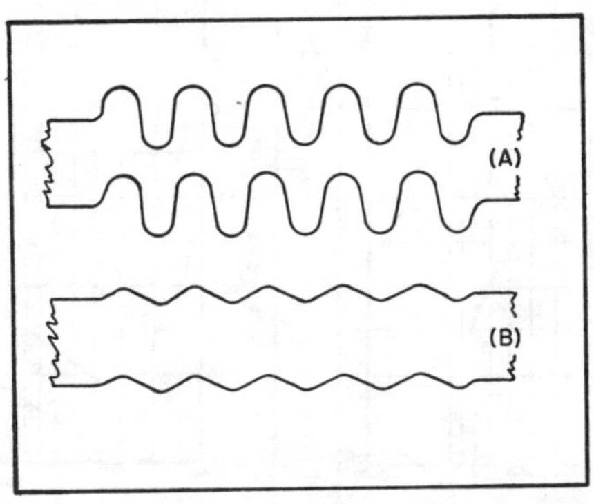

Fig. 5. (A) High-frequency signal recorded at 100% modulation on constant-amplitude basis. (B) Same signal recorded on constant-velocity basis.

Figure 6 shows two usual ways of providing the required low-end boost to compensate for this characteristic. (A) consists of a losser circuit placed between two high-gain triode stages. This type introduces a loss which increases gradually from zero at zero frequency up to a maximum at the frequency where the reactance of the capacitor becomes a small portion of the total impedance of the shunt circuit across the line. Above this frequency the loss is practically constant. The output voltage for zero frequency is essentially equal to the input voltage; at high fre-

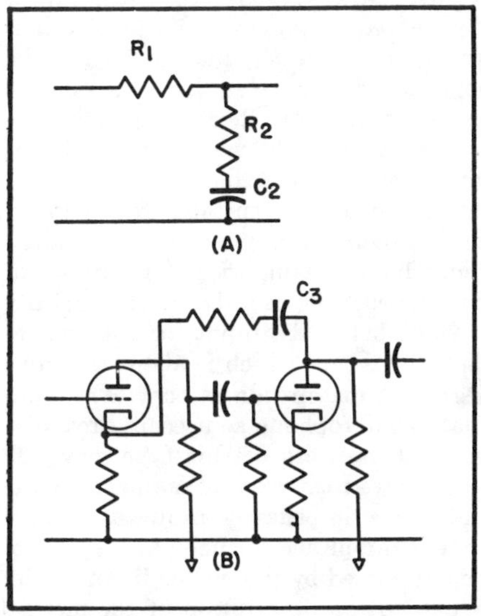

Fig. 6. Losser circuit to provide low-frequency boost. This circuit is usually placed between two high-mu triode stages. (B) Feedback circuit commonly used in preamplifiers to provide required low-frequency boost.

quencies it is approximately equal to $R_2/(R_1 + R_2)$ times the input voltage.

With one such equalizing circuit, consisting of the series resistor R_1, and the shunt resistor R_2 and capacitor C_1, the bend in the response curve at the turnover frequency is relatively rounded. Some equalizing amplifiers employ two such circuits, with an increase in the sharpness of the bend at the turnover frequency (see *Fig.* 7).

At (B) is shown a circuit which provides the required boost by varying the gain of the amplifier by means of a frequency-selective feedback circuit. In this circuit, the voltage fed back decreases as the frequency is lowered, with the result that the gain at low frequencies is greater than at high frequencies. Thus, the amplifier is suitable to compensate for the low-end droop employed in recording. The relative values of the capacitor and resistor in

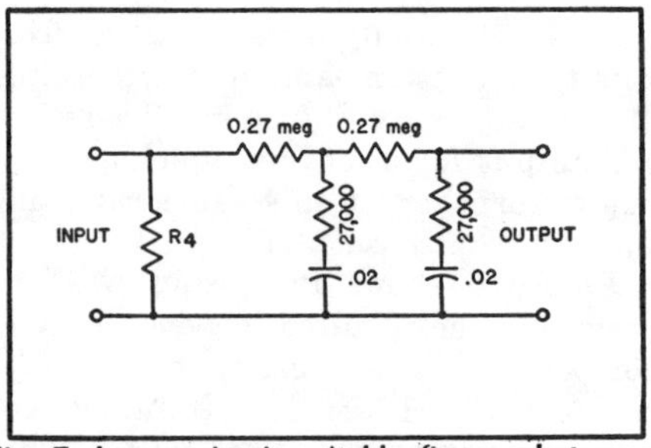

Fig. 7. Losser circuit suitable for use between magnetic pickup and high-gain, high-impedance input of P.A. system amplifier, such as normally used for crystal microphones.

the feedback circuit determine the turnover frequency.

In professional equipment, such as is used in broadcast stations, it is more common to employ a low-impedance equalizing circuit, such at that of *Fig.* 8. With this arrangement, it is possible to adjust the response for turnover at any desired frequency by suitable values for L_1, C_1, and R_1, and to adjust the slope below turnover by selection of suitable values for R_2 and C_2. This circuit is designed to work into a circuit with a nominal impedance of 250 to 600 ohms, and is thus suitable for connecting between the pickup and a standard microphone preamplifier. The output of this network is usually in the vicinity of 26 db below the input for frequencies above turnover, which is about the same degree of magnitude as that of a high-quality microphone as used in broadcast work. A network of this type may well be employed between a magnetic pickup and a low-impedance input of a P.A. system amplifier. Such an equalizer was discussed by the writer in an earlier issue[1], with a description of the method used to determine the component values.

Most P.A. amplifiers have only high-impedance inputs, so it is sometimes more convenient to connect an *RC* network of the type shown in *Fig.* 5 between a magn[illegible]kup and one of the high-gain microphone inputs of such an amplifier. The input resistor, R_4 is that normally employed for the termination for the pickup being used. The turnover frequency can be varied by changing the values of the two capacitors.

High-Frequency Equalization

Digressing for a moment, let us consider the power distribution in music with respect to frequency. A voluminous study of this problem has been reported by Sivian, Dunn, and White[2], and considerable work on speech has been reported by Fletcher[3,4].

These results are shown in the curves of *Fig.* 9, where (A) is the maximum power in ⅛-second intervals for male speech, and (B) represents maximum peak power in the same intervals for a 75-piece orchestra. While there is some reason to believe that higher efficiency in more modern microphones may modify these curves somewhat in the direction of more highs, they may still be used to show why it is possible to pre-emphasize the highs to some extent.

Assuming, for example, that the maximum probable power in a 10,000-cps tone occurring in orchestra music is 15 db below that for a 1000-cps tone, it is possible to record with a 15-db boost at 10,000 cps without exceeding the 1000-cps stylus velocity, since a constant velocity is assumed for flat frequency response. With such pre-emphasis, it is necessary that an equivalent de-emphasis be employed in the playback equipment, with a consequent reduction in surface-noise output, since noise is random and largely proportional to frequency.

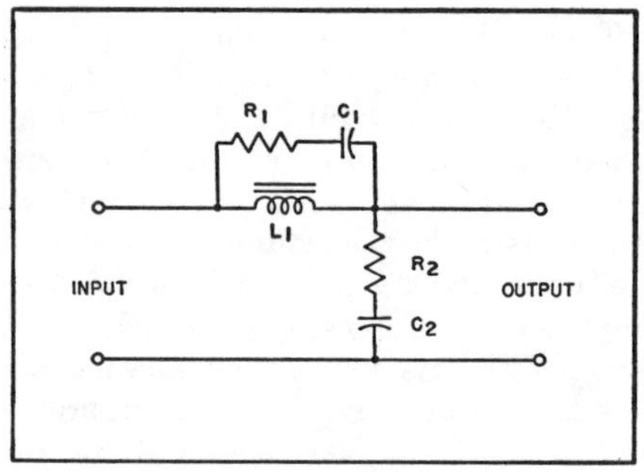

Fig. 8. Equalizer arrangement used between magnetic pickup and low-impedance input of microphone preamplifier.

Some pre-emphasis is employed in practically all commercial recordings made today. Typical curves are shown in *Fig.* 10, where (A) represents the NAB curve which is generally used with transcriptions and is standard for LP records. The curve used by RCA-Victor is believed to approximate that of (B), while the London (English Decca) curve is shown at (C). Undoubtedly there are some other curves in use, but these shown are commonly employed. If a phonograph amplifier is equipped with means to provide "rolloff" curves which are complementary to these recording curves, satisfactory results may be expected.

All of these rolloffs may be obtained by the use of RC networks, and many have been described in these pages[5], [6]. The reader is referred to these articles for details of the circuit arrangements. With correct rolloff curves, it is thought that little trouble should be encountered from surface noise, except in the case of badly worn records.

1 "Elements of residence radio systems, Part III." AUDIO ANTHOLOGY, page 113

2 "Absolute amplitudes and spectra of musical instruments and orchestras," *J. Acous. Soc. Am.*, Jan. 1931.

3 Speech and Hearing. New York: D. Van Nostrand Co., 1929.

4 "Physical Characteristics of Speech and Music," *Bell System Tech. J.*, July 1931.

5 St. George and Drisko, "Versatile phonograph preamplifier," AUDIO ENGINEERING, March, 1949. Reprinted in AUDIO ANTHOLOGY, page 58.

6 Howard Sterling, "Simplified preamplifier design," AUDIO ENGINEERING, Nov. 1949.

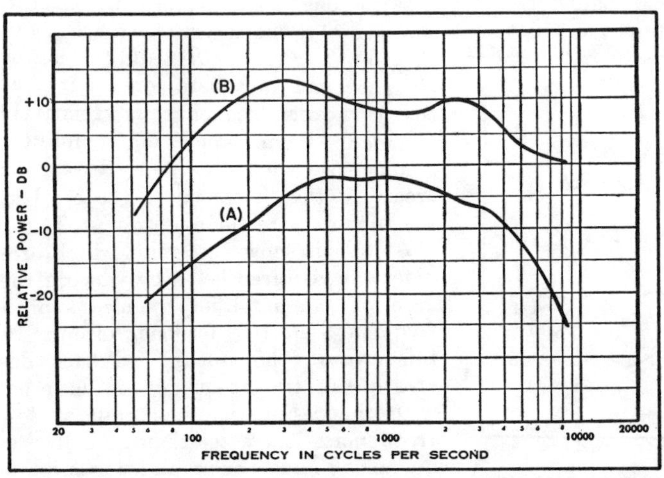

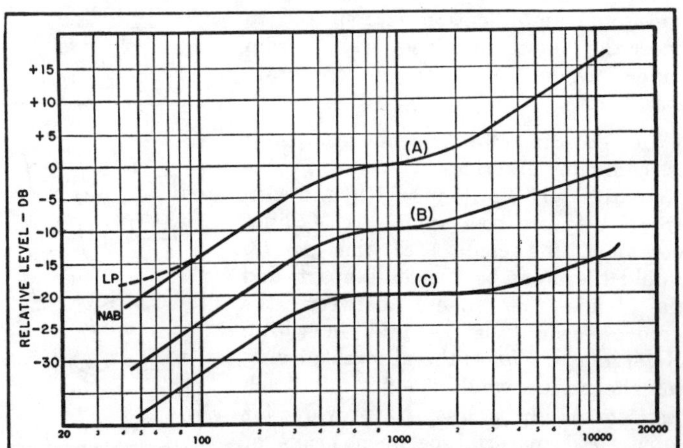

Fig. 9 (left). Curves showing maximum probable power in ⅛-second intervals in male speech (A), and in a 75-piece orchestra (B). Fig. 10 (right). Typical pre-emphasis curves: (A) NAB and LP; (B) presumed curve for older RCA-Victor 78-r.p.m. records; and (C) London ffrr.

Phonograph Reproduction

C. G. McPROUD

A discussion of the problems involved in designing a suitable preamplifier and control unit, and a practical circuit which provides sufficient flexibility for most home applications.

PHONOGRAPH REPRODUCTION can be enjoyable to a large extent only if is a fairly exact facsimile of the original performance. In spite of this, it is unusual to find that exactness in many systems because of a number of factors. One of these factors is the failure of the equipment designer to provide suitable electrical facilities to permit the control of de-emphasis to match the original recording. Another is the presence of annoying needle scratch, which can be extremely disagreeable in the case of worn records, many of which are sure to be highly prized by the owner. Still another factor is the tendency of the individual to overemphasize the high-frequency spectrum, as occasionally mentioned by Mr. Canby in his columns. The existence of "tweeters" in a speaker system creates a desire to "hear those highs" beyond the intent of composer or performer, with the result that the reproduced signal bears only a faint resemblance to the original.

Many of the preamplifier articles which have appeared in this compilation have said the same thing different ways, and many have provided means for achieving the required de-emphasis or "roll-off." None so far has provided for a sharper cutoff in the high-frequency response to reduce needle scratch to a reasonable level. It is the purpose of this article to describe the design steps taken to provide a suitable low-pass filter arrangement which can do the same work as a noise-suppressor circuit, but do it under the control of the operator rather than of the music itself.

Requirements

In this and the succeeding article, we shall attempt to outline the design procedure for a control amplifier designed to work with a power amplifier which is assumed to have no control except a simple gain control, preferably with discrete steps of the order of 5 db. Such an amplifier was described earlier[1] and in several additional forms and modifications in this volume, and it is for such an amplifier that the unit to be described was built.

[1] C. G. McProud, "Residence radio systems, Part III." AUDIO ANTHOLOGY, page 113.

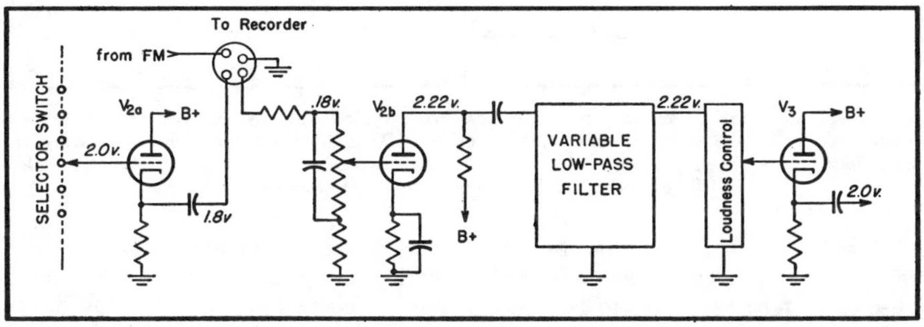

Fig. 1. Block diagram to show basic circuit before refinements of design provide complete schematic.

Before commencing any design problem, it is first necessary to set down the requirements so that they may be met successfully. Since this unit is to be used for a home entertainment system, it must provide for radio and phonograph, and for this particular application we shall add a wire recorder, just to make it more interesting. The writer finds it desirable to be able to record one radio program while listening to another. For this purpose a Webster Model 178 wire recorder is used because of its one-hour capacity and because it is simple to operate.

Therefore, with phonograph provision, assuming the use of one of the high-quality, low-level magnetic pickups, an equalized preamplifier is necessary. Certain modifications must be made in the design to accommodate the different makes of pickups because of the wide difference in output voltages. To play all types of records demands a number of different roll-off characteristics, since the de-emphasis required to play ffrr records correctly is hopelessly inadequate for LP's. An intermediate condition must be supplied to fit the Victor curve correctly.

It is also desired to have some control over the low-frequency response, and for this reason we shall incorporate a bass control which will give a number of curves which range from a slight droop up to a boost of about 10 db at 50 cps. We also wish to incorporate a loudness control, of course, and a series of low-pass filters. Last, in order to feed the main amplifier a few feet through a shielded cable, the output impedance of the control unit must be reduced to something of the order of 600 ohms, preferably without using a transformer. The remaining control to be furnished is a selector switch to permit choice between phonograph with any of three different turnover frequencies, AM or FM

Fig. 2. Circuit of bass-boost control (insert) and curves obtained when working into unloaded grid.

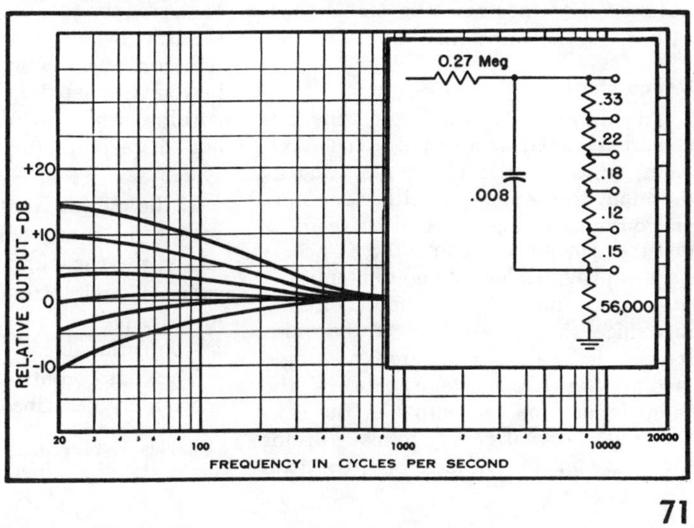

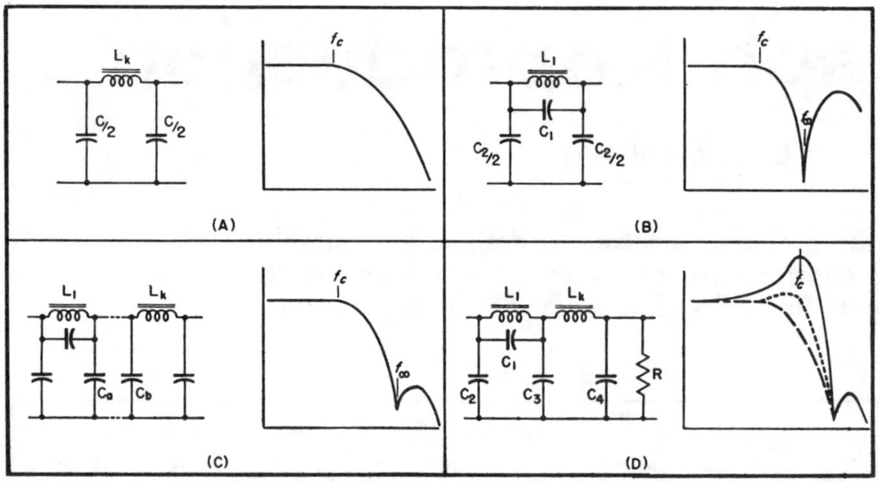

Fig. 3. (A) Constant-k low-pass filter section, with typical response curve; (B) m-derived section; (C) combined sections with resulting curve; (D) complete filter with terminating resistor. Solid curve shows response when R = 0.5 meg; dashed line is curve for tabular values of R; dotted curve for R higher than optimum values.

radio, and one additional position to work with a low-level output from a high-quality tape recorder, to be added at some time in the future. The wire recorder is to be connected only to the AM radio receiver, and any manipulation of the selector switch should not disturb this connection. For playback from the recorder, its own push-button switch shall open the circuits and feed the playback signal into the output circuit.

TABLE OF VALUES

f_c	L_1	L_k	C_1	C_2	C_3	C_4	R
11000	.75	1.25	90	100	450	350	56,000
9000	"	"	133	150	650	500	47,000
8000	"	"	170	200	820	620	39,000
6500	"	"	250	300	1300	1000	33,000
5000	"	"	400	500	2000	1600	27,000

Fig. 4. Component values for (D) of Fig. 3 for five different cutoff frequencies.

None of these requirements is difficult of attainment, and the design proceeds in a simple straightbackward manner—backward because the output signal must be approximately of a certain magnitude determined by the gain of the main amplifier. However, it is also necessary to accommodate the incoming signal magnitudes at the same time.

Design Steps

The output from the radio tuners is essentially fixed at about 2 volts maximum, which is also about the same as the input to the main amplifier for normal output. Because of the desire for a low-frequency boost control, it is necessary to provide about 20 db of amplification, since most of the boost circuits introduce that much loss at the mid-frequencies. This calls for a single medium-mu triode stage between the radio inputs and the output of the control unit, and that is only to make up for the bass control. Besides, both the bass control and the loudness control should work into an unloaded grid in order to operate correctly. To obtain the desired low-impedance output to feed the shielded cable to the main amplifier, a cathode follower can be used. This circuit has a slight loss, of the order of 10 per cent. *Figure* 1 is a block diagram of the system from selector switch to output, with approximate signal voltage indicated at various points in the circuit.

Following the circuit from the selector switch, we first encounter another cathode follower. This is used to provide a low-impedance line to feed the wire recorder, since it is to be connected to the control unit by a short shielded cable. The Webster 178 recorder has four push-button switches—microphone recording, radio recording, playback, and out-of-circuit. It is planned to channel the entire signal through the recorder, with the "out-of-circuit" switch (marked LISTEN 4) shorting the two leads so the signal goes straight through the recorder. In the playback position (LISTEN 3), the circuit is opened and the output from the recorder is fed to the control unit. The AM radio signal goes to the recorder through another lead, regardless of the setting of the selector switch on the control unit, and is recorded by depressing the RECORD 2 button. RECORD 1 is for microphone recording, which also may be done without disturbing the remainder of the system.

If the output from the selector switch were to be connected to the recorder directly, the capacitance of the connecting cables would introduce a droop in the high-frequency response which would be objectionable.

The bass control is one which is thoroughly described by Merchant,[2] and which is particularly useful for this application. *Figure* 2 shows the curves to be expected for the circuit diagrammed in the insert. Note the use of discrete steps in the tone control switch.

[2] Charles J. Merchant, "Simple RC Equalizer Networks," *Electronics,* Feb. 1944.

Low-Pass Filters

While the design of filters can be highly involved, for our purposes it is not necessarily so. To provide a reasonably sharp cutoff, it is customary to employ two sections—one of the constant-*k* type and one of the *m*-derived type. These terms are used to designate the type of filter section. *Figure* 3 shows these two types in pi configurations. (A) is a constant-*k* low-pass filter, with a response which cuts off gradually commencing at the cutoff frequency f_c as shown. (B) is an *m*-derived section with a sharper cutoff, but with a rise in the response above f_∞, the frequency of maximum attenuation. This frequency occurs at the point of resonance in the parallel circuit L_1 and C_1. By combining these two types of sections as at (C), a more suitable curve is obtained. It is desired to have a number of cutoff frequencies ranging from around 11 kc down to about 5 kc, with a suitable distribution being 11, 9, 8, 6.5, and 5 kc. This should cover practically all requirements, provided another switch position is so arranged as to cut out the filters altogether.

It is customary to design filters for a specific value of impedance. However, with five different cutoff frequencies, it would be necessary to employ a total of ten coils or to use tapped inductances which would have to be made specially to order, and at a high cost. To simplify the filter, let us consider the possibility of using only two coils and changing the impedance of the circuit. Eliminating some of the original calculations, let us choose two specific coil values from manufacturers' catalogs and proceed from that point.

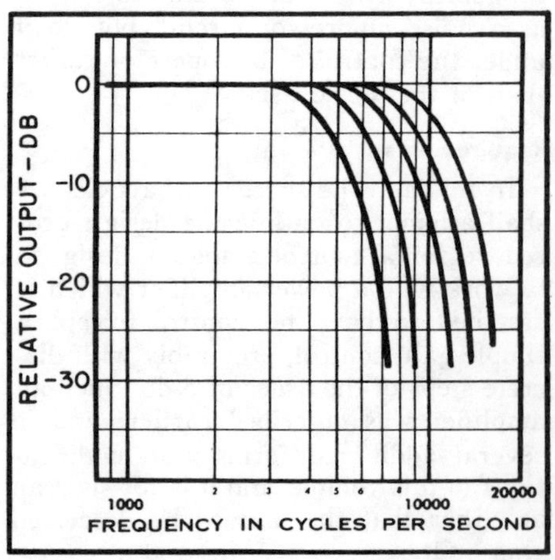

Fig. 5. Response curves obtained with filter switch in five cutoff positions.

These original calculations indicated that L_k should be approximately 1.25 h. for impedances of a suitable range. With the usual value of m at 0.6, L_1 would be 0.75 h., and both of these values are commercially available. The formulas for a constant-k section are

$$L_k = \frac{R}{\pi f_c} \qquad (1)$$

$$C_k = \frac{1}{\pi f_c R} \qquad (2)$$

Therefore, assuming a fixed value of 1.25 h. for L_k, R will vary with the cutoff frequency, f_c, since $R = \pi f_c L_k$. Calculating further, then, values of R for the five cutoff frequencies are shown to be 43,300, 35,400, 31,400, 25,600, and 19,700 ohms respectively, and the values for C_k may then be determined.

The formulas for the m-derived sections are

$$L_1 = m\, L_k \qquad (3)$$

$$C_1 = \frac{1 - m^2}{4m} C_k \qquad (4)$$

$$C_2 = m\, C_k \qquad (5)$$

With $m = 0.6$, the values for L_1, C_1, and C_2 may be calculated. We will consider later what to do with the varying impedance, R.

When the two sections are connected in series as at (C) of *Fig.* 3, the capacitors C_a and C_b may be combined since they are directly in parallel. Therefore, a table may be made up as shown in *Fig.* 4, with the values for the various capacitors rounded off. The values shown for R are approximately 1.25 times the nominal impedance of the filters, and are also rounded off to fit RMA preferred values. As mentioned before, the impedance of the filter changes with each switch position, yet the filters are being fed from a source of about 7500 ohms, which represents the impedance of the preceding triode tube. The performance of the filters under this condition may not fit the calculated curves exactly, but they will approximate the desired curves. If measuring facilities are available, the terminating resistor may be chosen on the basis of eliminating the peak which occurs just before the cutoff when the terminating impedance is too high. The tabular values for R will give approximately the same result, however, if no test instruments are at hand. These values were determined by measurement, and the filter as constructed gives the curves shown in *Fig.* 5.

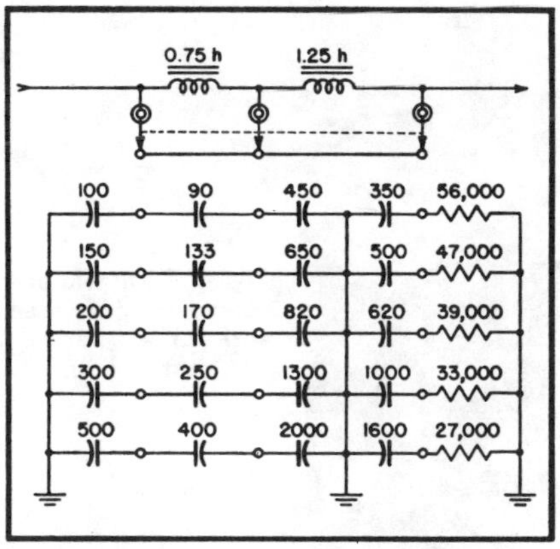

Fig. 6. Method of wiring 6-position, 3-circuit switch to make compact low-pass filter.

Filter Construction

Toroid coils are used for this filter because they are not very susceptible to hum pickup and because they have high values of Q. With low-Q coils, the filters would have a gradually decreasing transmission up to the cutoff frequency, instead of being flat up to that point. L_k is a 1.25 h. coil, such as Freed F-812T or UTC HQA-12; L_1 has an inductance of 0.75 h., such as Freed F-810T or UTC HQA-11. Toroids are somewhat expensive, but their freedom from hum pickup makes them particularly desirable for low-level circuits.

Figure 6 shows the configuration for the entire filter unit, with component values. The capacitors can be selected from most dealers' stocks of ceramic types, and while exact values are desirable, it must be remembered that the cutoff frequencies were chosen to give a variety rather than any exact curves. Except for the coils, the entire filter can be made up on a Mallory 3136J switch, which has three circuits of six positions each, allowing for one off position and five cutoff frequencies.

When combined with the balance of the control unit circuit, this filter arrangement makes it possible to eliminate undesirable needle scratch to a remarkable degree, although of course it also removes the high frequencies in the reproduced music. With the correct roll-off circuits as a part of the phonograph input equipment, however, trouble from needle scratch is minimized anyhow, and the filter will seldom be used in lower than the 8-kc position, except for particularly noisy records.

Phono Equalizer

The writer has a preference—determined by listening tests—for the low-impedance equalizer previously discussed by this writer [3]. This type of equalizer consists of an inductance and several capacitors and resistors, together with an input transformer and a pentode am-

[3] "Residence radio systems, Part III," AUDIO ANTHOLOGY, page 113.

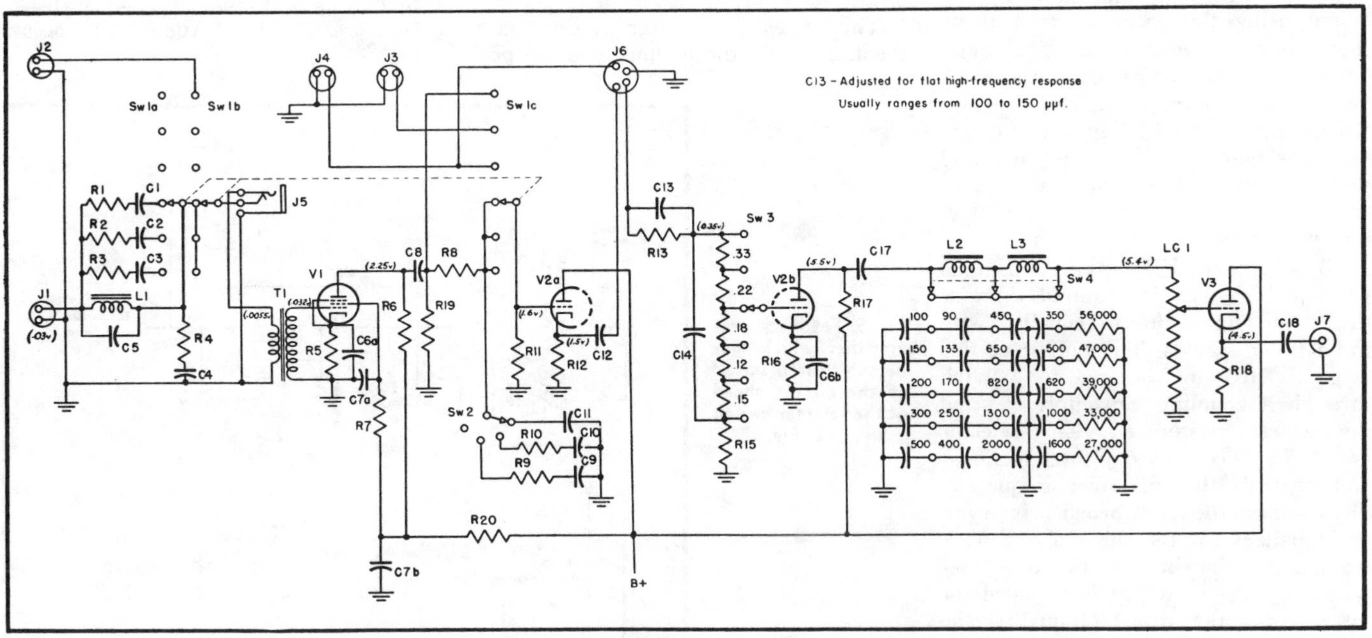

Fig. 7. Over-all schematic of control unit described.

Fig. 8. Turntable and pickup assembly used with the author's control unit.

plifier stage. The output of this combination is approximately the same as that of the more commonly used dual triode with feedback. Referring to the over-all schematic, *Fig. 7*, the series inductance L_1 is tuned by one of the three *RC* circuits R_1C_1, R_2C_2, or R_3C_3, depending on the turnover frequency. These values are best determined by measurement with the pickup to be used, although any of the presently available magnetic pickups give comparable frequency response with the circuit shown. The resonant frequency of the capacitance-inductance is usually at approximately 0.8 times the desired turnover frequency, and empirically, the value of the associated resistor is approximately equal to the reactance of the inductance at the turnover frequency.

The *RC* circuit R_4C_4 determines the slope of the curve below turnover, and experimental determination of values indicates that a relatively large capacitance is required. Calculations for this equalizer circuit are extremely complicated and well beyond the scope of this series. With the constants shown in *Fig. 7* and using the specified components should ensure similar results. The inductance L_1 should most certainly be a toroid to avoid hum pickup and to obtain a sufficiently high Q. The input transformer should be of high quality, and should be well shielded magnetically.

Amplifier Stage

To provide sufficient gain, a pentode preamplifier stage is required to raise the phono signal up to equality with radio tuner signals. The output of this stage is fed to the selector switch through a coupling capacitor, C_8, and the series resistor R_8. The selector switch Sw_1 is composed of three sections: (a) controls the turn-over frequency; (b) connects the transformer primary to the equalizer for the three phono positions, and in the sixth position to a separate input circuit which is intended for future connection to a high-quality tape recorder playback head; and (c) connects the following stage of the unit to the preamplifier through R_8 for the three phono positions or to C_8 directly in the sixth position. Positions 4 and 5 are for FM and AM radio tuners. The switch end of R_8 is connected to the arm of Sw_2, which is the roll-off control. This switch is a Centralab 1461 which has been altered to permit the rotor to turn over four positions instead of the three normally permitted. This alteration consists of filing the slot in the frame so the rotor turns a full 90 deg. In one position, the circuit is open, resulting in essentially flat response—this position being the new one created by altering the stop. Actually, due to the capacitance in Sw_1, the response is down 3 db at 10,000 cps. In the second position, R_9 and C_9 cause a roll-off of 5 db at 10,000 cps, which is suitable for ffrr and most European 78 recordings. R_{10} and C_{10} give a roll-off of 8 db at 10,000 cps for most domestic 78's. C_{11}, with no series resistor, gives a roll-off which is correct for LP's. By actual check, using a pre-equalized test record, this circuit plays LP's within ±0.5 db of flat from 50 to 10,000 cps.

This type of equalizer is essentially identical with those employed in professional equipment, having been practically copied from the Pickering 163 Equalizer. When used with a high-quality pickup and a good turntable, the reproduction is excellent. The importance of a level, well balanced and *constant* speed turntable is not generally recognized. The unit pictured in *Fig. 8* used as a test setup for various pickups and motors, as well as for regular phonograph reproduction in the home, has shown the importance of turntable quality. For example, the writer previously used a two-speed, gear-driven turntable with a 12-pound platter—presumably heavy enough to iron out any irregularities of speed. However, when the transition was made to the rim-drive unit shown, a definite improvement in quality was observed—particularly with respect to "flutter." Flutter may be considered a high-frequency speed variation as contrasted to "wow," which is a low-frequency variation. It affects, principally, the "cleanness" of the reproduction of high-frequency tones, such as those of a violin. The importance of turntable steadiness and speed constancy must not be underestimated.

Optional Preamplifier

Figure 9 is the schematic of a preamplifier which can be substituted for the one shown in *Fig. 7*, using the same switching arrangement for turnover frequency. The roll-off control is placed at the input to the preamplifier, after the manner of St. George and Drisko.[4] This preamplifier is somewhat less expensive to construct and much easier to adjust unless measuring facilities are available. It should give equally satisfactory results in most applications.

[4] St. George & Drisko, "Versatile phonograph preamplifier," AUDIO ANTHOLOGY, page 58.

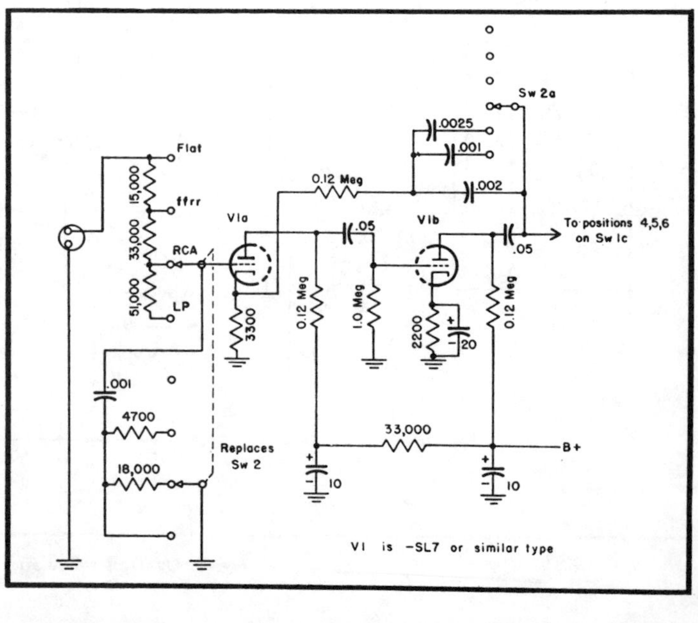

Fig. 9. Feedback preamplifier which may be substituted for all of the circuit ahead of the selector switch Sw_{1c} of Fig. 7.

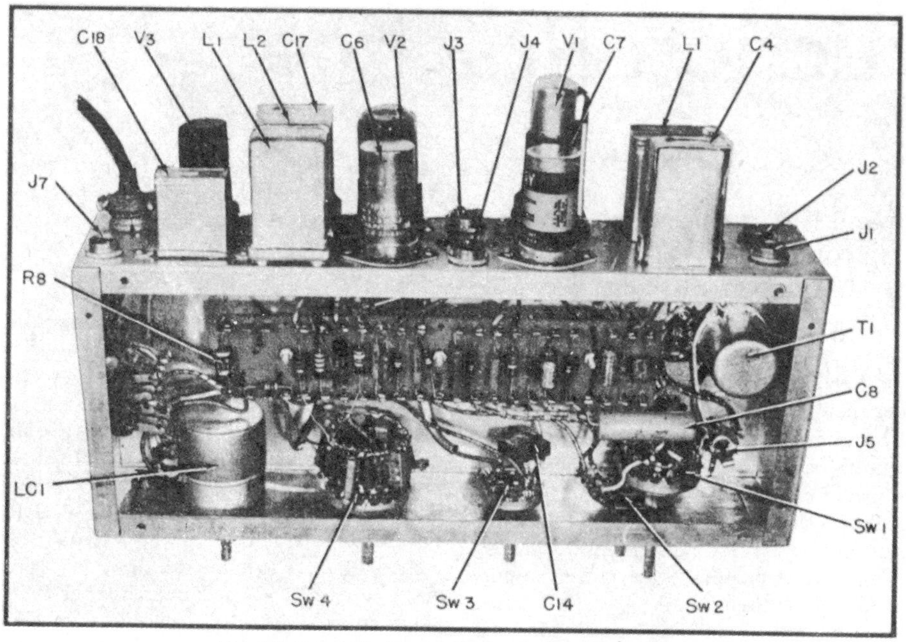

Fig. 10. Internal view of the control-unit chassis.

Construction

The writer still favors a construction, devised some time ago, which places the tubes and other above-chassis components on one side of a chassis, with the controls on the opposite side, as shown in *Fig.* 10. The major parts are labeled in the figure, with a first model of the capacitor assembly mounted on the low-pass filter switch, Sw_4.

A few brief words of explanation may be necessary. All normal input circuits are on Amphenol 80PC2F receptacles. The jack J_5 gives direct access to the primary of T_1 for microphone use or for testing. The filament circuit is designed to work from a 12-volt d.c. supply, with V_1 and V_3 in series and with a 12-volt tube for V_2. V_1 is a 6J7 because of the lower hum possibility with the control grid connection being on the cap in case the unit should be operated with a.c. heater supply. Furthermore, if an absolute minimum of noise is required, a 1620 could be substituted immediately. C_{17} and C_{18} are oil-filled units, mounted above the chassis. The loudness control requires that some gain adjustment be provided so that the control will be operated at its normal position for average room volume. This control is a tap switch mounting a number of resistors providing 5-db steps of gain variation, and is located on the main amplifier chassis. The entire control unit is constructed on a 6 × 14 × 3 chassis, which is sufficiently large to avoid crowding of the components.

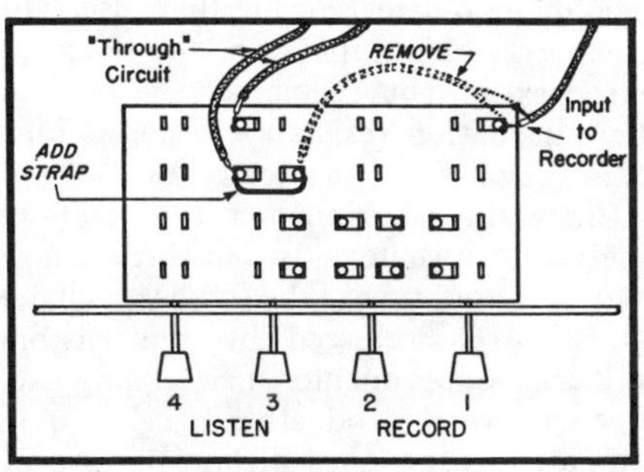

Fig. 11. Changes required on switch in Webster 178 recorder to enable switching operations described.

	V1	V2a	V2b	V3
Ep	96	225	82	225
Esg	40			
Ek	0.96	10.0	2.9	11.4
Ebb	194	*		

Measured with zero-current voltmeter

Fig. 12. Operating potential chart.

As mentioned previously, the AM radio output was permanently connected to the recorder input. Further consideration of this connection indicates the desirability of having the recorder connection made to the AM radio to take advantage of the better quality of FM for direct listening. However, the change can be made simply by interchanging the connections to the two input jacks and relearning the switch positions. The radio-recorder connection is ahead of the selector switch, so no operation of the switch can disturb the program being recorded. The Webster 178 recorder has a 4-pin plug at the rear to which connections are made. Normally, one lead is used both for input and playback output; one is a grounded shield, and the other two are used to open an r.f. cathode circuit in the radio, thus cutting the radio off during playback.

The circuit employed in the unit being described requires three shielded leads—one for the incoming signal, and the other two for a loop through the recorder switch. Thus, it is desirable to rewire the cable with three lengths of shielded wire and to shield the two leads from the pin plug to the switch in the recorder itself. Also necessary is a change in the wiring on the bottom of the switch to cause the playback output to feed the desired lead. The shielded jumper shown in *Fig.* 11 should be removed, and a strap added between two of the switch contacts. This will open the loop circuit and feed the playback output to the loop lead when the LISTEN 3 button is depressed. This will provide the operation desired.

Operating Conditions

The d.c. operating potentials are shown in *Fig.* 12 with a regulated supply of 225 volts. The signal voltages are shown on the schematic for a typical input from a magnetic pickup, with the selector switch in the 300-cps turnover position, at a frequency of 1000 cps, and with the loudness control at maximum volume.

PARTS LIST

C_1 0.13 µf (0.1 + .03) paper, 200 v.
C_2 .08 µf, paper, 200 v.
C_3 .05 µf, paper, 200 v.
C_4, C_{17} 1.0 µf, 600 v. oil-filled, Aerovox 618B
C_5 500 µµf, Erie Ceramicon
C_6 40–40/25 electrolytic
C_7 10–10/450 electrolytic
C_8 0.1 µf, 600 v. molded
C_9 100 µµf, mica
C_{10} 1000 µµf, mica
C_{11} 2000 µµf, mica
C_{12} .05 µf, 600 v. molded
C_{13} 120 µµf, mica
C_{14} .008 µf, mica
C_{18} 0.25 µf, 400 v. oil-filled, Aerovox 418B
J_1, J_2, J_3, J_4 Amphenol 80PC2F connectors
J_5 closed-circuit jack, Switchcraft 12A
L_1 3.0 h; Freed F819T, UTC HQA-14
L_2 0.75 h; Freed F810T, UTC HQA-11
L_3 1.25 h; Freed F812T, UTC HQA-12
LC_1 Livingston MB Loudness Control
R_1 8200 ohms, ½ watt
R_2 10,000 ohms, ½ watt
R_3 18,000 ohms, ½ watt
R_4 1800 ohms, ½ watt
R_5 820 ohms, 1 watt
R_6, R_{17} 0.1 meg, IRC DCF
R_7 0.56 meg, 1 watt
R_8 0.1 meg, ½ watt
R_9 0.22 meg, ½ watt
R_{10} 68,000 ohms, ½ watt
R_{11} 1.0 meg, ½ watt
R_{12} 3900 ohms, 2 watt
R_{13} 0.27 meg, ½ watt
R_{14} Five ½-watt units assembled on Sw_3
R_{15} 56,000 ohms, ½ watt
R_{16} 1800 ohms, 1 watt
R_{18} 4700 ohms, 2 watt
R_{19} 2.2 megs, ½ watt
R_{20} 33,000 ohms, 2 watt
Sw_1, Sw_4 Mallory 3136J
Sw_2 Centralab 1461 modified (see text)
Sw_3 Mallory 3126J
T_1 500/50,000 input transformer, shielded. Langevin 401B, UTC A-11, Triad JO-1
V_1 6J7 (or 1620)
V_2 12SN7
V_3 6SJ7

Construction Practice

C. G. McPROUD

Regardless of the design and layout, the final performance of an electronic unit depends upon its components and the way they are put together.

Audio equipment—like any other—must be built before it can be used, and there are almost as many different styles of construction as there are constructors. Fortunately, performance is not nearly as dependent on the actual physical layout and arrangement in the audio spectrum as it is in some others. High-frequency techniques are quite specific, and short leads, socket-mounting of resistors and capacitors, and point-to-point wiring are mandatory if good results are to be obtained. Anyone who has assembled a television receiver from a kit will remember how parts are placed, and how little similarity there is to standard audio practices.

In this series, we shall discuss the physical design and layout of audio apparatus, choice of components, wiring practices, cabling, and many other points involved in the construction of amplifiers and power supplies. This series will not discuss circuit design—that will be assumed complete—but will carry on from that point, considering each step of the work necessary to convert a schematic into a finished piece of apparatus. Circuit design is an entirely different problem.

Initial Steps

In most instances, the builder will have an idea of the size and shape he wishes a piece of apparatus to have. Obviously, the size will be governed by the components required, particularly with respect to transformers, chokes, and filter capacitors. In general, however, these components will rarely need to be replaced, and can be mounted quite close together, with capacitors taking up the space under transformers if necessary.

Assuming that a circuit is complete, the first step is to select the components. Many of these will be indicated by the circuit specifications, particularly transformers—power and audio—and chokes. The use of half-shell power transformers is not recommended for high-gain audio equipment because the chassis—if steel—is likely to have an a.c. field induced in it because it is in direct contact with the transformer core. The mounting with the core perpendicular to the chassis is more suitable. For the finest equipment, completely cased and potted transformers and chokes are desirable, and rarely give any trouble. In humid climates they are almost essential, to avoid failure due to moisture in the windings.

As readers may infer, the writer is a confirmed experimenter, and many of these suggestions may be tempered with a consideration of the possibility of rebuilding and the saving occasioned by re-use of certain of the components, particularly the more costly units such as transformers. For this reason, the transformer types preferred are those having terminals rather than leads, because once cut to a certain length for one chassis, leads may not reach the sockets or terminals in a new one.

The same suggestions apply to power-supply chokes, although space may be at a premium and open-frame chokes can often be mounted under a chassis. Chokes are less expensive than transformers, however, and the need for re-use may be lessened in the interests of space-saving or of economy.

Let it be said firmly, however, that a saving in cost of transformers is precarious—the best obtainable is usually the cheapest in the long run. Transformers have an unusually long life—we have never known a *good* one to fail—and their characteristics remain constant over a long period. Once built into an amplifier or power supply, they may be expected to function perfectly for years.

Capacitors

For the highest class of equipment, it is usual practice to employ oil-filled capacitors for filters. Except for the surplus units that are now available, these units are quite expensive, but their power factor is considerably better, they are more efficient in filter circuits, and less capacitance is needed for equivalent results. However, when operated well within ratings, good electrolytics serve quite satisfactorily, and are smaller and less costly. It is recommended strongly that an oil-filled unit be employed as the first capacitor in a power-supply filter (of the capacitor-input type) since the ripple is high at this point, and a high ripple voltage means a high current—a.c.—through the capacitor with a resulting reduction in life.

Capacitors used in power-supply filters preferably should be of the single-section type, to reduce the voltage gradient within the same case, although multi-section units always seem to work satisfactorily with different voltages on the sections. For cathode bypass use in audio circuits, the high-capacitance, low-voltage electrolytics are ideal.

Coupling capacitors are necessary in most amplifier circuits. The tubular molded types are compact and well suited for this use, for they are easy to mount and have no capacitance to the case. The grounded foil of tubular units should be connected to the plate of the preceding stage where the impedance is lower than usual in grid circuits, since there is less chance of hum pickup with this connection. Many designers prefer to employ metal-cased capacitors for coupling use. However, the capacitance to case must be considered, and while some types are satisfactory, others have too much capacitance for plate-to-grid coupling circuits, resulting in an attenuation of high frequencies. In shunt-fed transformer-coupled stages they are usable provided the capacitor is located in the ground leg of the circuit. The "bathtub" capacitor usually has a high capacitance to case; for coupling use the upright, oil-filled types such as Aerovox -16CT and CB, -18CT and CB, Sprague CN- and CA-, or General Electric styles 60, 62, 64, 66, and 68 have low capacitance to case, and are easy to mount and neat in appearance.

In most layouts, there is usually more space above the chassis than below, and the upright models with bottom terminals which extend through the chassis reduce the number of components under the chassis. These models provide complete shielding for the capacitors, and make for a professional-looking amplifier.

For filter and equalizer use, mica capacitors are most suitable. They do not normally have high voltage ratings, but in most instances there will be relatively low voltages across these units, so this should be no deterrence to their use. The use of ceramic capacitors for coupling purposes is not recommended because their insulation resistance is not as high as in the oil-filled or mica types.

Since the additional cost of 600- or 1000-volt capacitors is but little more than 400-volt types, the higher voltage ratings are preferred by the custom builder or experimenter for coupling use. In commercial production of electronic equipment, economics dictate the use of the lower-cost items, but the additional cost is of little importance to a person who is making but a single unit.

Variable Capacitors

The use of variable capacitors in audio work is becoming more frequent, and they bear some discussion. They are used in variable-frequency oscillators and tunable filters. Capacitors for this service must be precision built on rigid metal or ceramic frames and should have the plates soldered into carefully machined slots, or assembled on the shaft with precision spacers and locked in place. Among the better capacitors for this service are the National PW series and the Cardwell PL 24,050. Also these capacitors must have a drive system with a minimum of backlash, and the dial or scale should be carefully hand engraved or machine divided.

With the advent of magnetic recording equipment the audio man is often faced with the selection of small trimmer capacitors. These should be of the mica spaced, ceramic mounted type because they have lower losses and less tendency to change value under varying conditions of temperature and humidity. Care must be used when soldering to mica trimmers to prevent solder or flux from flowing between the plates and causing erratic operation. It is not good practice to reuse trimmers; in fact it is economically unsound.

Resistors

For general use, the ordinary type of metallized or molded resistor in the ½-, 1-, and 2-watt sizes are commonly used, and are completely satisfactory for most applications when used within their power ratings. It is good policy to allow a factor of safety of around 100 per cent for resistors in plate and cathode circuits. In standard sized assemblies, it is preferred to use 1- or 2-watt resistors for any application where there is a current flowing. Half-watt units are satisfactory for grid resistors or for tone correcting circuits. In low-level stages, some care must be exercised in the choice of plate-load resistors. Precision wire-wound types or the new deposited-carbon resistors (such as IRC type DCF) have lower noise level than most commonly used resistors, and are preferred for the input stage of amplifiers used for phonograph reproduction or with microphones.

When higher power must be dissipated in bleeder or voltage-dropping resistors, vitreous enamel wire-wound types are most satisfactory. They may be used almost up to their dissipation rating with a much lower factor of safety than usual in resistor selection.

Variable potentiometers are commonly used in audio work for gain and tone compensation controls. They are available in three general groups, carbon, molded carbon, and wire-wound. The most popular before the war was the carbon element potentiometer in which a rotating contact arm carried a sliding contact over a circular card on which a layer of carbon had been deposited. Since the end of the war, the molded carbon element used in the same mechanical case, but having a ring of plastic into which the carbon has been molded, has become extremely popular. These two types find application in the grid circuit of vacuum tube equipment where little or no current passes through the unit. For low level use the molded element gives more satisfactory service because it introduces less noise. Wire-wound potentiometers are used in plate, cathode, and screen circuits where they are required to dissipate up to four or five watts of power. Under these conditions, the carbon elements would either burn out or generate excessive amounts of noise. The rating of the molded element is two watts, and that of the usual small carbon element is one watt. Where a potentiometer is required to fit into a small space, a special series may be obtained from most manufacturers as stock items, but they have reduced power and life ratings. Large quantities of the molded element and wire-wound potentiometers have been available since the end of the war, on the surplus market.

Switching Devices

Switches, outside of the a.c. power switch, are used to change circuit conditions in a positive-acting manner. This covers changes in input source and output transducer, and changes in equalizaton, operating impedance, and gain when it is desired to go from one fixed known value or condition to another. The most common switches for these applications are the wafer-type rotary switches with laminated plastic decks and silvered contacts. High-power switching requires mica-filled or plastic decks and for applications where leakage is important, ceramic decks may be used. In cases where only two or three switch positions are needed, lever-type switches can be used. They are particularly useful where the operating conditions must be known without reference to small dial indications, since their position may be seen at a glance. These switches are also made with an anti-capacitance feature where low switch capacitance is necessary, but these are more expensive. When more than three positions are to be used and rapid visual indication of the operating point is necessary, pushbutton types, with interlocking buttons that release any depressed button when another is pressed, are similar in electrical characteristics to the laminated decked rotary switches. For broadcast and long life service, pushbutton switches using jack leaf springs are preferred. All of these switches are self cleaning and may be had with either shorting (make-before-break) or non-shorting (break-before-make) contacts, and with up to twelve positions in stock styles.

For a.c. power control, toggle switches are suitable and are made with both laminated and plastic cases. The latter are more desirable but are higher in price. In addition to all of the above considerations in selecting switches, no switch should be used in a circuit where its rated current-carrying capacity will be exceeded. Disregarding this practice leads to burned contacts, poor connections, and sometimes causes the switch contacts to weld.

Relays are electrically operated switches, and have been popular in audio equipment for remote operation by switch, clock, voice, and photocell actuation. In general the considerations governing the selection of switches also apply to relays, with a few additional features found only in relays. Among them are their operate and release time, bounce characteristic, and coil characteristics. It is necessary when specifying a relay to give its resistance and its operating and release currents or the operating voltage and power consumption. These will be determined by the circuit in which it is to be used. Relays can be provided with latching and release mechanisms for use where is is undesirable to have the actuating current flowing continuously, or where there is only a short pulse of energy to actuate the relay. Mechanically looking relays come with a variety of contact and frame arrangements, but the electrical switching capabilities are the same as the switches already discussed. The stepping relay, so common in telephone work, provides the same type of contact arrangements as the multideck rotary or pushbutton switch.

A practice often suitable with other components—the purchase of a unit at a reduced price and the redesign of the circuit to make it possible to use the unit—is not too sound when applied to relays. They should be selected to fit the simplest conventional circuit which may be used or adapted for the purpose. When extreme sensitivity is needed for differential current or battery operation, "sensitive relays" can be obtained from several manufacturers as stock items.

In mounting both relays and switches, consideration should be given to lead lengths and the relation to adjacent parts. For convenience in mounting and locating relays many are now packaged in octal-based cans providing for easy installation and maintenance. Where practical they should be mounted close to the other components which they control, and when used in low-level circuits they should be mounted away from power supply components. Both of these conditions are particularly important when the application is to low-level stages.

Inductors

Inductors are not commonly thought of as being an audio component. However from the earliest days of audio they have been used in impedance-coupled amplifiers and as components in loading, equalizing, and filter circuits. Both the core style and inductance determine the case size and shape, but where expense is not a primary consideration, toroid coils are available in a wide range of values and in small cases. The toroid coil has several advantages over the conventional laminated, rectangular-core inductor in that it provides greater coupling between turns and between separate windings, and since there are no corners or gaps in the core, the flux distribution outside of the core is extremely small. Another desirable feature is the low sensitivity of the toroid to stray hum and noise fields which recommend its use in low-level equalizers and filters. These last two considerations make it possible to put two toroid coils in close proximity on a chassis without the danger of interaction or, when used on different signal circuits, of crosstalk.

Care should be exercised in the mounting of air- or iron-core inductors, so that they will not be within the strong fields that surround power supply chokes, transformers, and tubes. When used in low-level circuits such as an input playback equalizer, it is best to mount them as far from the power supply as possible, and where practicable the equalizer should be mounted on a separate chassis. In some cases it may be necessary to determine the best mounting position for inductors, to obtain minimum pickup, after the rest of the unit is built. The approximate position may then be explored with the coil connected to an oscilloscope.

Plugs and Connectors

Interconnecting plugs and sockets are available to the audio engineer and hobbyist from a wide variety of sources, and their selection is governed by the general service application as well as the electrical characteristics. For broadcast type equipment, it is necessary that the microphone and other low level input connectors have small physical size combined with positive electrical contact, low contact resistance, cable clamping, and plug locking devices that still permit quick disconnect. Another important feature of the connector is its pin and socket insert style. The best connectors use solid pins and split cylindrical tubes or rings as the mating pieces, each with a small "solder pot" as the wiring facility. Most AN type plugs are of this construction, as are the connectors manufactured by Cannon Electric Development Co. and Winchester Electronic Co. In addition to these features the latter may be obtained with spring loaded pins making it possible to separate a twenty or thirty pin connector with almost no effort.

For transferring power from one unit to another a different set of requirements must be met. The usual contact resistance and noise requirements should be met, but most important is the size of the contact area. This determines the current handling capacity, which affects the transmission in filament and low-impedance power circuits. Plugs for this service may be of the tube-base-and-socket variety, knife blade style, or in the same style as the microphone connectors.

The first of the styles mentioned above is typified by the Amphenol series 86 and 78 having from four to eleven contacts. An often used knife blade connector is the Cinch-Jones series 300, 400, or 500 with up to 33 contacts and with a variety of shells and chassis mounting brackets. Pin-style connectors for power use are made by Cannon and Winchester, and, in addition to the regular power transfer pins may be equipped with pairs of signal connectors inside a shield for use with two-conductor shielded wire or twinax cable. Connectors of this type are the Cannon DP series. Miniature and battery connectors are not recommended for audio design because of their poor electrical and mechanical characteristics, except for the very expensive military series. For very high current carrying capacity both Hubbell and Russel & Stoll make suitable connectors in a variety of cases and mountings including waterproof assemblies.

Contact arrangement for a particular application is dependent on the selection of arrangements listed by the manufacturer. Microphone connectors are made with one to four contacts, and some styles of low-level signal connectors—although not specifically designed for microphone service—have more. Among them are the Cannon types K and P. It is seldom advisable to use single-pin connectors except at the chassis end of a microphone line. Otherwise it will be impossible to avoid capacitive or magnetic hum pickup or ground loops. In general the three-circuit connector is best for this service since it carries not only the signal leads through on the pins but can also carry the ground lead or shield without depending on the dubious electrical capabilities of the shell assembly. For interconnection of plug-in chassis assemblies, either the knife blade connectors or the Cannon DP series may be used. They should have sufficient contacts to provide several spares for future modification without having to be changed and the mounting hole enlarged. Also they must be mounted "floating" so that the guide pins will pick up the mating connector and carry out their guide function. If rigidly fastened it is easy to break off the pins. When it is necessary to provide isolation, shielded pins may be used or a ring of pins may be grounded. For low leakage a ring or row of pins can usually be removed.

Although most manufacturers number the pins on plugs and sockets, there is as yet no industry standard. It has been common practice on microphone connectors to have pin #1 as ground and pin #3 as the *hot* lead, and some manufacturers make the circuit of pin #1 close before the other two in order that the shield or ground connection is completed before the signal circuit. On larger connectors it is almost impossible to follow any consistent system, and the designer should choose his own system and *stick with it.* At least be consistent within any given group of units designed to work together. Where all pins are of the same size an acceptable numbering method is to use the numbers in relation to increasing potential. However, even this method is not completely satisfactory.

The rules governing the use of male and female connectors is simple. In low-level circuits the male is placed on the source side of the line and the female on the load side or amplifier input. This protects the amplifier from overload caused by hum, clicks and pops which occur when open input terminals are touched. In *all* other applications the male should be placed on the load side of the line and the female on the source. The reason is obvious. Nobody wants to pick up a pair of high-voltage contacts or to have a good transformer ruined by being shorted at the connector terminals.

Input plugs should, if space permits, be placed close to the input circuit elements and as far away from the power supply components as possible. Where a single plug carries all input, power, and output leads the input leads should be kept as far from the a.c. and output leads as possible or brought through on a twinax pair. Otherwise a.c. and output leads should be placed for the operator's convenience. In rack mounted units this would usually be along the rear edge of the chassis. In some home audio systems, the amplifier is on the floor of a cabinet in which case the plugs are brought out on the top of the chassis.

Wherever possible, connectors should be mounted on the chassis in a position that is clear of tube sockets and terminal boards. Otherwise whenever maintenance is required the connector will have to be dismounted from the chassis to get at the parts that are hidden.

Miscellaneous Components

The total number of component types is too great to be considered all in one article. However, some mention may be made of a few of them. Many different types of tube sockets are available, and their selection is of some importance. For general use, the socket which is mounted in a keyed hole by means of a wavy spring ring is most easily installed, but a special punch is required to prevent the socket from turning. Punches for these sockets are easy to use, the Pioneer Ham-R-Press being one of the most useful for chassis construction. The moulded-in-plate sockets are reliable and make good contact. For output stages, where the potentials across the socket may be quite high on peaks, ceramic sockets are desirable to reduce the possibility of flashover with attendant damage to transformers. For miniature tubes, the mica-filled sockets are preferred. Low-level stages should be cushion mounted to reduce microphonics, and several manufacturers make sockets which provide for such cushioning.

Electrolytic capacitors of the fabricated or etched plate type are commonly mounted on metal or Bakelite wafers. For lowest hum level in high-gain amplifiers, Bakelite wafers should be used, with separate leads connecting the shell of the capacitor to a common ground point.

Chassis Selection

In choosing a chassis for a given unit of audio equipment, there are usually three conflicting requirements: appearance, available space, and functionality. When the unit must fit into a predetermined space, smaller than the functional optimum, care must be exercised in the layout, and some sacrifice in appearance will usually result. When this condition is encountered, it is best to place the large parts transformers, chokes, relays, etc.) on the chassis and move them about until tentative positions are found that will still leave enough room for tubes and capacitors. Also the signal level at which the various tubes and transformers are to operate must be kept in mind in order that no low-level components will be placed near a.c. or power output parts. To facilitate layout it is often desirable to cut out small pieces of cardboard having the shapes of the bases of the various parts. These may be shifted about with ease and with no regard to lugs or leads which would otherwise be in the way.

When space is not a limitation, function and appearance should govern layout. When possible, parts should be in rows at right angles to one another merely for the sake of good appearance. However, in no case should quality suffer at the expense of appearance. Slight

Fig. 1. Special dual terminal strip made up for an experimental equalizer circuit.

variations in positioning and alignment of parts may frequently be hidden by careful finishing. An easy way to determine the chassis size is to lay the parts out on a piece of paper which has only two adjacent sides drawn on it. After the parts have been positioned approximately, the commercial chassis size that will be suitable may be selected from a catalog.

The material of which the chassis is made is important in two respects. Where magnetic shielding of sub-chassis components is desirable steel should be used, but if shielding is not required and ease of drilling and punching is desired, aluminum should be selected. Mumetal and stainless steel are not good chassis metals because they are too hard to drill or punch.

Parts Layout

After the parts are laid out on paper, the chassis may be marked using a good adjustable square and a sharp scribe. Another method that simplifies work is to make a spare chassis drawing to full scale, including the front, back, and side edges. This should be done with a sharp hard pencil. It may then be fastened to the chassis with tape, and all the center punching, drilling, and punching done through the paper. In any case, care must be exercised in making the drawings, laying out the lines on the chassis, and in centerpunching for drilling. Care here will save many headaches later. After drilling and punching, the holes should be carefully scraped to remove burrs, and the transformers, chokes and other cased parts mounted. It is at this point that the chassis may be painted a uniform color. The three most popular commercially available colors are telephone black, gray, and aluminum. By spraying the entire unit, a very professional appearance may be obtained. Parts to be protected from paint may be covered with masking tape. The choice of the color must take into consideration the temperature ratings of the components.

Physics tells us that black bodies absorb and radiate heat better than light colored ones. Therefore, transformers are usually painted black, capacitors light. If everything is to be painted the same color, then, for light colored units, all parts that get warm should be overrated, while dark-painted capacitors should have high ratings. Here again care in the selection of components and planning of layout for specific temperature conditions will solve the problems. After the chassis finish has dried, all of the projecting components should be wrapped in heavy brown paper and taped. After this is done, the tube sockets, plugs, switches, and those mechanical sub-assemblies that require wiring may be fastened in place. The brown paper will protect the finish on the projecting parts while the smaller pieces are being assembled and the wiring is done. Large sub-assemblies that do not require wiring should not be mounted until they are needed for dressing leads around them or fitting to other parts; otherwise they will only serve to the chassis heavier and more unwieldy.

The Wiring Layout

Before any wiring layout can be made it is necessary to determine the position of the numerous small resistors and capacitors used in audio circuits. Point-to-

Fig. 2. Example of terminal board construction in a finished professional unit. (Courtesy E. R. P. Div'n, Western Electric Co.)

point wiring is sometimes used, in which case the parts may be laid out directly on the wiring layout and drawn in. However, for ease of construction and servicing and for neatness, terminal boards are recommended. *Figure* 1 shows a type of construction used in an experimental equalizer, and *Fig. 2* is a professional example of a completed unit employing the terminal-board form of construction. Terminal boards can be obtained from several manufacturers in made-up form, or lugs and terminals having various sizes and shapes may be purchased and mounted on sheets of electrical laminate to meet the builder's own specification. A few of the styles are illustrated in *Fig.* 3. In planning the layout of terminal boards, the parts should be mounted so that they will be reasonably close to the other components with which they must work, and that the wiring and connections on the strip itself permit the majority of the external connections to be brought out along one edge of the strip. This will facilitate assembly of the final units and future maintenance. Where only one or two isolated parts are to be mounted, tie points are quite satisfactory. They should have laminate of sufficient quality and thickness so that they will not warp or change position once they are installed. Tie points are also useful where bias and erase circuits in tape equipment must be built close to the tube socket.

When the chassis layout is being drawn it is a good idea to draw a bottom view also. This drawing can then be used to make the wiring layout. The positions of the wires and the color of the insulation can be indicated by drawing them in with colored pencils. In laying out the wiring, long grid or plate leads should be avoided, since they increase the grid-cathode and plate-cathode capacitance. Also the leads within the low-level stages should be as short as possible and all a.c. and power output leads kept as far away from them as possible. D.c. power supply leads may be grouped together and neatly laced after they are soldered in place or laced in harnesses made up to be dropped into place for wiring.

Fig. 4. Example of wiring showing the use of a cabled form or harness. (Courtesy E. R. P. Div'n, Western Electric Co.)

After the parts are mounted and the wiring layout is complete, the wiring may be started. If harnesses are used, they may be made by transferring the bottom view to a sheet of ply wood, placing finishing nails at corners and bends and laying the wires along the paths thus indicated. After the wires are in place the lacing may be done. A typical wiring job using a harness is shown in *Fig.* 4. If harnesses are not used, the heater circuits should be wired first and then the short leads put in, socket to socket, and terminal board to socket, and finally the long leads. By leaving the long leads to last it will be easier to check the major portion of the wiring before they are installed.

Wire Selection

The choice of wire is determined by circuit considerations and insulation. Acetate-coated rayon insulation or wax impregnated cotton are the two commonest types used, although plastic, asbestos, and glass insulation are becoming more popular. All of these types are supplied in a variety of colors, and they should be used in accordance with a standard color coding system such as the RMA color code shown in Table I. With the exception of filament, a.c., power output, and shielded leads, the wire size generally used is #20. For higher current-carrying capacity, #18 or #16 should be used. In case of doubt consult a wire chart. The big and as yet unanswered question is whether to use solid or stranded wire. This writer has used both with equally good results. For laced harnesses, stranded wire is usually best, but otherwise there is no definite rule. Solid wire makes a neat, well-dressed job, but some engineers feel that stranded wire is less apt to break at the joints. The reader can try both and take his choice.

When the wiring is done, the unit checked and ready to put in the cabinet, it should be equipped with a tightly fitting bottom plate. This will prevent dust, mice, and bugs from getting into the wiring and causing poor operation.

This information does not begin to cover the entire field of construction, but should serve as an introduction to to the problems involved. Relatively little information on this subject has heretofore been published—the newcomer usually has to learn by experience.

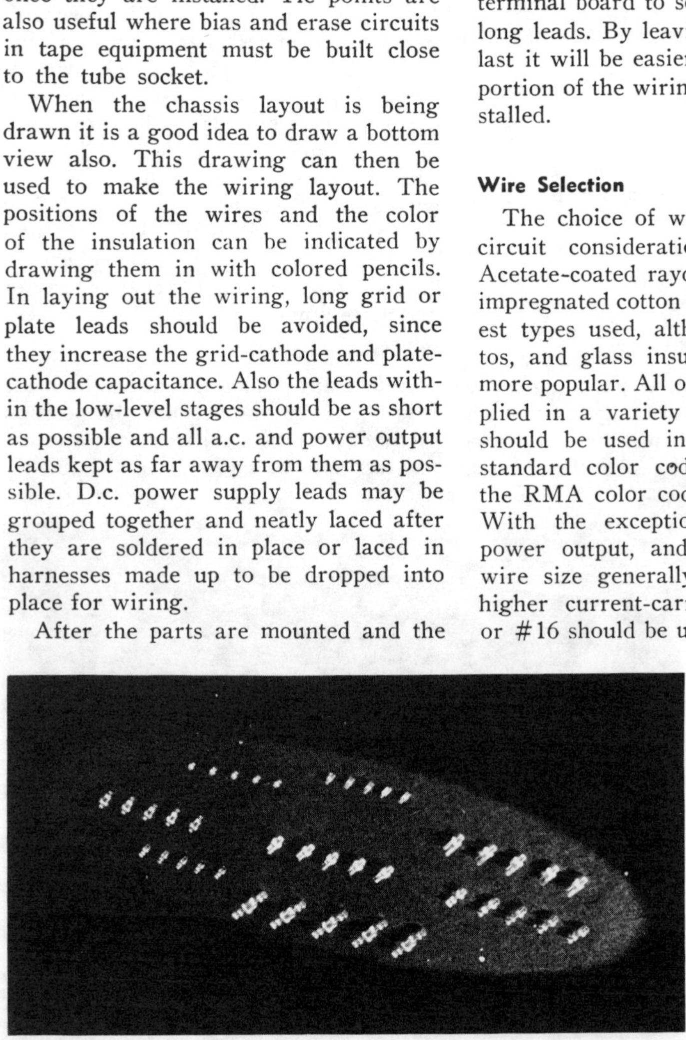

Fig. 3. Typical terminals for use on mounting strips. (Courtesy Cambridge Thermionic Corp.)

TABLE I—RMA COLOR CODES

Circuit	Color
Grid circuits	Green
Plate circuits	Blue
B – (maximum)	Red
B – (intermediate)	Orange
Ground	Black*
Power Transformers	
Primary leads, no tap	Black
Primary leads, tapped	
Common	Black
Tap	Black & Yellow
Finish	Black & Red
Rectifier Plates	Red
Center tap	Red & Yellow
Rectifier Filament	Yellow
Center tap	Yellow & Blue
Filament Winding #1	Green
Center tap	Green & Yellow
Filament Winding #2	Brown
Center tap	Brown & Yellow
Filament Winding #3	Slate
Center tap	Slate & Yellow

* Some manufacturers use Yellow for Ground, Black for Cathode circuits.

Resonant Loudspeaker Enclosures

BOB H. SMITH

THIS CHART is based upon the assumptions that the dimensions of the enclosure are small compared to wavelength, that the thickness of the port is negligible, and that the amount of air moving in the port is equal to the three halves power of the area of the port. Thus, the inertance of the port is:

$$M = \frac{\rho}{\sqrt{A}}$$

where A is the area of the port and ρ is the density of air. The compliance of the enclosure is:

$$C_a = \frac{V}{\rho c^2}$$

where V is the volume and c is the velocity of sound. The resonant frequency of a Helmholtz resonator is:

$$f_r = \frac{1}{2\pi\sqrt{MC_a}} = \frac{c}{2\pi}\,\frac{A^{\frac{1}{4}}}{V^{\frac{1}{2}}}$$

This expression seems to agree well with experiment. In a typical case the error is 5 per cent.

It is customary to choose the resonant frequency of the enclosure equal to the anti-resonant frequency of the moving system of the speaker. Since the impedance of the resonator is purely resistive at resonance and represents a very heavy load, no oscillatory transient occurs at this frequency. However, a new transient of higher frequency appears. It is caused by the mass of the moving system going into anti-resonance with the compliance of the box. Thus, the resonant enclosure does not completely eliminate the production of oscillatory transients but usually the new transient will be of shorter duration than the one which would have occurred without a resonant enclosure.

The radiation resistance is proportional to the area of the port and is usually too low for efficient energy transfer. Thus, the amount of power radiated by the port is approximately proportional to the area. In practice, this means that for a given resonant frequency the larger the enclosure the larger will be the port and the more effective the resonator. Usually, one does not use a port area less than half nor more than twice the cone area.

The importance of sturdy construction of the cabinet can not be overemphasized. Motion of the walls absorb energy which would otherwise have been radiated from the port. It is good practice to stiffen the back and front of the cabinet with 2 x 4's. The force produced by the voice coil might be as high as 5 lbs. at the resonant frequency and the enclosure must be able to stand this force without rattling.

In order to eliminate radiation from the port at higher frequencies the interior of the enclosure is usually lined with some absorbing material such as Celotex. The material used should have high absorption at the higher frequencies (500 cps up for example) but negligible absorption at the resonant frequency. The objection to simultaneous radiation from the port and cone is that destructive interference occurs at certain frequencies causing large dips in the response curve. This difficulty does not occur at the resonant frequency because here the radiation from the cone is negligible compared to the radiation from the port.

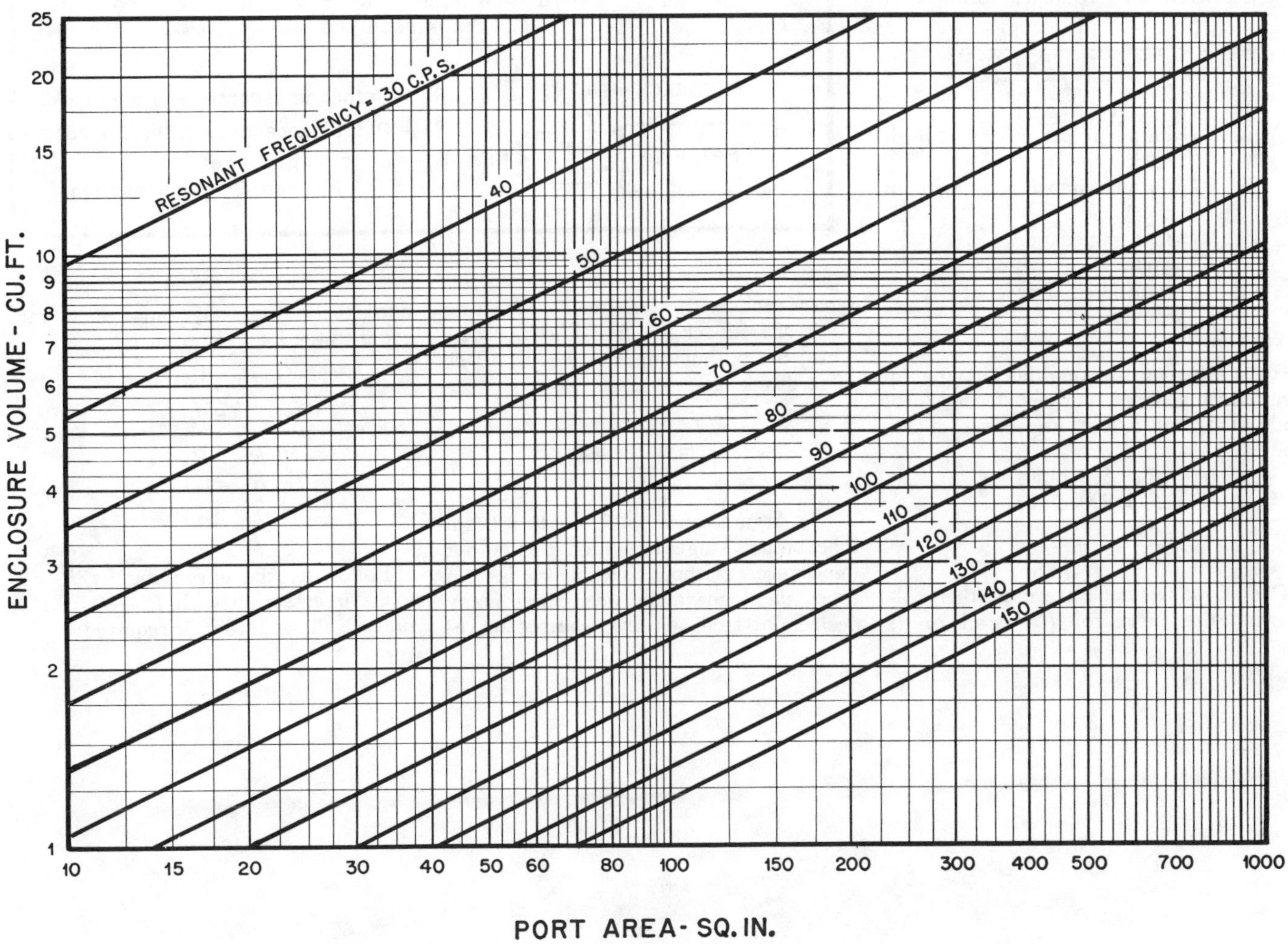

Chart Showing Reduction in Output Impedance Obtained with Negative Feedback

WILLARD F. MEEKER

An important property of voltage-controlled negative feedback in audio amplifiers is the reduction of the output impedance which it produces. The amount of this reduction is ordinarily obtained in terms of the fraction of the output voltage which is fed back and the amplification factor of the amplifier (gain with output open-circuited). However, the amount of feedback in an audio amplifier is often expressed as the number of decibels reduction in gain produced by the feedback (db of feedback). Consequently, it would be convenient to have the reduction in output impedance expressed in terms of the reduction in gain. This can be done if the output impedance without feedback and the load impedance are known. The chart shows these relations for negative, voltage-controlled feedback and non-reactive output and load impedances. The chart is based upon the following relation, which is derived from conventional feedback formulas:

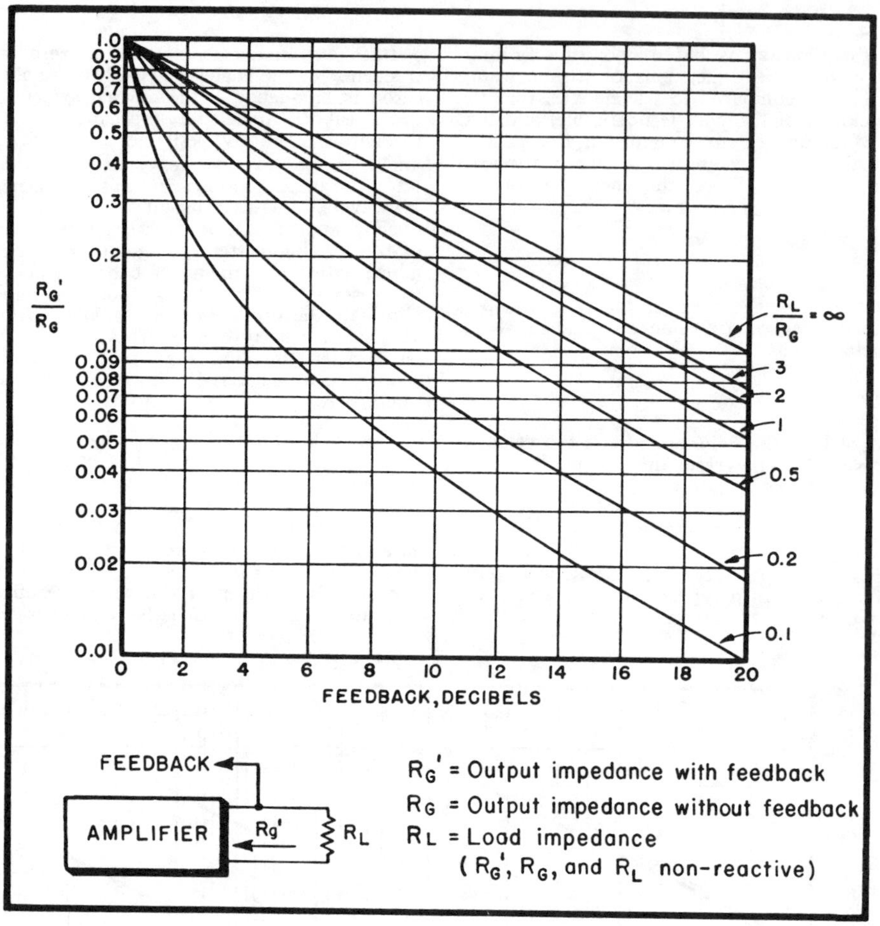

$$R'_g = \frac{1}{1-(1-Q)\dfrac{R_g+R_l}{R_l}}R_g$$

where R'_g = output impedance with feedback

R_g = output impedance without feedback

R_l = load impedance

Q = factor by which gain is multiplied if feedback is removed.

Examples:

(1) Assume an amplifier having an output impedance without feedback of 500 ohms and a load impedance of 500 ohms.

$$R_g = 500 \text{ ohms}$$
$$R_l = 500 \text{ ohms}$$
$$\frac{R_l}{R_g} = 1$$

If 10 db of feedback are used, then following the curve for $R_l/R_g = 1$ to its intersection with the line representing 10 db of feedback, we find $R'_g/R_g = 0.19$. Thus, the output impedance with 10 db of feedback is

$$\begin{aligned} R'_g &= 0.19 \times R_g \\ &= 0.19 \times 500 \\ &= 95 \text{ ohms} \end{aligned}$$

(2) Assume an amplifier with a beam tube output stage in which the output impedance, without feedback, is ten times the load impedance. How many decibels of feedback are required to reduce the output impedance to a value equal to the load impedance?

$$R_g = 10\,R_l$$
$$\frac{R_l}{R_g} = 0.1$$
$$R'_g = R_l$$
$$R'_g = 0.1\,R_g$$
$$\frac{R'_g}{R_g} = 0.1$$

Following the curve for $R_l/R_g = 0.1$ to its intersection with $R'_g/R_g = 0.1$ we find 5.2 db of feedback required.

Adding Decibel-Expressed Quantities

ALFRED L. DiMATTIA and LLOYD R. JONES

The authors present a simple nomograph which reduces the work of adding levels to its simplest form.

WHEN QUANTITIES expressed in decibels are to be multiplied, it is an easy task to obtain the product (in decibels) by algebraic addition of the number of decibels.

A less common but important problem arises when two or more quantities expressed in the decibel notation are to be added. Ordinarily, this involves a tedious conversion of each quantity from decibels to the corresponding power ratio. Then the individual ratios must be added and the sum reconverted to decibels.

The nomograph eliminates the need for such conversions. The difference between any two decibel-expressed quantities is first determined by algebraic subtraction. This value is next found on scale A. The corresponding figure on scale B indicates the number of decibels to be added to the greater original quantity to yield the required answer. For example: supposing two powers expressed as 35.2 db and 37.0 db (relative to a common reference) are to be added. The difference is 1.8 db which, when located on scale A, corresponds to 2.2 db on scale B. This value is then added to 37.0 db to yield the resultant power of 39.2 db.

As another example, two voltages expressed as −2.0 db and +1.5 db have an algebraic difference of 3.5 db. The chart indicates that 1.6 db should be added to the greater original quantity, which in this case is +1.5 db; thus the answer is +3.1 db.

Problems involving the addition of more than two decibel-expressed quantities also may be solved. Any two quantities are chosen and added by means of the nomograph. The result is then added to any one of the remaining quantities by repeating the operation; thus, each step reduces the number of quantities by one. A succession of such operations will yield the desired answer.

The nomograph can be applied equally well to expressions of powers, voltages, currents, sound pressure levels, and components of noise or distortion. By reversing the procedure, it is possible to evaluate the contribution of either of two added quantities to the total, when the other quantity and the total are known, if the difference between the total and the known quantities is equal to or less than 3.01 db.

When the difference between two added quantities is greater than 10 db, the contribution of the smaller quantity is generally neglected. However, the scale has been drawn to accommodate a difference of 20 db in order to satisfy more exacting demands for accuracy. Disregarding the smaller quantity when the difference exceeds 20 db produces an error of less than one per cent.

The nomograph scales are based on the formula:

$$B = \left[10 \log_{10}\left(1 + \log_{10}^{-1}\frac{A}{10}\right) \right] - A$$

where A and B correspond to points on the respective scales.

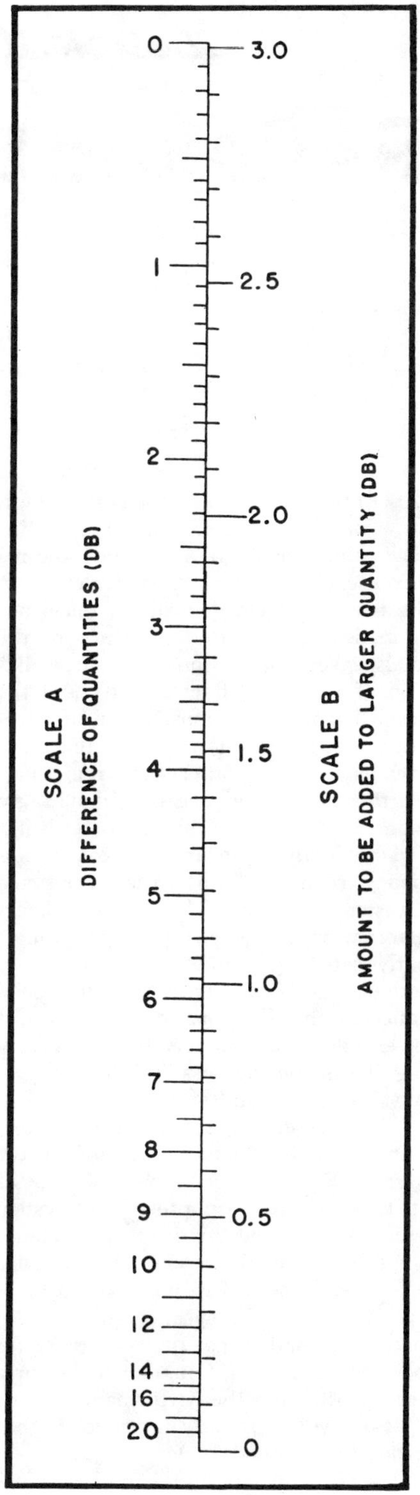

Design, Construction and Adjustment of

Reflexed Loudspeaker Enclosures

DAVID W. WORDEN

Practical procedure in the planning for a reflexed speaker cabinet, with constructional hints which will simplify the work of building.

UNLIKE THE EXPONENTIAL HORN, multiple speaker, and large or "infinite" baffle arrangements, the reflexed enclosure is a resonant device. If the resonant frequency of the enclosure is made equal to the frequency of the loudspeaker cone resonance, a cancellation of resonant effects occurs and the result is smooth response down to a frequency somewhat lower than the loudspeaker would otherwise reproduce. Furthermore, the speaker diaphragm works into a favorable acoustical impedance, which means increased efficiency, reduced distortion and improved transient response. The damping characteristics of this enclosure are inherently rather poor, but a liberal use of sound absorbents—necessary for good cancellation of resonant peaks—results in excellent damping. A further advantage is its compactness and flexibility of physical shape and size.

The speaker resonant frequency determines the low-frequency cutoff of the system, since there is naturally a limit as to how far the response of the system will extend below this frequency. Hence the speaker should be chosen which has a low resonance; if response down to 30 cps is desired, the speaker should resonate at around 60 cps or less. Hence for best results 12- or 15-in. drivers are preferred, although the performance of any speaker will be greatly improved with a properly designed enclosure.

The reflexed enclosure is nothing more nor less than a cavity resonator of the type developed by H. Helmholtz. Referring to *Fig.* 1, it consists of an enclosed volume of air V coupled to the outside by means of a mass of air M in an open tube, or port. The magnitudes of V and M determine the resonant frequency. The operation is analogous to that of a parallel tuned circuit.

The volume, V, and the mass, M, of air in the exhibit acoustical reactance (capacitive and inductive, respectively) just as do their electrical counterparts. Also, similarly, the Q of the circuit is determined by the amount of resistance present, the acoustical resistance being supplied by sound absorbent lining inside the box and by curtains of burlap or similar material stretched across the port. The impedance of such a parallel tuned circuit is maximum at resonance.

The speaker is also a resonant device. The moving parts (cone and voice coil) and their suspension are mechanically equivalent to a weight acted upon by a spring. Such a system behaves like a series resonant circuit, which shows minimum impedance at resonance. Hence if the two systems be connected together and adjusted to resonate at the same frequency, the impedance "peak" of one fills the "valley" of the other and the combination tends toward constant impedance over a broad range of frequencies. If the resistive element, Q, of one of these circuits is adjustable, the cancellation of resonant effects can be brought about more closely.

The simplified equivalent circuit of the combination of speaker and enclosure is essentially as shown in *Fig.* 2.

Phase Effects

The question is often raised as to the phase of the signal issuing from the port relative to that from the speaker. The popular belief that phase shift in the reflexed box is due to internal reflections, and hence to greater path length, must be discounted in view of the fact that—with the usual box dimensions—path length could not possibly account for more than 20 deg. or so around 60 cps. The phase shift is due rather to the nature of the resonator, which may be considered as a closed organ pipe with lumped constants; the enclosed air, V, must always be a node and the air, M, a loop. (Note that no overtones are possible, in contrast with the organ pipe.) One quarter wavelength then must exist between node and loop, which means 90 deg. phase shift. This is sufficient to give an additive component, even if the signals from port and cone are equal; however, near resonance, the stiffness of the enclosure limits the cone amplitude to a very small value, and the radiation is almost entirely from the port. Furthermore, the output of a loudspeaker contains, near resonance, a strong component at 90 deg. with respect to diaphragm velocity, which would be in phase with the enclosure output. Thus the phase relations are favorable regardless of the shape or proportions of the box.

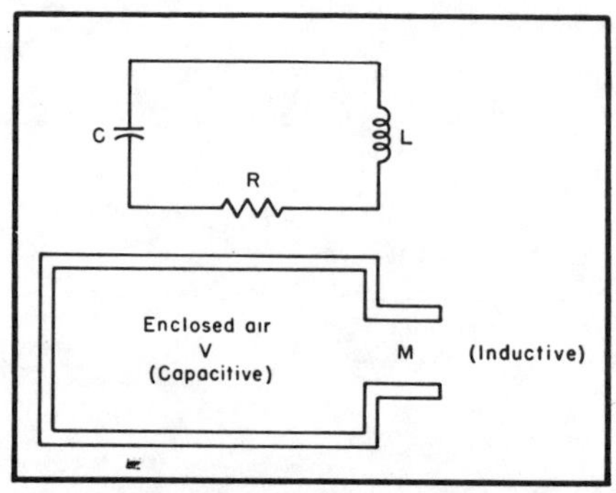

Fig. 1. Reflexed enclosure reduced to Helmholtz resonator equivalent, with electrical circuits corresponding to the acoustic network.

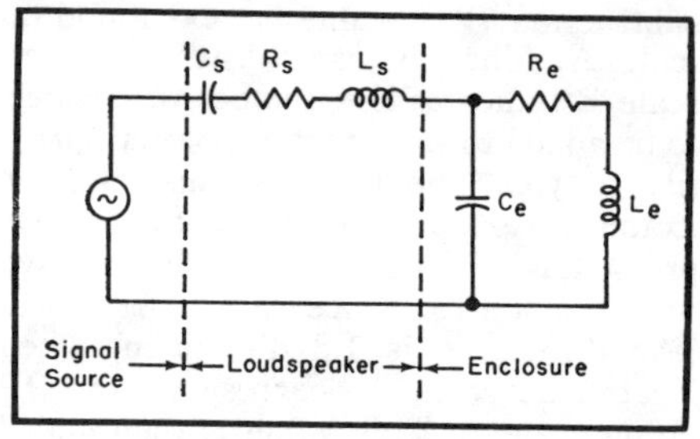

Fig. 2. Simplified electrical equivalent of loudspeaker mounted in reflexed cabinet.

The large reduction in loudspeaker-generated distortion is due to the restriction of cone amplitude mentioned above. Henney[1] shows the maximum distortion in an open-back cabinet of 43 per cent to be reduced to a maximum of 12 per cent in a reflexed enclosure. This represents a reduction in distortion of over 72 per cent.

Design Procedure

Lord Rayleigh presents the following formula for the frequency of resonance

[1] "Radio Engineers' Handbook," 3rd. Ed. McGraw-Hill Book Co., New York.

(*Theory of Sound, Vol. II*), for an enclosure of the type shown in *Fig.* 3.

$$f = \frac{c}{2\pi}\sqrt{\frac{A}{V\,(L + \frac{1}{2}\sqrt{\pi A})}} \qquad (1)$$

where

- f = resonant frequency, cps.
- c = velocity of sound, in./sec.
- π = 3.1416
- A = area of cross section of port, sq. in.
- V = net internal volume of enclosure, excluding volume of port, speaker, sound absorbents, etc., cu. in.
- L = port length, in.

The velocity of sound is approximately 13,560 in./sec. at 70° F. Using this value, combining constants and solving for V, the formula becomes:

$$V = \frac{4.657 \times 10^6\,A}{f^2\,(L + .886\sqrt{A})} \qquad (2)$$

The first step is to determine the design frequency, f, which may be done in the following manner: with the loudspeaker in open air and connected to the output of an audio oscillator, vary the frequency slowly from about 30 to 150 cps. Note the frequency at which the cone amplitude is greatest. The peak may be rather broad; so run across it several times, noting the frequencies above and below the peak at which the diaphragm motion noticeably decreases, and average these two readings. Bits of paper torn up and placed on the cone may assist in observing the amplitude of the cone movement.

A better method, particularly with small speakers, is to isolate the signal generator and voice coil by means of a series resistance several times the nominal voice coil impedance, and read the voltage developed across the voice coil with a good a.c. rectifier-type voltmeter. These readings may be plotted against frequency and the resonant peak may be read accurately from the graph. This method is also the best for testing the completed enclosure.

Now that f is known, there remain an infinite number of combinations of V, A and L which would yield the desired result. A value for A may be arbitrarily chosen; it should be from one half to one times that of the speaker opening. Past practice seems to indicate this choice; at least, a number of successful enclosures have port areas within this range. The larger area is preferable since it radiates more sound energy, but, if it is too large the internal dimensions of the box may approach quarter wavelength, since the volume increases with port area. The area of the speaker cone may be computed from:

$$A' = \pi S \left(\frac{D + d}{2}\right) \qquad (3)$$

where

- A' = speaker cone area, square inches
- S = slant height of cone, inches
- D = diameter at outer edge of cone, inches (do not include corrugations)
- d = voice coil diameter, inches

From here on, the following procedure is suggested. Choose a value for A, say $A = A'$ to begin with. Set L equal to the thickness of the material of the box plus the absorbent lining, as this will be easiest to construct, then solve for V. Compare this computed volume with the space available, or cabinet size desired; and if it seems too large, either increase L or decrease A or both until a satisfactory compromise is obtained.

Box Shape

Now a word as to the shape of the box: the only restriction is that the inside lengths should be kept small in order to discourage air column resonances which may occur at frequencies where such dimensions are equal to a quarter wavelength. With the usual proportions, these resonant frequencies are high enough to be readily absorbed by the lining of the box, but if the enclosure were unusually long, trouble might be encountered. Also, the box will be inherently stronger and will require less material, the closer its shape is to a cube.

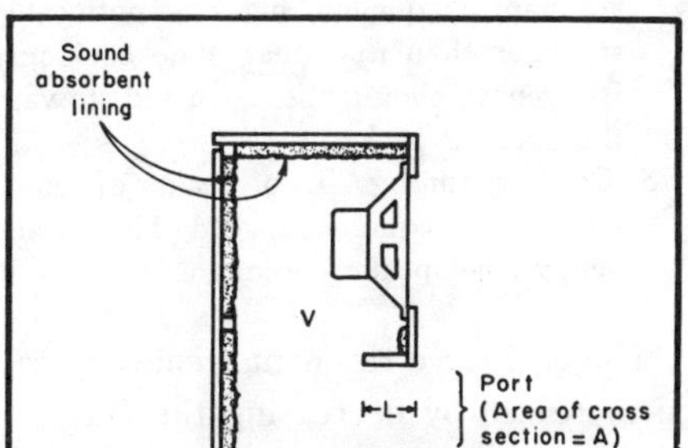

Fig. 3. Basic arrangement of ported cabinet using a port of finite length.

To the calculated net volume, V, must be added the volumes of the speaker itself, of that portion of the port which projects within the cavity, and of any other objects to be located within the box. The dimensions corresponding to the resulting gross volume will be inside measurements, and the thicknesses of the enclosure walls and lining must be added in order to obtain the overall dimensions. The volume displaced by the speaker may be estimated by computing the volumes of the cone plus a cylinder enclosing the magnetic structure. A table of approximate values is given below for convenience; however, individual speakers vary greatly, and actual measurements should be used whenever possible.

Nominal Speaker Size (In.)	Approximate Displacement (Cu. In.)	Approximate Cone Area (Sq. In.)	Approximate Resonant Frequency (cps)
6	10– 20	20	150–200
8	30– 60	38	100–150
10	70–140	60	70–100
12	100–200	85	60– 85
15	200–400	115	40– 65

The actual shape of the enclosure, if not dictated otherwise, usually develops in this fashion: the front face area is made large enough to accommodate the speaker and port comfortably, and its area computed. The gross inside volume divided by the (inside) area of the front face gives the required depth. The frontal area, port depth, or port area may be changed, if necessary, to adjust the depth to a satisfactory value.

An example may help to clarify the foregoing. Suppose a reflexed enclosure is to be designed around a 12-in. speaker which shows cone resonance at 70 cps. The cone area [A' in. Eq. (*3*)] is computed to be 85 sq. in., and this value will be used for A in Eq. (*2*). Now, assuming ¾-in. plywood for the box and ¾-in. lining, L = 1½ in. for the first trial. The net volume as computed from Eq. (*2*) is 8355 cu. in. Adding 200 cu. in. for the speaker gives a gross volume of 8555 cu. in.

The area of the front is estimated as

follows: the long side of the port may be made approximately equal to the diameter of the speaker cutout; this is economical in space and balances the appearance. The diameter of the speaker opening (and one side of the port rectangle) is $10\frac{1}{2}$ in. The other side of the port opening is then $85/10\frac{1}{2} = 8.1$ in., or approximately $8\frac{1}{8}$ in. (Great accuracy is not necessary, since the resonant frequency varies as the square root of the volumes, areas, etc., [See Eq. (*1*)]. Allowing 3 in. edge clearance and 2 in. between the speaker and port, the inside dimensions of the front will be $(3 + 8\frac{1}{8} + 2 + 12 + 3) = 28\frac{1}{8}$ in. long by $(3 + 12 + 3) = 18$ in. wide (speaker diameter is 12 in.). The frontal area is $(18)(28\frac{1}{8}) = 506$ sq. in. and the depth, then, must be $8555/506 = 16.9$ in. inside. Allowing $1\frac{1}{2}$ in. for the wall thickness including lining, the outside dimensions become $31\frac{1}{8} \times 21 \times 19.9$ inches. Suppose, now, that the front dimensions are satisfactory but the depth is too great. The port length L may be arbitrarily increased, say to 4 in. The volume, Eq. (*2*), now becomes 6639 cu. in. net; adding 200 cu. in. for the speaker and 289 cu. in. for that portion of the port tube projecting into the enclosure, measured as shown in *Fig.* 4, the gross internal volume equals 7128 cu. in. The inside depth, then, is $7128/506 = 14.1$ in. or 17.1 in. outside.

Construction Notes

The box should be very rigid in order to resist vibration. All joints, corners, etc. should be strong and tight, preferably reinforced with strips, and large panels should be braced. The back should be attached with a liberal number of screws so that it may be removed to give access to the interior. If a pliable material such as hair felt is used for lining the enclosure, it may be attached to furring strips, thus spacing the lining away from the wood and increasing the low-frequency absorption. Take precautions against air leakage; the speaker gasket should seat against the wood, wiring should be brought out through a bulkhead type of plug or receptacle which may be mounted securely with screws, and the removable back panel should fit snugly. A good method for attaching the back is shown in *Fig.* 5.

Items of equipment may be located within the box provided they are not affected by the high pressures developed insided the resonator. Output transformers and dividing networks may be mounted in the box, but amplifying stages, for instance, might be subject to acoustical feedback if placed inside. Tweeter mechanisms should be well protected from this pressure.

Adjustment Procedure

Install the speaker in the completed enclosure and screw the back into place. Connect an audio oscillator to the loudspeaker input, and adjust the signal to a comfortable level. Now vary the frequency through the range below 200 cps, noting the frequencies of any peaks which may appear in the output. One of

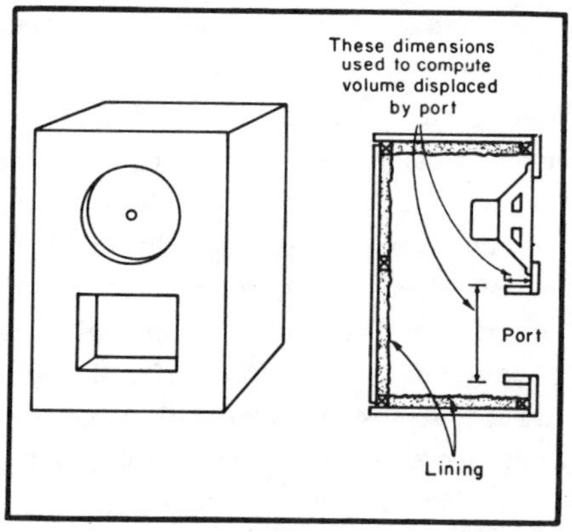

Fig. 4. Reflexed cabinet appearance, showing measurements to compute port displacement.

three conditions is likely to be encountered, as follows:

1. Enclosure frequency too high or too low. Two large peaks appear; one at loudspeaker resonance and the other at enclosure resonance.
2. Enclosure frequency slightly too high or too low. Two peaks appear, equally spaced above and below loudspeaker resonant frequency, but one noticeably stronger than the other. The enclosure frequency should be adjusted toward the smaller peak.
3. Correct tuning. Two peaks of equal amplitude, equally spaced above and below loudspeaker resonance.

The enclosure resonant frequency can be increased by decreasing the enclosed volume, V. A simple method for doing this is to place wooden blocks, such as might be cut from 4×4 material, inside the enclosure. These may simply be tossed in through the port while adjust-

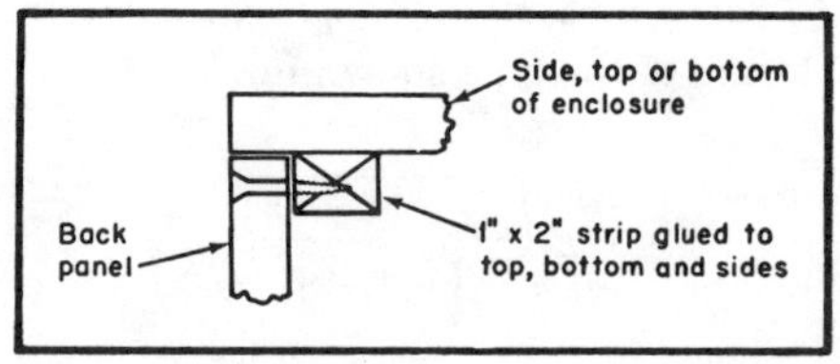

Fig. 5. Construction detail for corners of speaker cabinet.

ments are being made, and fastened down later. The easiest way to lower the frequency is to decrease the port area, A, which may be done with strips of wood cut to fit along one side of the port, the width being equal to the port depth, L. These strips may be fastened in place with screws.

When the frequency has been correctly adjusted, the damping may be increased by stretching porous cloth material (burlap, etc.) over the port opening. Experiment with various weights and layers of cloth until the two peaks just disappear. Too much damping will cause the single peak at speaker resonant frequency to appear.

As an alternative to merely detecting the resonant peaks by ear, a voltmeter may be used as described previously, and the response curve plotted.

It may be advisable to recheck the frequency adjustment after the system has been in use for some time. Loudspeakers, particularly when new or recently reconed, tend to show a lower value of resonant frequency after a period of time due to the cone suspension becoming more pliable with use. The loudspeaker resonance may always be found by blocking off the port and exploring with the audio oscillator. The only peak which shows up with the port blocked is that due to the loudspeaker cone resonance.

One more requirement is that the amplifying equipment used to drive this speaker system be capable of good frequency response, low distortion, and low output impedance. Then, a correctly adjusted and well damped reflexed enclosure will add greatly to the "presence" effect by providing extended bass response that sounds full and true without the usual resonant "boom" or "rain barrel" effect. Percussion instruments, plucked strings and other signals with high transient components come through clean and sharp because of the excellent response to such signals. In short, the improvement in over-all performance due to the enclosure is great enough to more than justify the labor and expense involved.

Loudspeaker Enclosures

DANIEL J. PLACH and

PHILIP B. WILLIAMS

Any evaluation of a speaker system must include the cabinet in which the loudspeaker is mounted; the authors analyze commonly used enclosures with respect to size and performance.

ENCLOSURES must be considered as an integral part of an acoustic radiating system. The low-frequency performance of a loudspeaker system depends to a large extent on the enclosure—in many cases is governed by it. It is the purpose of this article to discuss the features of the more generally used types of enclosures, and factors that must be considered in order to obtain the best possible low-frequency performance. Charts and equations are given which will enable the constructor to design enclosures of various types with reasonable assurance of obtaining an optimum design.

As an aid to understanding the function and operation of enclosures, a review of the principal types of enclosures is helpful. These may be divided into five main types:

1. Flat baffle
2. Open back cabinet
3. Enclosed cabinet
4. Horn loaded
5. Bass Reflex

No one type will fit all purposes and uses. It must be decided from a study of the characteristics of the various types which enclosure or which combination of enclosures will be most desirable for the application.

Flat Baffle

To describe the behavior of a flat baffle, it is necessary to consider the concept of doublet and simple sources of sound. A doublet source is one in which two point sources of equal strength and opposite phase are separated by a small distance. This is exemplified by a loudspeaker cone operating at low frequencies without a baffle.

The power radiated by the acoustic doublet is proportional to the fourth power of frequency and to the square of the velocity of the diaphragm.

A direct-radiator loudspeaker is essentially mass controlled above its resonant frequency so the velocity of the diaphragm is inversely proportional to frequency. The radiated power is therefore proportional to frequency, and increases 6 db per octave. Below resonance the speaker is stiffness controlled and the velocity is proportional to frequency so that the radiated power is proportional to the sixth power of frequency and falls off 18 db per octave. By comparison, the output of a simple source radiating into a semi-infinite medium—as in the case of a speaker cone in a large or infinite baffle—is proportional to the square of both velocity and frequency. Since the velocity is inversely proportional to frequency, the power output is independent of frequency as long as the diaphragm behaves as a simple piston. Below resonance, the response is proportional to the fourth power of frequency, falling off at the rate of 12 db per octave.

The radiation impedance of the diaphragm in the infinite baffle is considered as a simple mass in series with a re-

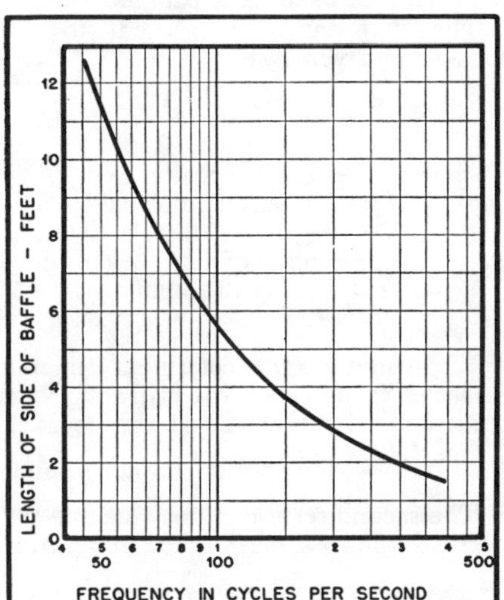

Fig. 1. Relation between low-frequency response and baffle size.

sistance whose magnitude is a function of frequency. In the region where $2\pi d/\lambda$ is less than *1,* this condition is valid. (d is diaphragm diameter and λ is wavelength.) In this region the magnitude of radiation mass M_A for one side is given by

$$M_A = .00658d^3$$

where d is the effective diameter of the speaker in inches, generally .8 to .85 of the speaker nominal diameter. The radiation resistance for one side is given by the expression

$$R_A = 5.6 \times 10^{-6} f^2 d^4$$

where R_A is in mechanical ohms and f is frequency. The radiation resistance is

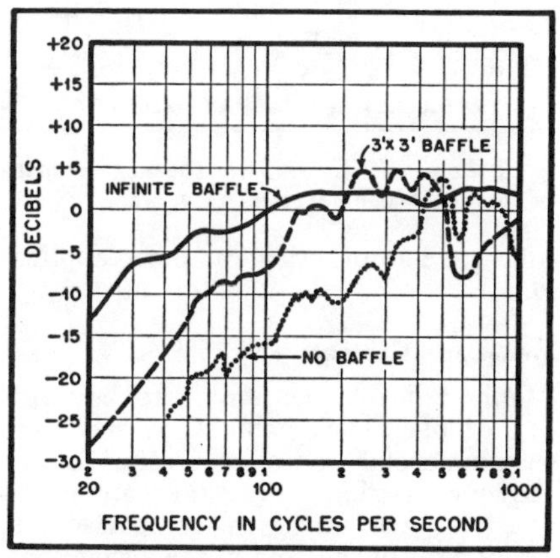

Fig. 2. Response of 15-in. speaker in different baffles.

of the magnitude of 360 ohms at 50 cps for a 15-in. speaker. This accounts for the poor damping and high distortion at resonance because of light speaker loading at low frequencies.

With an infinite baffle, good response may be maintained down to the resonant frequency of the speaker. The limiting factor in low-frequency performance is the resonant frequency of the speaker.

While a true infinite baffle does not

exist, it can be approximated for practical purposes by a baffle of dimensions equal to or larger than one-half wavelength at the lowest frequency to be reproduced. *Figure* 1 is a plot of baffle dimensions required for the lowest frequency to be reproduced. It can be seen from the chart that a baffle would have to be at least 11 ft. square for adequate reproduction at 50 cps. A good solution is to use the wall of a room as a baffle, permitting the speaker rear to radiate into an adjoining room. If the back of the speaker radiates into a relatively small volume, care must be taken to avoid enclosure resonances by addition of suitable damping materials. In addition, the volume of the enclosure has an effect upon speaker resonance, as described in the section on enclosed box operation.

If a flat baffle is used, the speaker should preferably be mounted off-center so that there is a variety of different path lengths for sound travel from front to back. This procedure avoids irregularities that may otherwise occur in the response curve.

Effect of baffle size on a 15-in. high-efficiency, low-resonance speaker is shown in *Fig.* 2. Where economy in space, good low-frequency response and good transient response are needed, flat

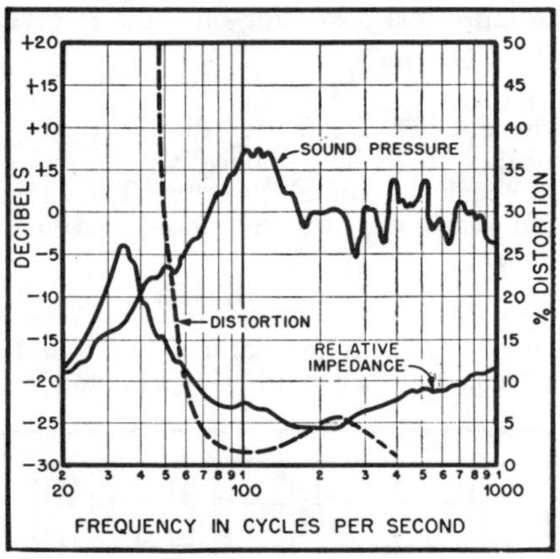

Fig. 3. Low-resonance 15-in. speaker in open-back cabinet 4 in. from wall.

baffles are impractical in most applications.

Open-Back Cabinet

Only a brief mention is to be made here of open-back cabinets widely used in commercial radio receivers. In the frequency region where the cabinet dimensions are large in comparison to wavelength of the sound, the system acts as a simple source. For a mass-controlled system, with constant driving force, the output is independent of frequency. Below this point, the transition occurs to the doublet source, and output falls off rapidly. Low-frequency performance is critically dependent upon positioning of the cabinet in the room. The action is that of a folded baffle. Typical response of a 15-in. speaker in an open-back 7-cu. ft. cabinet 4 in. from a wall is as shown in *Fig.* 3. The rise at 100 cps is caused by the cabinet acting as a resonant tube, and would be more pro-

Fig. 5. Increase in resonant frequency vs. cabinet volume.

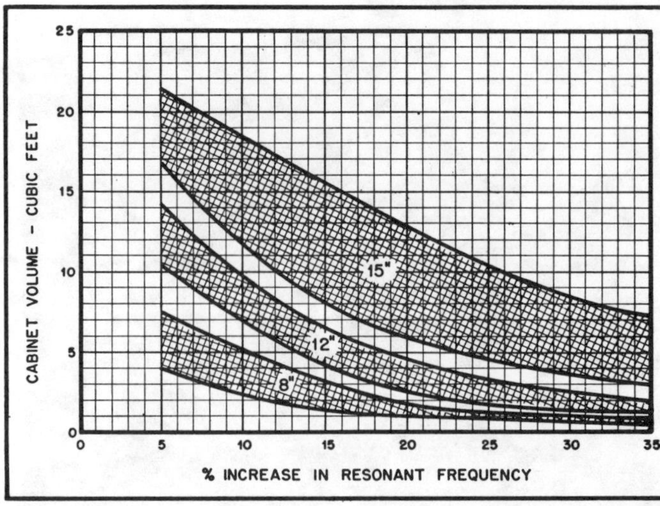

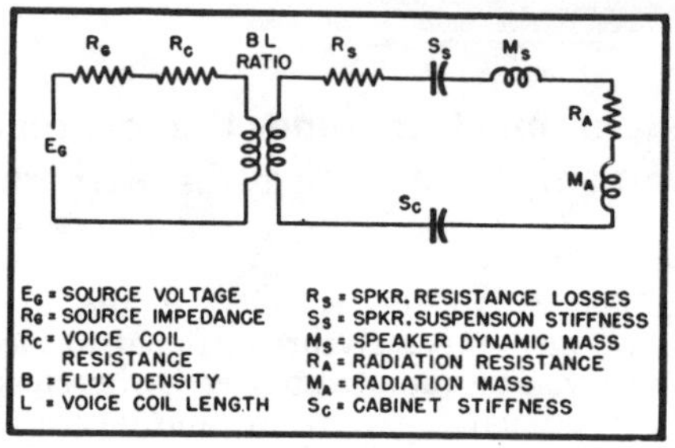

Fig. 4. Equivalent circuit of loudspeaker in closed-back cabinet.

nounced with a speaker of lower efficiency or if the cabinet frequency were placed at the resonant frequency of the speaker. The exaggerated output in this region may cause undesirable "boominess." This type of cabinet is not commonly used in high-quality reproducing systems today.

Enclosed Cabinet

The use of an enclosed cabinet with adequate volume makes it possible to attain satisfactory low-end performance, since a loudspeaker system with this type of enclosure operates essentially as a simple source. The output falls off much more slowly at low frequencies than with an open-back cabinet.

The impedance presented to the speaker is given by:

$$Z = -j\rho c \frac{A_s^2}{A_c} \cot \frac{2\pi L}{\lambda}$$

where A_s = effective speaker area
A_c = cabinet area normal to L
L = length of side normal to A_c
ρ = density of air
c = velocity of sound
$j = \sqrt{-1}$

Expansion of the cotangent function in series form yields the stiffness S_v as contributed by the box to the speaker:

$S_v = \frac{\rho c^2 A_s^2}{V}$ where V is cabinet volume.

The second term is a positive one that corresponds to a mass M of magnitude:

$$M = \frac{1}{3} \frac{A_s^2}{A_c} \rho L$$

Below the point where L equals 1/4 wavelength, the box volume introduces stiffness into the mechanical mesh of the speaker. Above this point, the box acts as a mass until L equals 1/2 wavelength, at which point the first normal mode of the box occurs. These modes or resonances have the effect of reducing the stiffness in the limited region where L is less than 1/4 wavelength. They occur at integral multiples of π and are points of high reactance as presented to the speaker. These resonances have the effect of introducing irregularities in the response of the loudspeakers. Since they generally occur at higher frequencies, it is possible to reduce their effect by addition of absorbent material in the box. This treatment—consisting of heavy felt, cellulose, glass fiber mat, or other damping material—is applied to at least one of each pair of parallel surfaces. Damping material makes the box appear as a resistance at high frequency and adds to the speaker damping.

A study of the cotangent function

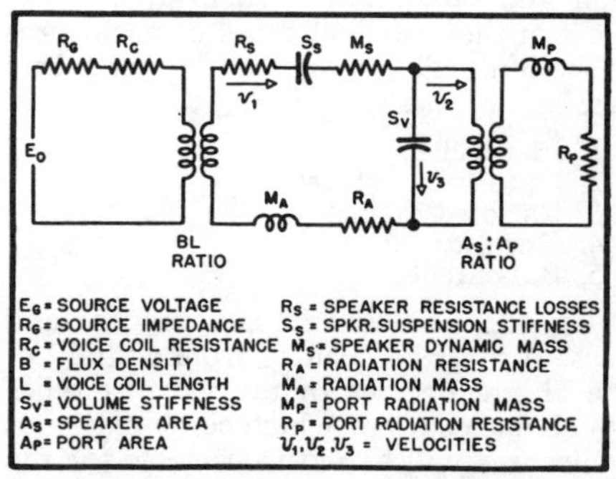

Fig. 6. Equivalent circuit of loudspeaker in typical bass-reflex enclosure.

shows that the box does not act as a simple stiffness at all frequencies. The box differs from this simple stiffness behavior even for values of L somewhat less than 1/4 wavelength. However, if the linear dimensions of the box are less than 1/8 wavelength, the error results in stiffness values somewhat smaller than expected. When the speaker resonance occurs at a frequency high enough so that the box dimensions are larger than 1/8 wavelength, the more exact expression for stiffness must be used in calculations.

For most design purposes, the stiff-

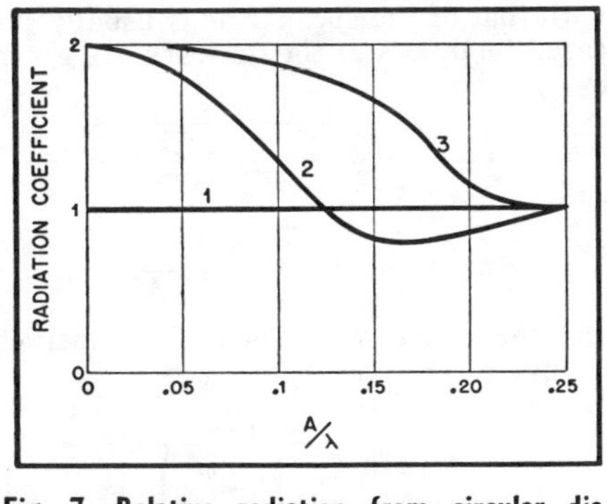

Fig. 7. Relative radiation from circular diaphragm combination. (1) single diaphragm; (2) two diaphragms 4A apart; (3) two diaphragms, in contact.

ness contribution due to the box as seen by the speaker is given approximately by:

$$S_v = \frac{2.26 \times 10^6 d^4}{V}$$

V is in cubic inches and S_v is in dynes per centimenter. This stiffness acts in series with the mechanical mesh of the loudspeaker as in the equivalent circuit in *Fig.* 4.

The effect of the compliance is to raise the resonant frequency of the loudspeaker above that which would exist when mounted in an infinite baffle. The effective system stiffness S_e resulting from the volume stiffness addition to the speaker then becomes:

$$S_e = S_s + S_v$$

where S_s is the stiffness of the loudspeaker vibrating system. The speaker resonant frequency f_c, in the enclosed cabinet, has the relationship to f_b, resonant frequency in infinite baffle:

$$f_c = f_b \sqrt{\frac{S_s + S_v}{.S_s}}$$

So as the volume is made large, S_v approaches zero and for practical purposes the conditions of an infinite baffle are attained.

From these equations, it is possible to calculate the volume required to limit the resonance shift to a prescribed value. *Figure* 5 gives the relationship between enclosure volume and frequency shift in terms of speaker size. The chart is based upon suspension compliances of Jensen loudspeakers as listed by nominal diameter sizes. From a practical standpoint, the chart is adequate for use with most speakers of the nominal sizes listed. A resonance shift of 5 or 10 per cent is not excessive, if one considers that a shift of 10 per cent at 50 cps is only 5 cps. For

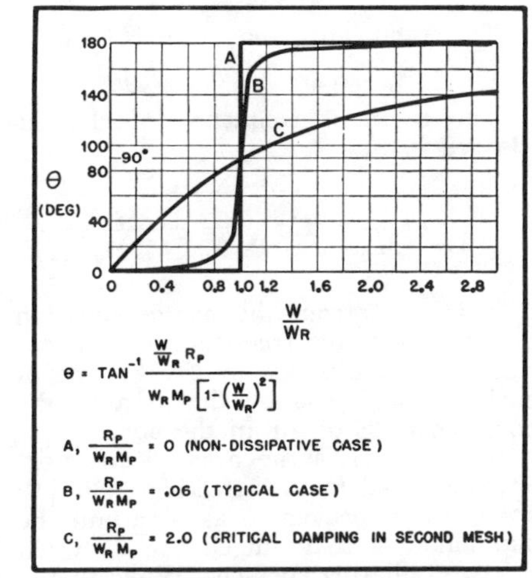

Fig. 8. Relative phase between currents (velocities) in two mechanical meshes.

each nominal speaker size, a shaded area is given. The upper limit of the area corresponds to speakers of least stiffness, and the lower limit corresponds to speakers with the highest stiffness. Information on the actual stiffness of loudspeakers is not usually available, so the upper limit of the shaded area should be used whenever possible to insure sufficient enclosure volume for all speakers. The geometry of the cabinet has not been found to be overly critical, if the longest side is not more than ⅛ wavelength in the range under consideration. The stiffness contribution of the cabinet will, however, be somewhat dependent on the cabinet geometry.

In addition to precautions to be taken in placement of absorbing material, the cabinet wall conditions must be considered. High pressures exist in the enclosure in operation. It is desirable to use material at least ½ in. thick for the enclosure walls to reduce vibration during operation, and special bracing may be necessary. Vibrating walls introduce a variable compliance and add dissipation, tending to produce irregularities in the response, as well as possible objectionable rattles. This phenomenon often shows up as irregularity in the impedance curve.

While satisfactory response can be attained with this type of enclosure in adequate volume, lower distortion and better damping characteristics can be achieved with other types of enclosures.

Horn Loaded

Due to some restrictive limitations, horn loading for direct-radiator speakers

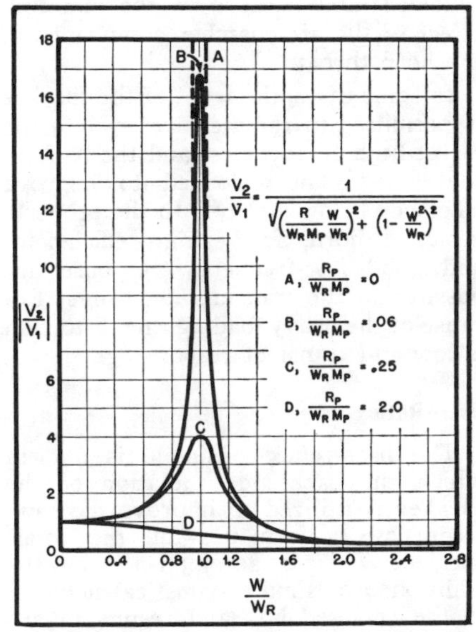

Fig. 9. Magnitude of the ratio of port velocity to diaphragm velocity.

is not used extensively, despite considerable advantages in performance. There are two main ways of horn-loading the speaker. One is to load the front end of the cone with a horn. A horn of the

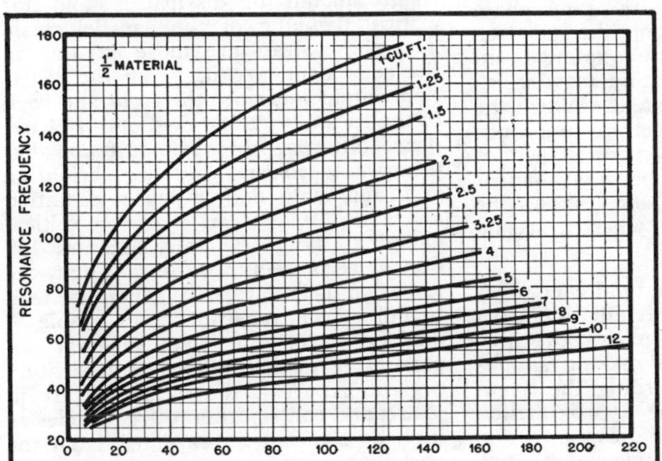

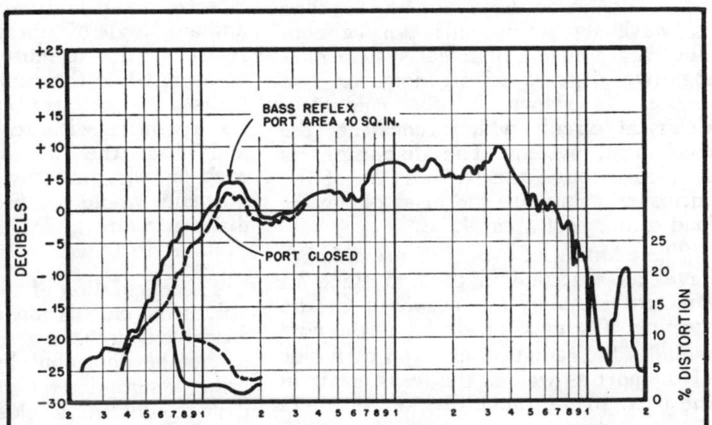

Fig 10. (left). Port area vs. resonant frequency in bass-reflex cabinet. Fig. 11, (right). Sound pressure response and r.m.s. distortion of 8-in. speaker in 1 cu. ft. enclosure.

necessary dimensions is quite bulky, and to conserve space it is usually necessary to fold or otherwise change the geometry. The mouth diameter required is determined by the lowest frequency that must be passed, so it must be equal to or greater than 1/3 wavelength at the lowest frequency. If mouth shape is other than circular, the corresponding area may be used. This gives the relationship that minimum diameter of mouth, in inches, is equal to *4000/f*, where f is in cps. Thus at 50 cps, mouth diameter must be 80 in. In some cases, the walls of the room are utilized as a continuation of the horn, thus affording very large mouth diameter. High-frequency performance is usually limited to frequencies of the order of 400 cps, because high frequencies are lost in the folded horn due to the tortuous path followed, and because of the shunting effect of the air chamber compliance at the horn throat.

In some cases, the front of the speaker is permitted to radiate directly into the air at high frequencies, and the back of the cone is horn loaded to improve efficiency below 400 cps. In the properly designed horn, fairly high efficiencies and good low-frequency response are possible in the transmission range. Because of the heavy loading, the distortion is low and transient response good.

Bass Reflex

The bass-reflex principle is one in which the back side radiation of the speaker is utilized to improve the low-frequency performance of the loudspeaker. This is accomplished by the addition of a simple acoustical network to the enclosed box and results in another degree of freedom for the equivalent circuit of the speaker and cabinet. *Figure* 6 shows the typical bass-reflex enclosure and its equivalent circuit.

It is seen that an additional degree of freedom is attained by placing a slit or port—for the back side radiation—at the front of the cabinet and near the speaker. At frequencies for which the maximum linear dimension is less than ¼ wavelength this unit can be considered a simple dynamical system having two degrees of freedom, or two meshes as shown in the equivalent electrical circuit which represents the mechanical system. The diaphragm is coupled through the stiffness S_v of the enclosure volume to the mass of the air load of the port area A_p.

The addition of a port of area A_p behaves as a second diaphragm since an effective mass of air oscillates in the opening. The mass of air M_p in the port includes the radiation mass on each side of the port as well as the mass of air in the port. In the range of interest where $2\pi a/\lambda$ is less than one-half, the radiation mass of the port is $M_A = \dfrac{16A_p\rho a}{3\pi}$ where a is the radius of an equivalent circular piston.

The radiation resistance of the port is

$$R_p = \frac{A_p^2\omega^2\rho}{2\pi c}$$

While these expressions apply rigorously in a circular piston, if the ratio of port length to port width does not exceed 2:1 the values calculated from formulas for a piston of equivalent diameter are a satisfactory approximation. The mass contribution due to the thickness t of the walls of the port opening is

$$M_t = \rho t A_p$$

The total port mass then becomes

$$M_p = A_p\rho(t + .96\sqrt{A_p})$$

The resonant frequency of the enclosure is then

$$f_r = \frac{1}{2\pi}\sqrt{\frac{S_v}{M_p}} = 2155\sqrt{\frac{A_p}{V(t+.96\sqrt{A_p})}}$$

It is seen from the above equation that the resonant frequency of the enclosure is determined by the volume of enclosure and also by the area of the port and mass of air in the port.

In many cases the area of the port is made equal to the effective radiating area of the speaker so as to attain the maximum mutual impedance between the two radiating surfaces. When this is done, in order to make the volume V of reasonable value for a given resonant frequency, the length t is varied by the use of ducts. If a duct of volume V_d is used, the previous equation must be modified to the following extent:

$$f_r = 2155\sqrt{\frac{A_p}{(V-V_d)(t+.96\sqrt{A_p})}}$$

With the values of port radiation resistance and mass defined, the equivalent circuit can be simplified by referring these parameters to the mechanical mesh of the speaker by multiplying the port impedance by A_s^2/A_p^2.

The exact analysis of the equivalent circuit is complicated by the fact that the mutual radiation impedance between the diaphragm and port is a function of the size and spacing of the radiating surfaces. If two surfaces of equal area are closely spaced and have the same phase and amplitude of vibration, the radiation resistance due to mutual coupling is increased, while the radiation mass is increased to a lesser extent. When the pistons differ with respect to amplitude and phase the radiation resistance of each diaphragm may be less than it normally would be. *Figure* 7 shows the difference in radiation when a second radiator is added.

The calculation of the total energy involves the calculation of radiation from a double source. In the general case this cannot be done, but by making simplifying assumptions regarding the boundary condition, a close approximation may be obtained in special cases.

The modes of the network with two degrees of freedom can be calculated or obtained by use of Mohr's circle.

If the resonant frequency f_r of the enclosure and port is chosen to coincide with that of the speaker as is usually the case, the modes of the network are given by

$$f = f_r\sqrt{\frac{(a+2) \pm \sqrt{a^2+4a}}{2}}$$

$$\text{where } a = \sqrt{\frac{S_v M_p}{S_s(M_A+M_s)}}$$

and the phase θ between the velocities v_1 and v_2 is given by

$$\theta = \tan^{-1}\frac{\omega R_p}{S_v\left(1-\frac{\omega^2}{\omega_r^2}\right)}$$

When $f = f_r$, analysis of the equivalent circuit shows that the impedance is resistive and has a maximum at this point so the excursion of the speaker diaphragm is very much less than it would be in an infinite baffle or type of enclosure other than the properly adjusted bass-reflex enclosure. At this frequency f_r, the radiation from the port predominates and is in quadrature with the diaphragm radiation. There are two frequencies f_1 and f_2 which are below and above, respectively, frequency f_r, and are determined by the quantity a previously defined. The mechanical impedance at these frequencies is a minimum and will produce an impedance maximum on the electrical side. It may be mentioned at this point that even though resonance occurs at frequency f_2 higher than that of the same speaker in a large closed cabinet, damping is better. This can be attributed to the increased mutual radiation impedance between port and diaphragm and a larger value of radiation resistance at the higher frequency. In most cases, however, the damping is largely dependent upon the magnetic energy of the speaker and the source impedance.

Below f_r, when dissipation is small the phase shifts very rapidly so that the radiation from port and diaphragm are out of phase by 180 deg. while above f_r the two radiating surfaces are in phase. *Figure* 8 shows the phase relationship existing between the velocities v_1 and v_2 in the two meshes for no dissipation, small amount of dissipation and for critical damping of the parallel mesh. The phase shifts are those existing in the equivalent circuit. From the radiation standpoint, since the front and back of the speaker are 180 deg. out of phase, phase difference between port and diaphragm approaches 180 deg. below ω_r and approach the in-phase condition above ω_r. It is seen that for critical damping, the phase shift becomes 180 deg. at zero frequency. As a result, the radiation below f_r will generally tend to fall off in transition from simple to doublet source, the exact behavior depending upon the amount of dissipation present. Above f_r the radiations are in phase, and contribution to frequencies as high as $2\omega_r$ can be expected from the port.

Figure 9 shows the ratio of the magnitude of port velocity to that of diaphragm velocity as a function of frequency for various values of damping

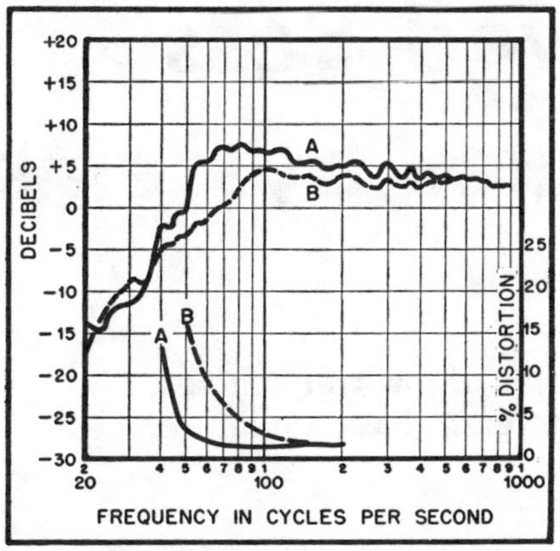

Fig. 12. Sound pressure and distortion, bass-reflex and enclosed cabinets of 7.3 cu. ft. volume, with 15-in. speaker. (A) solid curve, 66 sq. in. port. (B) dotted curve, no port.

in the added mesh. At higher frequencies the response from the port is reduced by the shunting effect of the cabinet volume and losses in the absorbing material in the cabinet.

In the application of the bass-reflex principle to loudspeakers, the speaker of the largest possible diameter should be chosen because the radiation resistance is proportional to the fourth power of effective diameter; the amplitude of motion of the diaphragm for a given amount of radiated power is less with a larger diaphragm, thereby reducing distortion that arises from non-linearities in the speaker.

The dimensions of the cabinet have not been found to be overly critical. Design charts given here hold very well for cabinets in which the longest side is not more than one eighth wavelength.

The port should be placed close to the speaker in order to take advantage of mutual impedance increase resulting from this close coupling. While port area is not critical, the port must be at least one quarter the effective speaker area. If smaller, the cabinet tends to act as a closed cabinet. Port area may be larger than the speaker area where the larger area gives the proper volume-port resonance.

With a port area in this range, and with resonant frequency of the speaker known, the volume of the enclosure can be found from *Fig.* 10. The correct port size can then be determined by observing experimentally the frequency placement of the modes for a given trial port size. These modes are always on opposite sides of the blocked port mode, and move in the same direction as the port area is changed. In practice, it has been found that the scalar impedance of the two resulting modes should be approximately equal in magnitude. If the port area becomes too small, a duct can be used to tune the enclosure to a lower frequency for the same port area.

A loudspeaker enclosure can be a bulky and even an expensive piece of equipment. In the case where economy and need for space dictate the use of relatively small enclosure volume, a properly tuned bass-reflex system helps to offset the effect of the small volume. In this case, where cabinet volume is small, it is best to tune the port exactly to the speaker resonance. *Figure* 11 shows the response and distortion characteristics of an 8-in. speaker in a one cubic foot cabinet, with and without port. This volume is small for this size speaker.

In the case where the cabinet volume allowable is more generous, a bass-reflex enclosure offers considerable gain in output at the extreme low end. *Figure* 12 shows the response and distortion characteristics of a low-resonance 15-in. speaker in a 7.3-cu. ft. cabinet with 66-sq. in. port, and the same speaker and cabinet without port. *Figure* 13 shows the effect of mistuning the port. Curves are shown for tuning to frequencies above and below the speaker resonance. It is seen that in this case, where enclosure volume is neither large nor small, the port should be tuned to the speaker resonance or slightly below.

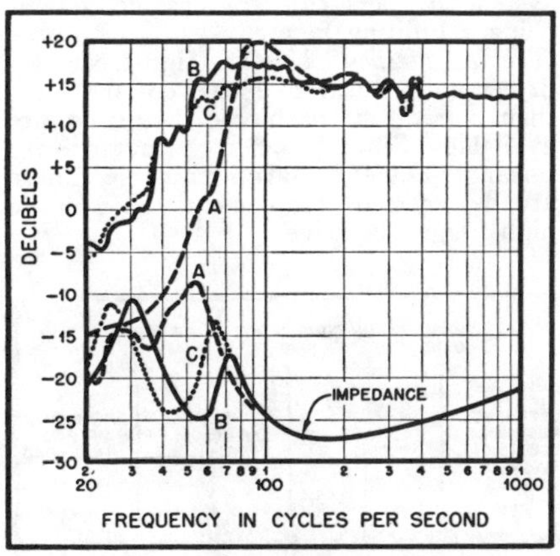

Fig. 13. 15-in. speaker in 7.3 cu. ft. cabinet. Port areas—(A) dashed, 132 sq. in. (B) solid, 66 sq. in. (C) dotted, 32 sq. in.

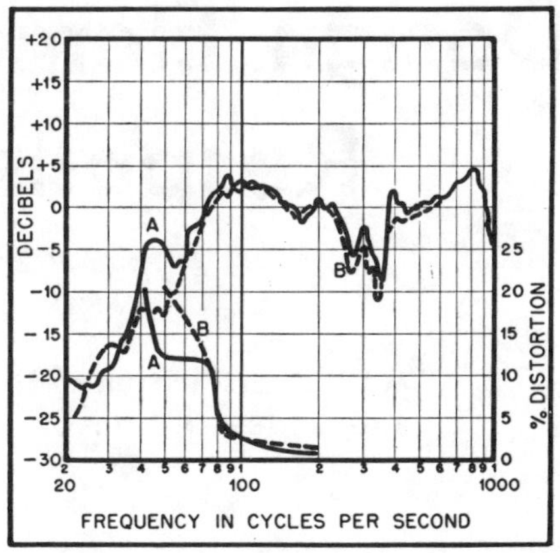

Fig. 14. 8-in. speaker in 12 cu. ft. cabinet. (A) solid curves, 60 sq. in. port; (B) dotted curves, no port.

There may be some instances where a cabinet to be used has a volume which is larger than necessary. If the volume is very large, it becomes difficult or impossible to tune the system to the speaker resonant frequency without a port of excessive size. In this case, the best adjustment is to set the port size at about 1½ to 2 times the effective speaker area, and allow the resonant frequency of the cabinet to be lower than the speaker resonance. There will be little effect on response or distortion at speaker resonance, but there will be improvement in both these factors in the region of the cabinet resonance.

Figure 14 shows what happens when an 8-in. speaker is used with a 12-cu. ft. cabinet with a port of 60 sq. in. The speaker has a resonance of about 100 cps, the cabinet of about 40 cps.

The ratio of port area to speaker area can be used to determine when the cabinet volume should be considered large or small. If the port area required to tune the cabinet to speaker resonance is less than one-half the speaker area, the volume is small. If the port area is more than one and one-half times the speaker area, the volume is large.

The photograph on page 87 illustrates a modern bass-reflex enclosure with a removable front panel for ease of installation and adjustment.

Theoretical considerations and corresponding experimental work, combined with extensive listening tests, lead us to believe that the properly adjusted bass-reflex cabinet is the most generally suitable enclosure for loudspeakers.

Design Elements for Improved Bass Response in Loudspeaker Systems

HOWARD T. SOUTHER

A frank discussion of the factors which influence the reproduction of low frequencies, and a description of means which may be employed to increase bass radiation.

IT HAS BEEN SAID that music is the only form of dissipation which may be engaged in to any degree by the individual without harm to his physical well being.

Recognition of this postulate helps to explain the accelerating interest in quality sound reproduction as evidenced by the growing ranks of high-fidelity enthusiasts. This enthusiasm constitutes a profitable market for the manufacturer. Accordingly, acoustic research has been spurred recently to satisfy this demand. The listening public, lacking the objectivity of the engineer, has found it difficult to articulate its needs, thus making it hard for the engineer to direct his efforts with any degree of certainty towards a problem the solution of which is primarily subjective. As in the formative stages of any art, much of what has been accomplished in high-fidelity reproduction was through trial and error methods. Exploratory tests conducted by two well-known audio authorities may suggest the objective criteria so necessary to the engineer, and disclose some important points which previously have been overlooked by the lay listener. It is the purpose in what follows to indicate that the incorporation *of improved bass response in reproduced sound is probably the most important step* in the further development of the art.

Listener Reaction Tests

Howard Chinn of CBS conducted a series of listener tests using reproduced music. The general conclusion from this test was that listeners prefer a narrow, or at least a restricted band of frequencies.

This conclusion violated every aim of the idealist who instinctively believed his goal to be duplication of reality—including the reproduction of all frequencies between 16 and 16,000 cps.

In refutation of the previous test conclusions, Olson of RCA employed a live orchestra with acoustic low-pass filters to test listener range preference. *The choice was incontrovertibly for the widest-range response possible.* The discrepancy in test results was shown to be principally the result of odd-order distortions introduced by the electro-acoustical system. Narrow band reproduction subdued this distortion, making the sound more tolerable.

An Indicated New Approach—The rather conclusive tests by Olson show that the frequencies from 16 to 16,000 cps are necessary for ultimate listening. A concomitant conclusion is that the widest frequency range must be accompanied by very low odd-order distortion or the band must be restricted to be tolerated.

This last compromise cannot be reconciled by the progressive electro-acoustic engineer or he denies his reason for being. For sometime he has had virtually perfect wide-range reproduction in the laboratory, and has awaited only sufficient arguments to warrant presenting it to the public. The way to this presentation is paved by these facts:

1. *Public Acceptance*—The determination that the public desires high quality sound reproduction has been made. Note that this desire includes all the audible frequencies without unbalance from the bass range.
2. *Wide-Range Source Material*—New wide-range microphones, low-distortion amplifiers, vinylite records, and magnetic tape will allow the wide-range speaker system to perform, sources no longer being a limiting factor.
3. *Treble Driver Units*—High-frequency driver units linear to 10,000 cps with less than 2 per cent harmonic distortion are available. "Super" high-frequency drivers linear to 17,000 cps are obtainable. These are designed to supplement the average high-frequency driver.
4. *Low Speaker-System Distortion*—The principle of separating the driver units for reproducing only their respective portions of the spectrum reduces the transducer distortion to the region of acceptable limits.

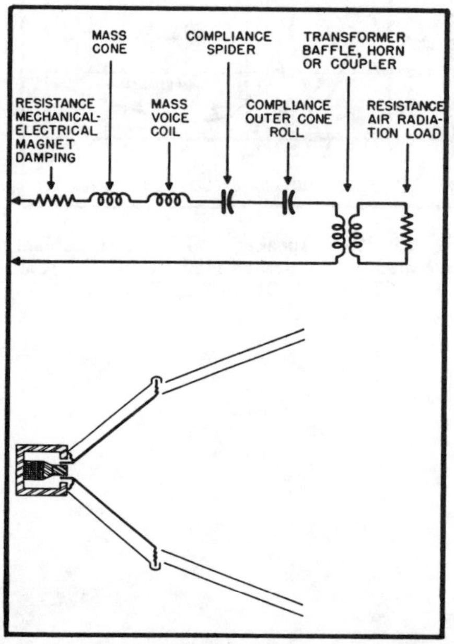

Fig. 1. Typical loudspeaker and its equivalent circuit for investigation of design requirements.

The notable exception to the listing above is adequate bass in the reproducer response. The goal of good listening in the home cannot be achieved without full consideration being given to the reproduction of the first three octaves up to 130 cps.

Consider for a moment that at 50 cps, a quarter wavelength is 80 in. In order to reproduce down to this frequency only, *just slightly below the third octave,* a horn 80 in. across the mouth is necessary. Thus, the usual claim of response to 40 cps would appear specious, to say the least, much less the sometimes added phrase of "usable response to 20 cps."

Further need for better bass response lies in the fact that the system is generally called upon to reproduce program music in the home at a much lower level than that at which it was originally recorded. Reference to the ear sensitivity curves of Fletcher and Munson show that the usual playback level of around 80 db leaves the bass attenuated by about 20 db. The best commercial amplifiers seldom incorporate more than 12 or 15 db of bass boost, for this equalization is costly of final over-all power output, especially at the extremely low frequencies where the amplifier ordinarily finds some difficulty in meeting its specifications.

It can be incontrovertibly shown that good extreme bass response is necessary for music reproduction, even though the particular fundamentals involved are not included in the passages being reproduced.

This is explained through the fact that music consists to a very large degree of transient signals which seldom approach a steady state. Thus, staccato passages on a flute with a fundamental of 500 cps will not be reproduced with the proper envelope shape if the low-frequency interruptions of the fundamental tone are below the pass band of the speaker system. Moreover, the mathematics involved will show that for perfect reproduction of this same staccato flute passage the band should be *infinitely* low *and* high. For practical purposes, however, it can be assumed that distortion will be minimized, even

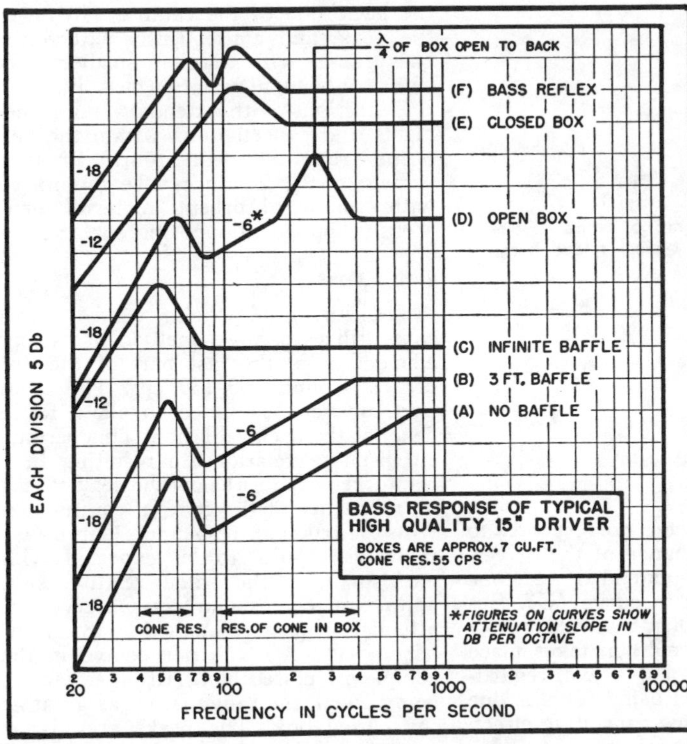

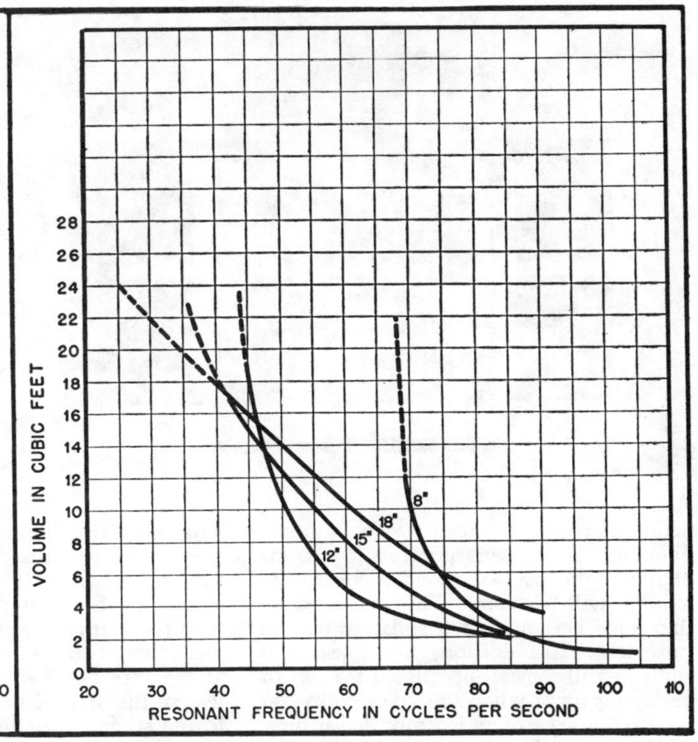

Fig. 2 (left). Comparison low-end response obtainable from various types of speaker mountings. Fig. 3 (right). Optimum size of speaker enclosure for speakers of various diameters and resonant frequencies.

on narrow band sources, if the pass band of the speaker is as wide as possible.

It has been previously concluded that the widest reproducing range is important, and it may be assumed that the extreme low end of the spectrum in reproduced sound has been neglected. It is remarkable in view of the obvious and overwhelming importance of these low frequencies that so little has been done to accomplish this reproduction.

It is the purpose of the balance of this writing to disclose what *can* be done, and perhaps to point the way to the complete accomplishment of adequate bass.

The Reproducing System

A proper understanding of the reproducing system cannot be achieved unless we consider the elements with which we have to work. In *Fig.* 1, we observe these elements and the equivalent electrical circuit. For best comprehension of what is to follow let us discuss the items shown in the circuit one by one.

(1) Magnet Damping (Mechanical-Electrical Resistance)—In an electrical circuit, resistance will lower the *"Q"* and broaden the transmission band. Exactly the same thing transpires in the acoustic circuit. High magnetic damping results in fine broad-band efficiency and subdues the natural modes of the vibratory system. There are other forms of damping which could be grouped along with this item. For instance, a shorted turn in the voice coil would provide this same sort of smoothing action, but at the expense of efficiency. Low internal impedance in the associated amplifier is of great value in providing this desirable damping action and accomplishes this without lowering over-all speaker efficiencies.

(2) Mass of the Voice Coil—A short analysis of the equivalent circuit will disclose that voice-coil weight becomes increasingly important as the frequency goes up. The circuit becomes mass controlled, with compliances having less effect. Copper wire is used universally in low-frequency drivers because of increased conductivity irrespective of the weight disparity between it and aluminum. The higher resistance of aluminum is a disadvantage at low-frequencies, whereas, above 1000 cps the lighter mass offsets the disadvantage by a factor of some 35 per cent. At low frequencies, the mass of the cone and air load is little affected proportionately by the addition of 5 or 6 grams, due to the copper, with its attribute of high current capacity. It is in the extreme bass range that the current assumes high proportions. At 15 watts input, the current at 50 cps is on the order of 2 amperes in the voice coil.

(3) Cone Mass—Observation of the inductive element represented by the cone mass will lead the electrical theorist to assume that increasing the cone weight will lower the resonance and increase the bass range. This is correct, but an increase in cone weight of double the usual 25 grams will lower the resonance by only a few cps and decrease the efficiency by over 75 per cent. In a driver of top design using 5 lbs. of Alnico V, the iron of the structure is working at its economical limit, and this loss of efficiency cannot be made up in any practical way.

(4) Compliance—The next point of attack is logically the suspension system or capacitive element. The usual low-frequency driver has fairly stiff outer compliance rolls, as well as a stiff inner suspension, or spider. The reasons for this are quite practical. The resonance will naturally become lower if the compliances are increased. But here the

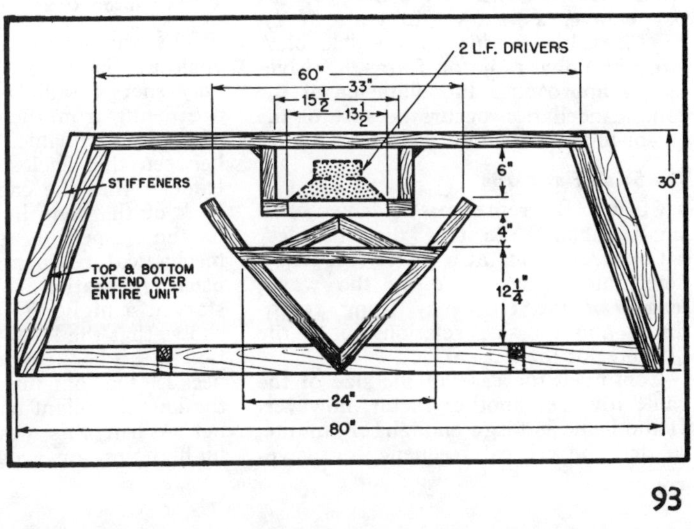

Fig. 4. Cross-section of theatre-type folded horn—the C-5 "bin."

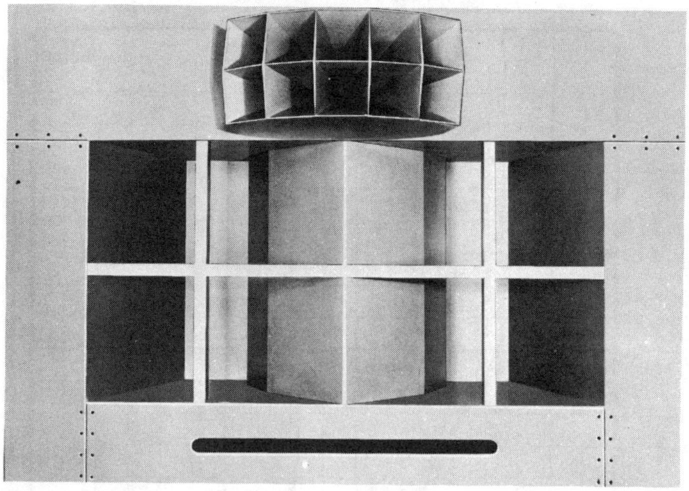

Fig. 5. Front view of typical folded horn.

manufacturer is faced with another dilemma. If he incorporates low compliances, the cone without adequate air loading will "bottom." This is caused by the voice coil hitting the flange on the nosepiece. In addition, non-linear response will ensue because of lack of restoring force when the voice coil rides out of the area of uniformly dense flux. A 30-watt rating for a driver becomes less than 3 watts when the suspensions are made more compliant without proportionately increasing the air load. But this, along with increased air loading, points the direction towards increased bass range and efficiency.

The average good 15-in. low-frequency driver has a free resonance of about 50 to 75 cps. Increasing the compliances reduces this resonance to about 35–50 cps. One 18 in. low-frequency driver on the market has a resonance guaranteed below 30 cps.

(5) The Acoustic Transformer—(4) above suggests that the operation of a driver unit cannot be treated on without consideration of its acoustic load. This load of very low impedance, i.e., the open air, must be coupled to the region of cone movement, an area of very high acoustic impedance. Considering the size of the wavelengths to be reproduced, we find that the "transformer" must deal in large surfaces and volumes. Hence, we must work with a baffle, box, and horn concept which is fundamentally architectural in nature. With no baffle at all a 15-in. speaker will begin to lose level from 800 cps down at the rate of 6 db per octave, as shown in curve (A), *Fig.* 2. This is due to the "doublet" effect of a dual radiator, for as the wavelength approaches the diameter of the cone, cancellation occurs because of the out-of-phase condition.

The Small Flat Baffle

Curve (B) represents a driver on a small baffle. Note that the "doublet" action takes place at a lower frequency, determined once more by the wavelength of the frequency being reproduced and its size relationship to the diameter of the flat surface.

Continued increase in the size of the baffle involves another factor, however. If the baffle is large enough, or infinite, so that the critical frequency or wavelength of the diameter corresponds to the mechanical resonance of the driver, range response is extended to cone resonance, as shown in curve (C). But past the critical resonant wavelength of the driver cone, attenuation takes place at the rate of 12 db per octave, regardless of the size of the baffle. This is due to the stiffness, or low capacitive effect, of the driver cone mechanism, and indicates the value of a driver with low free-space resonance. From the practical aspect, it can be assumed that an 8-in. speaker with a free-space cone resonance of 62 cps will not benefit with a baffle larger than 5 feet in diameter.

The Open-Back Box

Operating like the infinite baffle when the proportions are large, the open box is quite adequate provided that the dimensions prevent cancellation of the front by the back wave. However, in the usual dimensions suitable for a living room, say 6 to 12 cubic feet, the resonance of the box itself causes a violent peak of perhaps 10 to 17 db around 175 cps, resulting in a disagreeable boominess characteristic of commercial radio sets, juke-boxes and the like. As shown in curve (D), attenuation below resonance is at the usual rate of 12 db per octave plus 6 db per octave due to the "doublet" or phasing out effect of the back radiation. In general, any open-back box is considerably improved by closing off the rear.

The Enclosed Box

The simplest form of good speaker enclosure is the enclosed box, with the only energy radiated into the listening area being from the front, as in the case of the infinite baffle. The chief difference between this enclosure and the infinite baffle is that the capacitive load on the back of the cone increases the stiffness of the suspension system, raising the mechanical resonance. This causes the usual attenuation of 12 db per octave to start at a higher frequency. Curve (E) shows that the level just before attenuation is higher, or peaked. This is true because the cone mass is resonating with the low compliant mass within the cabinet. When this peak is objectionably high, holes, or acoustic resistances in the back side of the cabinet will lower the peak, and consequently extend the bass range with another smaller peak before violent attenuation sets in.

Reflections with attendant reinforcements and cancellations within the box occur from about the fourth or fifth octave on up. These can be minimized with sound absorbent material generously applied to opposing surfaces.

Reflex Boxes

The currently popular vented enclosure utilizes a novel method of raising efficiencies in the last part of the 3rd octave around 100–130 cps. The front port, or vent, is tuned to allow for a delay of the back radiation. This permits an in-phase relationship resulting in a reinforcement just below the peak caused by resonance of the cone in conjunction with the true box resonance. Curve (F) indicates results to be expected with enclosures of the same relative size, approximately 7 cubic feet. Observe that attenuation past resonance is 18 db per octave versus 12 db per octave in the completely closed box. All other factors being equal, the range is not as great as the closed box. The peaks are higher and the valleys lower. Moreover, the characteristic boominess is objectionable to the discerning listener. But the fact remains that the added efficiency, in the more restricted spectrum that it does reproduce, finds considerable favor.

It is recommended as a guide that no less than 6 cubic feet be considered for satisfactory results. The depth of the box should be about one-half the length or height. Although formulas exist for computing the port size, the necessary data on cone weights, compliance, cone resonance, and other parameters are seldom available to the constructor. The most practical and by far the most accurate method of determining optimum port size in a specific box for a particular speaker is the procedure as follows:

Place a voltmeter across the voice coil and start with a port of about 150 square inches. Scan the first three octaves with an audio oscillator playing through the

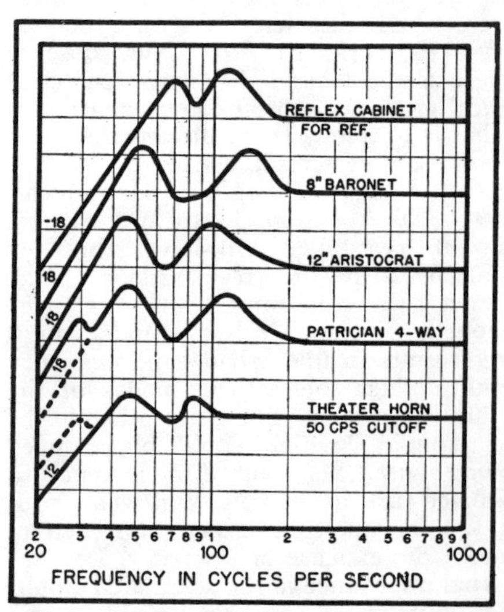

Fig. 6. Typical response curves of reflexed cabinet, several manufactured enclosures, and the folded horn.

speaker system, gradually decreasing the port size until two peaks are observed in the reading of the highest intensities and at the lowest frequencies possible.

As a guide to the optimum box size for top results with a speaker of given diameter, see *Fig.* 3. Note that for the best results—insofar as extended bass reproduction is concerned in a medium size cabinet (say up to 12 cubic feet)—the 12-in. cone has a considerable edge over the 15-in. model. The low-frequency driver of 18 in. is impractical and almost unuseable unless the volumetric capacity of the box is very large, or over 20 cubic feet.

Recapitulation

A study of the comparative designs offered above suggests that a rather poor job of bass reproduction is being effected unless large cavities, of the order of 12 cubic feet on up, are considered. The infinite baffle, combined with a cone resonance of 40–45 cps, does by far the best job if it is feasible to employ it. The vented box lends a "muddy" quality to the music and restricts the range, but is generally preferred over the plain closed box of equal size in spite of the added purity and extended range of the latter. This can be explained by reason of the foremost requirement of efficiency for the bass, even if of more restricted range.

Of the utmost importance in any of the designs above is low cone resonance, which also involves high compliances, or values of C in the equivalent circuit.

Towards Better Bass Response

Theater Horns—With space not being a limiting factor, the folded theater horn, is preferred. Response with this almost ideal enclosure is flat to 50 cps. Below this point, attenuation is fairly rapid. However, this horn delivers the smoothest, distortion-free response down to 50 cps that the art has been able to develop. The difficulty presented otherwise by this system is the need for dual drivers, and the requirement for a rather expensive cellular horn and treble driver above 400 cps. Observe the very low portion of the curve for this unit, shown in *Fig.* 6 with the curves of a number of commercial enclosures. This curve shows what can be done by enclosing the backs of the drivers with a cavity of about 16,000 cu. in. If the Klipsch-type units with high compliances are employed, it is possible to resonate this back cavity with the frontal air load mass and achieve added response down to 30 cps.

The vast majority of living rooms, however, have no provision for a system approaching the size of this rather ungainly device. The basic principles can be adapted, however, as disclosed in the more practical enclosures to follow.

As a rule folded horns are recommended for operation only up to 400 cps. At almost 500 cps and up, the higher frequencies find difficulty in following the circuitous path, ***unless supplemented by front radiation*** from the bass cone.

In several E-V designs, the folded horn loading is actuated from the rear of the cone while radiation of the higher frequencies takes place equally from the front. The important cavity directly behind the cone acts as an acoustic low-pass filter below 200 cps. This filter prevents phasing out of the front radiation by the back waves above this point. Some interference is measurable below 200 cps, but the generally smooth response to 65 cps overwhelmingly justifies the design. The sides of the room provide additional loading, acting as an extension of the mouth of the horn. This assists further in the formation of a spherical wave shape which promotes easy listening. Some radiation is effected, just barely usable, at 30 to 45 cps. The corner design proves itself an excellent substitute for the C-5 bin where the larger space required is not available.

Completely Horn Loaded Corner Design

The "Patrician" 4-way system employs a fully horn-loaded driver for utmost purity and efficiency, as shown in *Fig. 7*. In addition, the feature of resonating the back cavity with the front air load mass is utilized. The complete assembly is slightly less than 30-in. deep, allowing it to pass through a standard door.

To insure optimum results the bass horn design is slanted towards the lowest part of the spectrum, and a crossover at 220 cps transfers the task of reproducing the 200–600 cps range to a 12-in. low-frequency unit with an independent horn. To maintain optimum throat-to-mouth relationships, part of the face of each driver is masked off to the proper size. Closing down these throats effects attenuation only of the higher frequencies, well beyond the pass band of these two units. From 600 to 3500 cps, the treble driver exhausts through the large 2 × 5 horn. High-end attenuation for this unit is only 6 db per octave above 3500 cps. This allows a reinforcement in combination with the small Model T-10 super-tweeter of the critical 5000-cps band which so vitally effects the "presence" quality. Dramatic increase in total apparent efficiency is achieved with a 4 to 6 db rise at this particular point. The T-10 Sonax Driver carries the high end through 16,000 cps.

The back cavity of the large driver comprises some 8000 cu. in. to match the high compliance of the 18-in. cone. Note that the small triangular cavities, four in all, are fed to the larger area behind the cone by notches. This makes use of all the space in the bass section. It is essential that the back cavity be absolutely airtight. Horn assemblies are tested in production with a pressure gauge for leaks. The equivalent of a ½" hole will raise the lowest radiation peak by almost 5 cps. Moreover, the driver itself is sealed from front to back with a solid dome, incorporating no "breather."

The impedance characteristic of a loudspeaker rises at the extremes of the spectrum. To more nearly match a 16-ohm crossover network at the very low frequencies, the special Klipsch-type driver has a d.c. voice-coil resistance

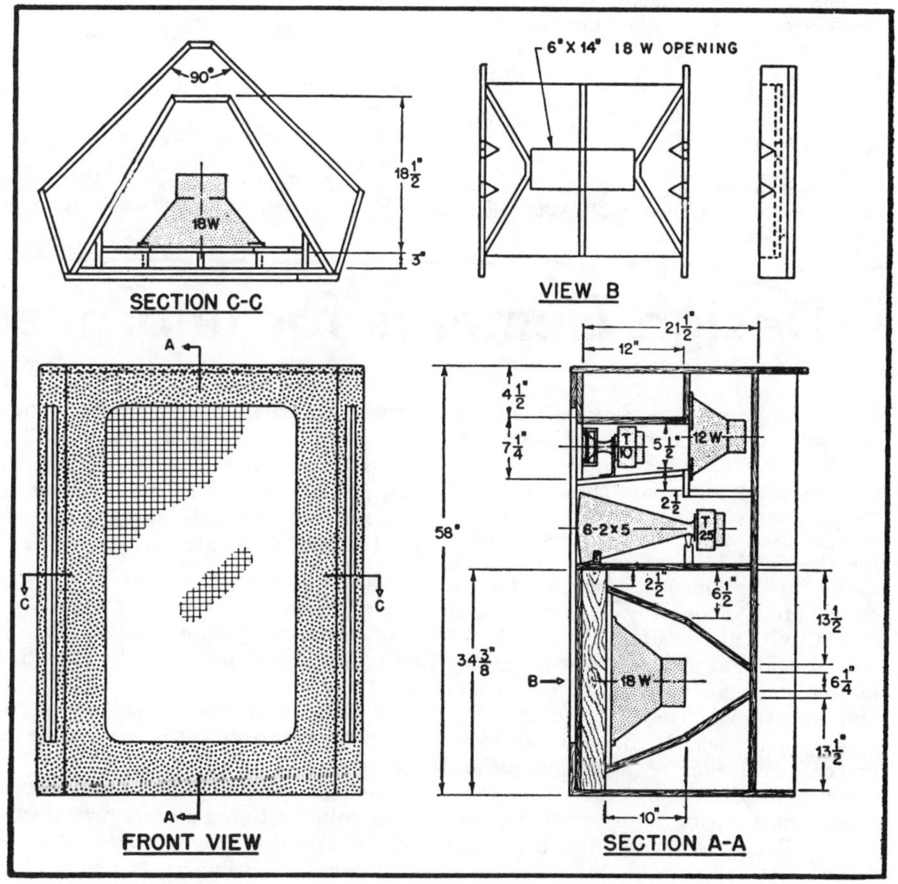

Fig. 7. Simplified constructional drawings for the "Patrician" four-way system.

Design Data for a Bass-Reflex Cabinet

J. A. YOUNGMARK

The charts presented by the author will simplify the selection of the dimensions for a satisfactory bass-reflex speaker cabinet.

If the full benefits of a high-quality audio system are to be realized, it is of course, vitally important to ensure that the loudspeaker system is equal to its appointed task. Fundamentally, this usually resolves itself into the provision of a suitable means of separating the radiation from the back and front of the loudspeaker cone, among certain other considerations, and the horn is notably efficient in such a connection. Unfortunately, the bulk of a horn such as would be required at the lower end of the audio range renders its use difficult, to say the least, and it is therefore obvious that some compromise must be made.

One such compromise between efficiency and size can be made by using the corner of a room as the outer section of the horn and a noteworthy example of this is the Klipschorn. This consists of a frequency-divided system in which both the high- and low-frequency units are horn loaded. It is, however, rather complex in design and quite expensive to produce, whilst another arrangement, the Pyramid Corner Speaker,[1] is not very suitable where only a single unit is used.

In the normal domestic installation there is much to recommend the bass-reflex design first described by Thuras.[2] Correctly designed, it gives increased power handling capacity, reduces the cone amplitude distortion at the resonant frequency, and at the same time gives additional output at this frequency. It may be built in a compact form and is relatively free from phasing difficulties. Although this type of loudspeaker system has never been so popular in Europe as in the U. S. it is now in fairly wide use there.

The theory of the reflex cabinet has been published, among others, by Hoekstra[3] and Smith.[4] The treatment which follows is based mainly upon their analysis in deriving an expression for the resonant frequency of any reflex cabinet and provides four sets of curves from which cabinets for the four most popular sizes of speaker (10, 12, 15, and 18 in.) can be designed without calculation.

The stiffness reactance of the air enclosed in a volume V is:

$$X_s = \frac{-j\rho C^2}{\omega V} \qquad (1)$$

where

ρ is the density of the air,
C is the velocity of sound,

and $\omega = 2\pi f$ (f being the frequency).

The cabinet is assumed to have a circular hole of radius R, provided with an internal tunnel of length L. The mass reactance of the air in such a tunnel mounted in an infinite plane baffle has been shown by Rayleigh[5] to be:

$$X_m = \frac{j\rho W}{\pi R^2}\left(\frac{K_1(2kR)}{R^2 k^3} + L\right) \qquad (2)$$

where

$K_1(2kR)$ is a Bessel Function of the first order

and $k = \frac{\omega}{c}$

Equation *(1)* represents the air load on the vent due to the air outside the cabinet, and equation *(2)* represents the mass of air in the tunnel. When the dimensions of the vent are small compared with a wavelength, this reduces to:

$$X_m = \frac{j\rho\omega}{\pi R^2}\left(\frac{16R}{3\pi} + L\right) \qquad (3)$$

When this cabinet is resonant,

$$-X_s = X_m$$

---

[1] W. E. Gibson and J. J. Andrea, "A symmetrical corner speaker," Audio Engineering, March 1950.

[2] A. L. Thuras, U. S. Patent 1,862,178 (1930).

[3] E. C. Hoekstra, "Vented loudspeaker enclosures," *Electronics,* March 1950.

[4] F. W. Smith, "Resonant loudspeaker enclosure design," *Communications,* August 1950.

[5] Rayleigh, *"Theory of Sound,"* Vol. 2, p. 306.

Design Elements for Improved Bass (Cont'd.)

of only 3.2 ohms, as opposed to the usual resistance of 11.6 ohms in a more conventional unit.

In listening to this system some rather startling effects are observed. When live tape is used as a source, it is possible for the spatial effect of the recording locale to be unconsciously evaluated, contributing materially to the illusion of reality. This is effected through the reproduction of extreme low-frequency effects, such as very low room noises and air rumble, more frequently felt rather than heard. Because all four drivers are loaded with columns of air, diaphragm excursions are held to a minimum, assuring high damping of the voice coils in the densest flux areas, and the smaller excursions minimize diaphragm breakup. Perhaps the most important observation is a negative one—the lessening of listener fatigue. This contribution may be ascribed principally to the elimination of harmonic distortion through the multiple division of the spectrum by disparate drivers.

Bass response in currently available loudspeaker systems and enclosures is poor. By considering the spatial requirements and utilizing the corner of the room as an extension of the necessary air-load on the driver cone a considerable improvement in bass range and efficiency can be achieved.

Though the foregoing offers no complete solution to the perfect reproduction of the first three bass octaves, it is felt that perhaps an important improvement of at least 100 per cent is offered from the standpoint of range extension. Efficiencies are improved greatly.

Improvements resulting from the horn load are an appreciable decrease in diaphragm excursions, with correspondingly greater power handling capacity of the driver, decreased intermodulation distortion, and vastly improved transient response.

Subjectively, it is found that apparent gains in realistic reproduction are considerably greater than the bare mathematics would indicate. The masking effect of the more adequate bass prevents extended high response from seeming shrill and thin. The resulting tonal balance achieves a more adequate feeling of listening satisfaction.

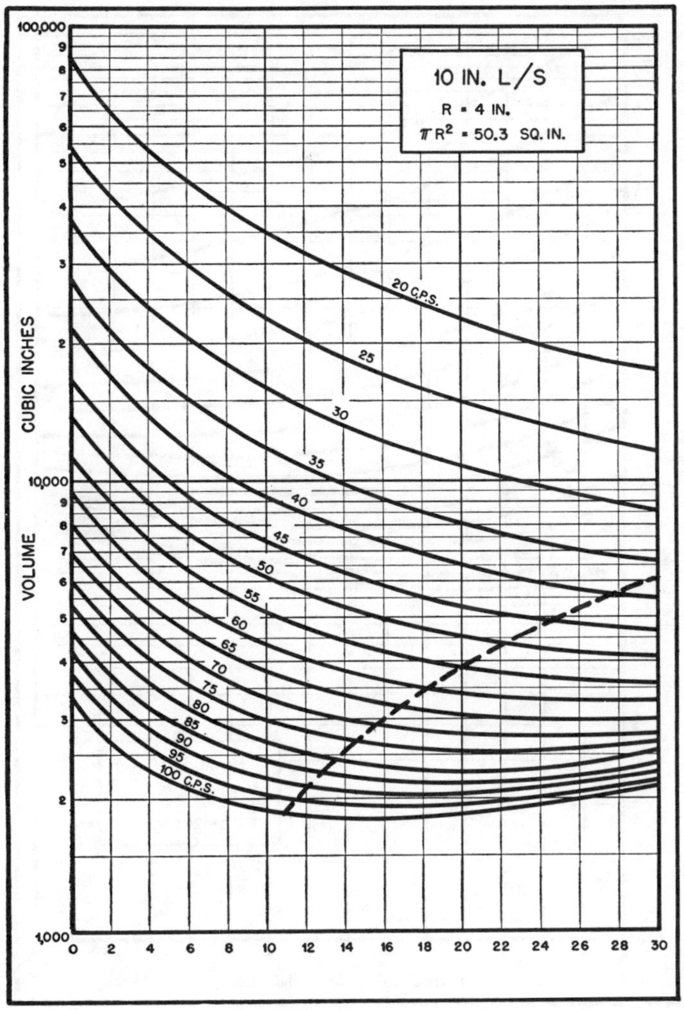

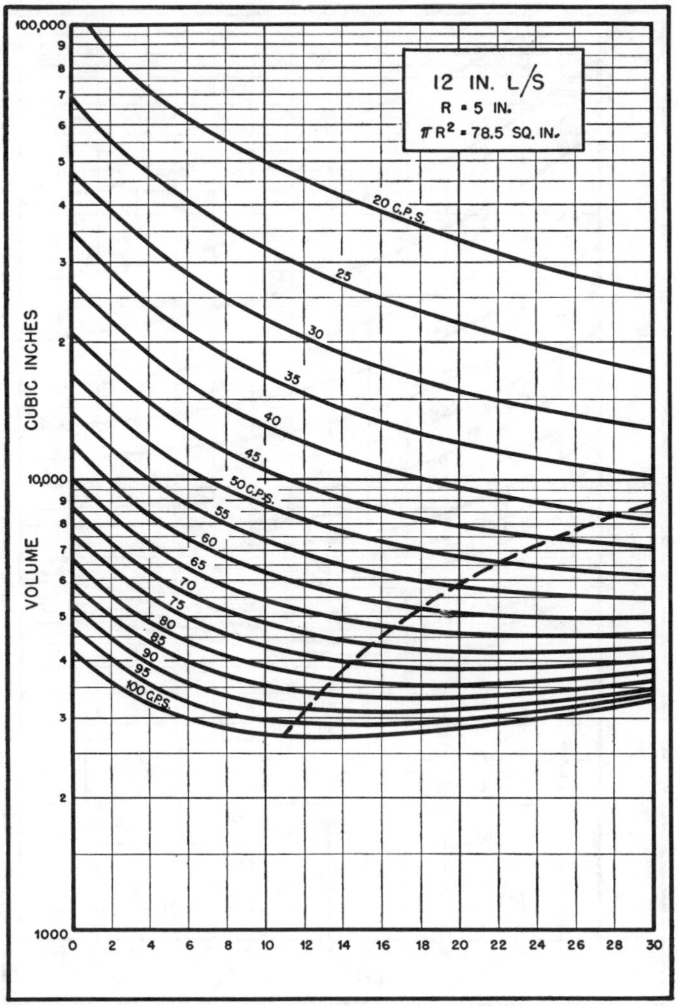

Fig. 1 (left). Curves showing tunnel length for 10-in. speakers resonant from 20 to 100 cps in a wide range of cabinet volumes. Fig. 2 (right). Curves for 12-in. speakers.

$$\frac{j\rho C^2}{\omega V} = \frac{j\rho\omega}{\pi R^2}\left(\frac{16R}{3\pi} + L\right) \qquad (4)$$

$$V = \pi R^2 \left(\frac{C^2}{\omega^2} \times \frac{1}{1.7R + L}\right)$$

At the resonant frequency given by this equation, the air in the tunnel moves vigorously, while a high impedance is presented to the walls of the enclosure where the speaker is mounted. It will be noted that the characteristics of the speaker itself do not enter directly into the equations at all, although for optimum results, ω should be equal to 2π times the resonant frequency of the speaker used and R should be equal to the radius of a piston equivalent to the speaker cone.

Tunnel Location

Practical cabinets have the tunnel inside the cabinet structure and the volume of the tunnel, R^2L, should be added to the volume V to obtain the total volume. In this case:

$$V' = \pi R^2 \left(\frac{C^2}{\omega^2} \times \frac{1}{1.7R + L} + L\right) \qquad (5)$$

The velocity of sound C is 1130 feet per second or 1.356×10^4 inches per second at a temperature of 20° C.

If all dimensions are expressed in inches

$$V' = \pi R^2 \left[\frac{1.84 \times 10^8}{\omega^2} \times \frac{1}{1.7R + L} + L\right] \qquad (6)$$

and this expression has been plotted for various practical values of R, L and f. *Figure* 1 shows V', L, and f when $R =$ 4 in. (10-in. speaker) and *Figs.* 2, 3, and 4 the corresponding values for 12-, 15-, and 18-in. speakers.

The volume occupied by the speaker itself is not included in V', and must be added to this factor or subtracted from the cabinet volume (if this is already fixed) to obtain V' before L and f are read off. There are two advantages in making L reasonably large: (1), the cabinet volume for a given resonant frequency becomes less as the tunnel is lengthened; and (2), as the mass of air in the tunnel is increased, the proportion of the mass reactance due to the outside air load on the vent is less, and the resonant frequency of the cabinet depends less on its position in the room. The assumption that the vent is radiating into semi-infinite space is usually incorrect and though the errors are not large the use of a long tunnel reduces them.

Tunnel Length

The length of the tunnel is limited by three factors. First, it must be short compared with a wavelength at the resonant frequency. Second, it must not approach the back wall of the cabinet so closely as to restrict the circulation of air (a distance equal to R is advisable). Third, after a certain point there is no advantage in increasing the tunnel length, since the decrease in unoccupied volume of the cabinet eventually causes the resonant frequency to rise again.

On differentiating equation *(6)* above, it is determined that the cabinet has a minimum volume when:

$$L = \frac{C}{\omega} - 1.7R. \qquad (7)$$

This becomes:

$$L = \frac{2160}{f} - 1.7R$$

if dimensions are expressed in inches.

In general it will be possible to use this optimum tunnel length only for large speakers with a high resonant frequency. In other cases the tunnel will not be short compared with a wavelength. The dotted line on each set of curves represents the point at which the length of the tunnel reaches 1/12 of a wavelength and points on the main curves to the right of this line should not be used.

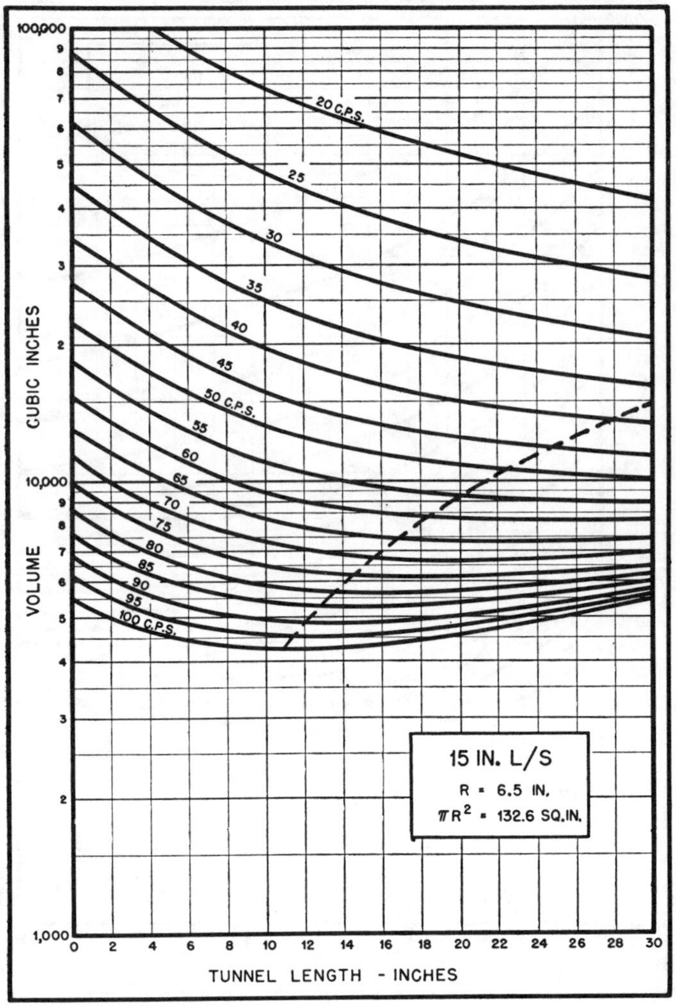

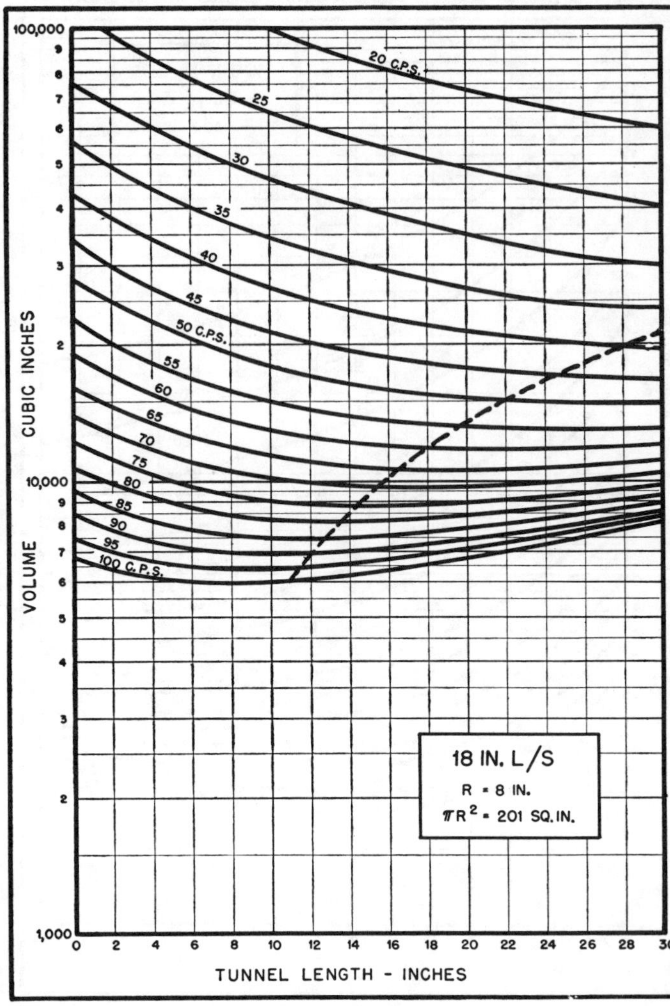

Fig. 3 (left). Curves of tunnel length for 15-in. speakers resonant from 20 to 100 cps in a wide range of cabinet volumes. Fig. 4 (right). Curves for 18-in. speakers.

The tunnel need not, of course, be round. Any shape will do if its cross-sectional area is πR^2.

Figure 5 shows elevation, plan, and sectional views of a correct reflex cabinet of about 8,000 cu.ins., designed for use with the Goodmans Axiom 12-in. speaker which has a bass resonance of 55 cps. The cabinet is constructed of 1-in. plywood and lined with soft felt 1-in. thick. It combines good performance with an appearance that is not out of place in the home.

Conclusion

The foregoing data has been prepared bearing in mind the over-all results which may be expected from the completed instrument and it is not out of place to stress the importance of the choice of speaker. This choice is affected not a little by the need for a unit with good top response, in the absence of a separate high-frequency speaker. It is of interest to note that, in such a correctly arranged system, the response to transients is considerably improved as a result of the increase of mechanical resistance due to the correct air-loading of the speaker.

In this latter connection it should be borne in mind that the actual reproduction of transients from any loudspeaker is largely affected by the room in which it is placed, quite apart from any other consideration. That is to say, the room usually has a period of reverberation which is very much longer than that of the speaker so that, in practice, quite large deficiencies in this respect on the part of the speaker may be of small moment, in unfavorable situations.

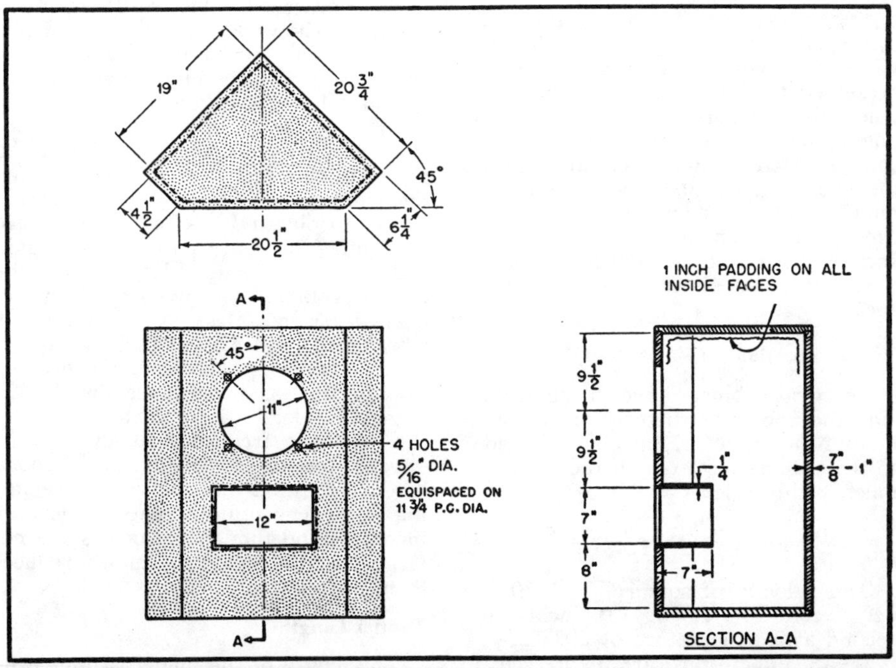

Fig. 5. Typical corner cabniet designed for a 12-in. Goodman speaker with a resonant frequency of 55 cps, which can be used with a speaker resonant at 75 cps by removing the tunnel.

Design and Construction of Horn-type Loudspeakers

WAYNE B. DENNY

In the never-ending search for good clean bass, builders are turning more and more to the exponential horn—not admirably suited to furniture designs. The two presented by the author should be pleasing, both acoustically and aesthetically.

Numerous requests for information about a speaker described by the writer in a previous article[1] have indicated that there are many audio enthusiasts who would like to construct horn speakers for their own use but who hesitate to do so without a better understanding of the theoretical and practical considerations which must be taken into account if results are to be satisfactory. Actually, there is nothing mysterious or esoteric about the design and construction of horn-type loudspeakers. The design of a satisfactory unit requires no great mathematical ability. The construction requires no more than a minimum of woodworking ability as far as acoustical performance is concerned. Of course, if appearance is a factor, the skills of a cabinetmaker may be employed for the visible part of the structure.

Theoretical Considerations

The acoustic horn can best be thought of as a transformer which accepts sound energy at its throat and delivers that same energy in somewhat different form at the mouth of the horn. The motion of the air in the constricted portion of the horn near the throat is characterized by relatively high velocity and displacement amplitudes. However, the amount, or mass, of air which moves in this manner is small because the cross-sectional area of the horn is small near the throat. In contrast to the motion of the air in the throat, the motion of the air at the mouth of the horn is characterized by relatively low velocity and displacement amplitudes but the mass of air moving in this region is much larger. An analogy is often made between the acoustic horn and the *impedance changing properties* of an electrical transformer. Thus, for example, a step-up transformer accepts electrical energy at high current and low voltage and delivers that same energy (minus losses, of course) at low current and high voltage. The fundamental purpose of the acoustic horn is to match the impedance of the cone or diaphragm of the driving unit to the impedance of the space into which the sound is to be propagated.

This analogy may be carried a bit further. A good audio frequency transformer must be designed so that its frequency characteristic is essentially flat over the entire audible range. An ideal horn would have a similar frequency characteristic. Electrical transformers are now available which are nearly perfect in this respect but it is found that it is difficult to cover the entire audible spectrum with a single horn if total space occupied by the horn is to be kept within acceptable limits. The physical shape is another factor which must be considered in the design of a practical horn. For these and other reasons it is convenient to use two or more separate horns, each one designed to cover but a portion of the audible spectrum.

The two design factors to be considered are shape and size. The shape will be considered first. Theory and experience[2] indicate that the most satisfactory shape is the so-called exponential horn. An exponential horn is best defined as one which obeys the following equation:

$$A_2 = A_1 \varepsilon^{kx} \qquad (1)$$

where A_1 = cross-sectional area at any point in the horn
A_2 = cross-sectional area at a distance x from the point where A_1 is measured. ($A_2 > A_1$)
k = flare constant
ε = 2.718, the base of natural logarithms

The application of this equation is shown in *Fig.* 1.

For design purposes it is convenient to let $\frac{A_2}{A_1} = \varepsilon^{kx} = 2$. From a table of exponentials

$$kx_d = 0.7 \qquad (2)$$

In Eq. (*2*), x_d represents the particular increment of horn length in which the area doubles. For a true exponential horn the area doubles for this increment of length no matter what particular value is chosen for A_1 (or A_2).

The value chosen for x_d depends on the lowest frequency for which the horn is to be used. This is called the cut-off frequency, f_c, where

$$4\pi f_c = kc \qquad (3)$$

where c = velocity of propagation of sound in air (about 13,200 inches/second)

After f_c is chosen, k may be determined from Eq. (*3*) and then x_d may be found from Eq. (*2*).

Example: Let $f_c = 40$ *cps*

Then $k = \frac{4\pi \times 40}{13{,}200} = .038$

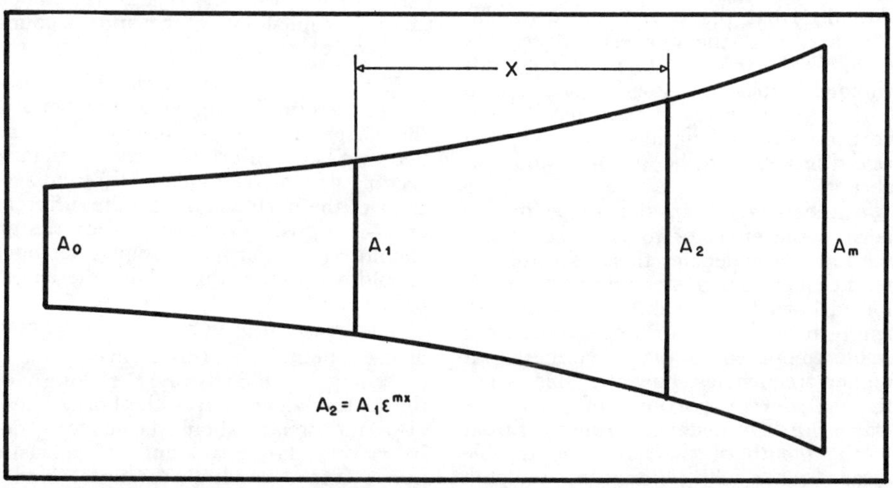

Fig. 1. The rate of expansion of an exponential horn, related to its mathematical expression.

---

[1] Wayne B. Denny, "For the discriminating listener: An audio input system." See page 49 *et seq.*

[2] Lawrence E. Kinsler and Austin R. Frey, *Fundamentals of Acoustics.* New York: John Wiley and Sons, Inc., 1950. Ch. 11.

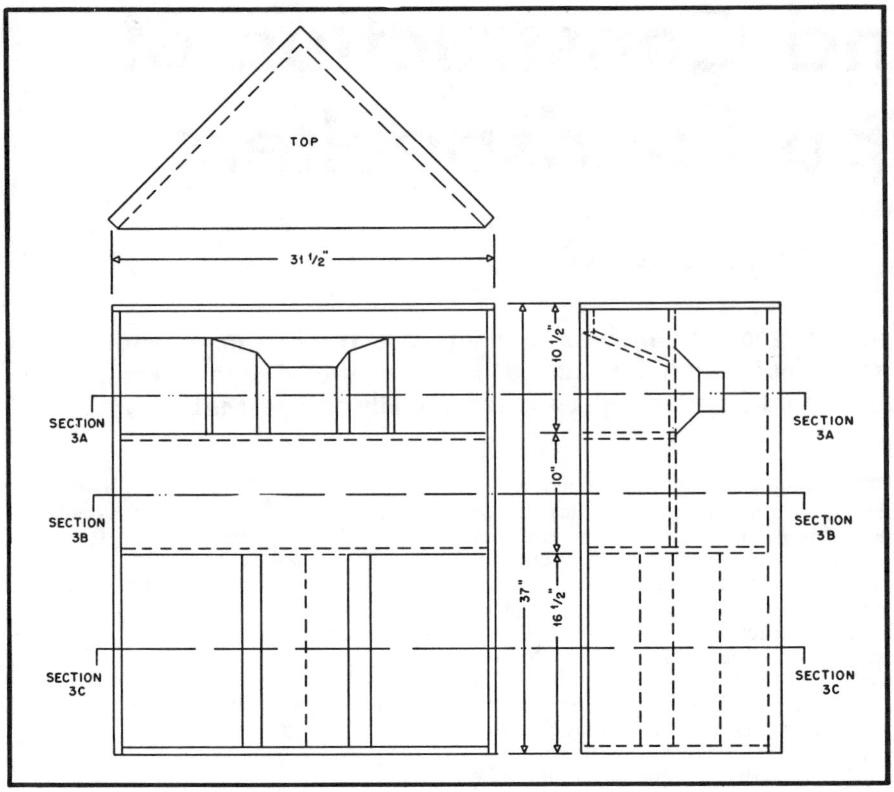

Fig. 2. Elevation of the corner speaker with duplex horn, with basic dimensions.

Finally, $x_a = \frac{0.7}{0.038} = 18$ inches, approximately.
Therefore the cross sectional area of an exponential horn should double every 18 inches in order to reproduce frequencies as low as 40 cps.

Other Considerations

There are three other considerations about horn shapes that should be mentioned. The first has to do with slight deviations from the exact rate of flare chosen. Experience indicates that for high-frequency horns (tweeters) the shape of the horn should conform as nearly as possible to the design criteria. Since high-frequency horns are small, this requirement presents no great problem. For low-frequency horns (woofers) considerable latitude seems permissible provided that the deviations are small compared with the shortest wavelengths to be used. Actually, there is some advantage in making a low-frequency horn with different rates of flare in different portions of its length. However, such deviations should be gradual and not abrupt.

Another factor in the shape of the horn is the effect of folds or bends. At the lower frequencies there is little difference between a straight horn and a folded horn because differences in path length in different parts of the horn are small compared to the wavelength. At higher frequencies, however, this is not true. Different portions of the wave travel different distances from the throat to the mouth of the horn and trouble occurs when this difference in path length approaches or exceeds one half wavelength. Under this condition it may be expected that there will be marked irregularities in the frequency response and in the directional characteristics of the horn. It is convenient, therefore, to use long, folded horns for the reproduction of low frequencies and to use short straight horns for the reproduction of the middle and higher frequencies. The situation can be summarized by saying that a folded horn has an upper limit to its frequency response as well as a low-frequency cut-off.

There are certain things that should be avoided in the design of folded horns even when they are to be used exclusively at low frequencies. In no case should the shape of the horn be favorable to the generation of standing waves, which are likely to occur whenever there is opportunity for sound waves to bounce back and forth between two parallel surfaces. This type of resonance can be avoided by taking into account the laws of reflection.

The last factor which should be considered about the shape of the horn is the shape of the cross-section. Horns need not be circular. They can be square, rectangular, or triangular. Different sections of the horn can have quite different cross-sections. However, the design should avoid abrupt changes: changes should be gradual throughout the length of the horn.

The question of horn size will be considered next. The throat area, A_o, is determined primarily by the design of the motor which drives the horn. However, no attempt should be made to deliver very large amounts of acoustic power into a small throat area. Under such conditions the variations in air pressure may approach the ambient pressure of the air and serious distortion can result. This is because the elastic properties of air are non-linear except for small amplitudes. Fortunately, for most applications this is not a design limitation. It is convenient, therefore, to give the throat an area whose value is somewhat less that the area of the cone for horns driven by conventional dynamic reproducers. In the case of tweeters the throat area is determined by the size of the coupler attached to the high frequency driver.

Next, the mouth area, A_m, must be determined on the basis of the lowest frequency which it is desired to reproduce. It is erroneous to conclude from Eq. (*3*) that the lowest frequency depends solely on the rate of flare. This is because, strictly speaking, Eq. (*3*) applies only to horns which are infinitely long and which have, as a direct consequence, infinitely large mouth apertures. A short review of some elementary acoustical theory will show the importance of the size of the mouth area.

Rate of Flare

According to Eq. (*3*) the less the rate of flare, the lower is k. To take an extreme case, k might be made zero in the hope that the horn will respond to extremely low frequencies. Then, according to Eq. (*1*), the horn degenerates into a tube of uniform cross section. But, as every physics student knows, a long open tube does *not* respond equally to all frequencies. Rather, it exhibits strong resonances at those frequencies for which the length of the tube is

$$l = n\left(\frac{\lambda}{2}\right)$$

where $n = 1, 2, 3, 4,$ etc.
and λ = the wavelength = c/f

Eq. (*4*) represents the condition for standing waves which are the resultant of two similar waves moving in *opposite* directions in the tube. The forward wave is the desired mode of propagation while the backward wave is produced by reflections which occur at the open end of the tube. In general, reflections occur wherever there exists a discontinuity in the air column unless the wavelength is very small compared with the diameter of the tube. For a uniform tube, reflections occur at the open end where the cross-sectional area changes abruptly from a small finite value to a large (quasi-infinite) value.

A long acoustic tube may be thought of as an acoustic transmission line with its own characteristic impedance. It is an elementary fact about finite transmission lines—be they acoustic or electric—that reflections will occur at the end of the line *unless the line is terminated in its own characteristic impedance.* The difference between the tube and the horn is this: the impedance of the tube is constant throughout its length while the impedance of the horn (due to its "transformer action" which has already been mentioned) varies throughout its length. At certain frequencies where the wavelength is not small compared with the dimensions of the mouth, the horn will exhibit reflections with consequent resonances and standing waves. The problem, then, is to *reduce* the magnitude of these resonances to a small value. They cannot be entirely *eliminated* because no horn of practical size can effect a perfect impedance match with the interior of the usual room.

Fortunately, this problem is not quite so serious as it seems. Most horns are used in rooms which themselves exhibit several resonant modes of vibration. The standing waves which result from room resonance are not ordinarily considered objectionable. Actually, horn resonances can be reduced to the point where they are masked by room resonances. When this is done the results are superior to those obtained from loudspeakers without horn loading.

There are three practical ways in which the effects of horn resonances can be reduced in loudspeaker systems. First, the horn structure can be made as large as practicable. Second, two or more rates of flare can be used in different sections of the horn. Third, the horn can be placed in the corner of the room so that the floor and walls will form a virtual horn which extends into the room itself. The last expedient is very effective in increasing the virtual mouth area. In most cases it provides the best possible impedance match between the generator (driver) and load (room).

Practical Considerations

The foregoing paragraphs have shown the most important theoretical considerations which govern the performance of horn-type loudspeakers *once the several design constants have been chosen.* However, it is usually impossible to assign numerical values to the design constants on the basis of theory alone. A loudspeaker is, among other things, an article of furniture and practical considerations of size and appearance must be taken into account as well as acoustical performance. Sometimes compromises must be made in order to fit equipment into available space. It is the purpose of this section to discuss the practical construction of horn speakers. This can best be done by describing the constructional features of two horn speakers built by the writer for which data are available. The two examples chosen for description are quite different in conception and will serve to illustrate the major requirements which must be met in any system. One of these horns is described in the following paragraphs, and the other follows next month.

Several years ago Olson and Massa[3] described in the literature a loudspeaker employing two horns and one motor. The low-frequency horn was coupled to the rear of the cone and the middle- and upper-frequency horn was coupled to the front of the cone. Later there appeared in the literature a description of the now familiar Klipsch horn which employed the corner of the room in order to extend the dimensions of the virtual horn. The speaker to be described is a modification of the Olson-Massa design and incorporates the room-corner feature. Although the speaker was built several years ago its performance is still considered excellent. It compares favorably with more recent designs which are vastly more expensive. Using the writer's speaker as a prototype, other constructors have built similar units and have reported excellent results.

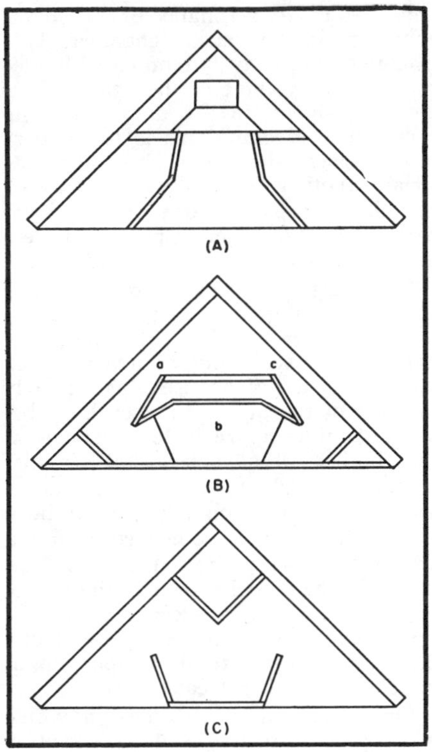

Fig. 3. Details of top (A), center (B), and bottom (C) sections of the duplex-horn speaker.

Figure 2 shows the main constructional features. The eight-inch cone is mounted near the top, secured to a vertical mounting board about half way back from the front of the structure. A short horn extends forward from the cone. Behind the speaker mounting board is the speaker chamber. The low-frequency horn connects to the lower portion of the speaker chamber and the mouth of the horn is at the front of the structure near the bottom.

For the sake of clarity some of the details are omitted from *Fig. 2* and are shown in *Fig.* 3. (A) in *Fig.* 3 shows the plan of the top section of the structure. This section includes the cone, the high-frequency horn, and the upper part of the chamber. (B) shows the plan view of the center section which comprises the lower part of the chamber, the throat, and part of the low-frequency horn. (C) shows the lower section of the structure, comprising the remainder of the low-frequency horn.

The functions of the various partitions can best be explained by tracing the path of the low-frequency horn. As shown in (B) of *Fig.* 3, the throat is in two parts: one half is at *a*, the other at *c*, The first portion of the horn consists of two parallel paths, from *a* to *b* and from *c* to *b*. The paths converge at *b* and immediately pass down into the bottom section through the hole at *b*. The horn then splits into two parallel paths, each of which opens separately at the front of the bottom section, as at (C) of *Fig.* 3.

The effective length of the low-frequency horn is about 45 inches. The total mouth area is about 360 square inches and the total throat area is about 38 square inches. Considerable liberty was taken in the rate of flare in the interest of simpler construction. However, the average rate of flare provides for a theoretical low frequency cut-off at about 52 cps. Eqs. (*1*) and (*3*). This should not be taken too seriously, however, since finite horns radiate appreciable energy at frequencies less than cut-off. This particular speaker, for example, produced fundamentals of 16 cps with considerable intensity. This was the lowest frequency available from the test oscillator. (No claim is made for "flat" frequency response at these low frequencies.)

Interference Effects

It was mentioned earlier that a folded horn does not radiate efficiently at higher frequencies where interference effects exist because of differences in path length. For this reason, the upper limit of the low-frequency horn should be determined before the other horn is designed and built. There is no simple way to compute what this upper limit is: it must be measured after the horn is in operation. A convenient method is to block off the radiation from the front of the cone with absorbing material (pillows, blankets, etc.) and check the operation of the low-frequency horn with an oscillator connected to the driver. Once the approximate upper frequency is determined, the high-frequency horn can be designed to complement the characteristics of the larger horn.

The two vertical boards which comprise the outside "vee," *Fig.* 3 are cut from ¾-in. plywood. The speaker mounting board is made of ½-in. plywood. All other partitions are ¼-in.

Fig. 4. The completed duplex-horn speaker. Barely visible is the protecting screen which hides the mechanism under normal illumination.

[3] Harry Olson and Frank Massa, *J. Acous. Soc. Am.* Vol. 8, No. 1, p. 48, 1936.

Harry Olson, *"Elements of Acoustical Engineering,"* 2nd ed., Van Nostrand, New York, 1947.

Fig. 5. Speaker chamber in the top section. The plywood partition below the speaker magnet indicates the reduction in chamber volume which was required to reduce cavity resonance.

plywood. All joints are nailed, glued, and reinforced with cleats. A portion of the top is secured by screws to permit access to the cone. The resulting structure is very sturdy. The inside horn surfaces were given several coats of thin shellac and sanded between coats. The resulting surface is very hard. The internal damping of the wood partitions is sufficient to eliminate all but a trace of vibration.

Experiments on this and other units have demonstrated conclusively that the use of thicker wood is no guarantee of reduced horn vibration. One speaker similar to the one described here was built of ¾-in. wood throughout. The vibrations of the horn walls were much more intense for the same electrical input signal. It was necessary to install heavy braces between partitions, and still the braces vibrated. Only by bracing the braces was the problem solved. A clue to the reason came when it was observed that the resonant frequencies of thick partitions were higher than those for the thinner vibrations, other factors being the same. Now the rigorous theory of vibrating plates is highly complicated but a simpler explanation appears plausible. Consider the following equation for the natural frequency of a simple oscillator:

$$f = \frac{1}{2\pi}\sqrt{\frac{s}{m}} \qquad (5)$$

where m = mass of the vibrating body
s = stiffness factor

In this case the "stiffness" is the reaction of the wood to bending. If we double the thickness of the wood the mass of the body is doubled. However, the stiffness of the wood increases much more rapidly than the mass. Consequently, the natural frequency increases: it does not decrease as might be expected. The only advantage gained by increasing the thickness of the wood is the consequent increase in mechanical resistance. Apparently, this advantage is sometimes outweighed by the increase in the resonant frequency *if* this shift means that the resonant frequency is raised from a sub-audible to a value which causes poor transient response in the audible range. It appears that the most effective way to damp out horn vibrations is to use relatively thin wood and to coat the outside of the horn with viscous damping material to lower the "Q" of the vibrating element.

The only difficulty experienced with this speaker when first constructed along the lines indicated by *Figs.* 2 and 3 was undesired cavity resonance of the Helmholtz type in the speaker chamber. This trouble was effectively eliminated by decreasing the volume of the chamber below that shown in the diagrams. One simple way of doing this is to fill part of this space with sound absorbing material. Another is to block off part of the chamber by the use of additional partitions. The writer employed the second alternative, as shown in *Fig.* 5.

The front of the completed unit can be covered by a grill to improve the appearance. *Figure* 4 shows a wood frame with bronze screening attached. Thick fabric should not be used, particularly in front of the high-frequency horn because it offers appreciable resistance at the higher frequencies, but Lumite plastic grill cloth may be considered satisfactory. The high-frequency horn shown in *Fig.* 4 is somewhat different from that shown in the diagrams of *Figs.* 2 and 3. The model shown in the photograph employs a three-section horn. The two additional partitions were found desirable to increase the dispersion at middle and higher frequencies.

One variation of this design which was brought to the writer's attention deserves mention. It uses a conventional two-way speaker with concentric tweeter in place of the single cone specified. Thus, no high-frequency horn need be constructed. Reports indicate that excellent performance is obtained in this manner. Since most two-way speakers use large low-frequency cones the total length of the low-frequency horn is lowered for the same mouth area. For dual speakers with relatively high crossover frequencies the action is threefold. The large horn is effective only at the lower frequencies, the middle range is taken from the front of the cone, and the "highs" are obtained from the tweeter.

Vertical Corner Horn

The second horn speaker to be described is depicted in *Figs.* 6 to 9, with the diagram and photographs showing the constructional features. The horn is vertical and opens into the *upper* corner of the room. This feature has the advantage that sound radiated from the horn avoids acoustic obstacles like chairs and other articles of furniture. Furthermore, the absorption by the ceiling is considerably less than that due to floors with carpeting. The space which is not used for the horn proper is used for shelves and this arrangement effectively hides the horn. The shelves add greatly to the rigidity of the structure.

A 12-inch driver is coupled directly to the throat of the horn and the speaker is entirely enclosed. Experiments over a period of weeks indicated that the low-frequency response was much smoother with the rear of the speaker entirely enclosed. Vents in the chamber

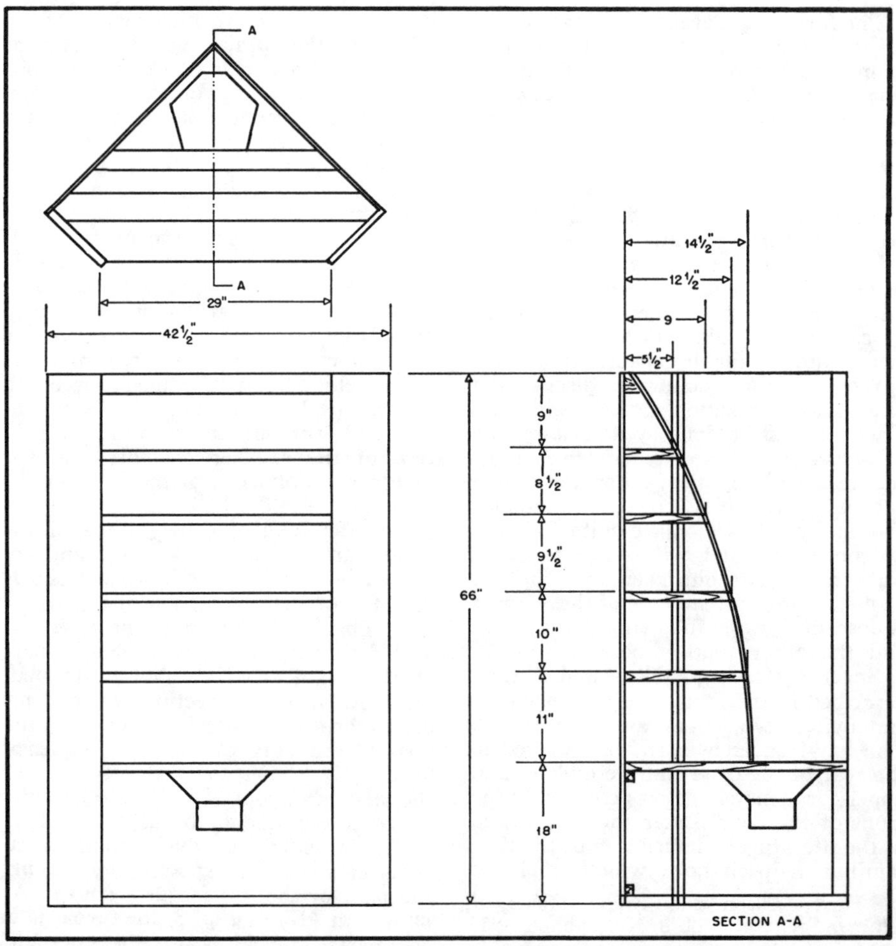

Fig. 6. Constructional details for the vertical corner horn speaker, showing top and front views, together with section through the center of the unit.

Fig. 7. Corner cabinet with vertical horn. The shelves serve to camouflage the horn and to add rigidity to the entire structure

resulted in marked resonances and "boom." Ozite and ordinary air filters of the type used in warm air heating systems are attached to the walls of this enclosure and effectively damp out undesirable cavity resonances.

The horn walls of this speaker are constructed entirely of ¼-in. plywood. The shelves, uprights, and the bottom are made of ½-in. plywood. (It is suggested that heavier wood be used for the uprights to avoid warping which has been observed over a period of fifteen months.) Plenty of bracing eliminates undesirable vibration. *Figure* 8 shows one of the two damping plates which were constructed to keep the outside walls from vibrating. They consist of a piece of ½-in. ozite glued to the outside of the walls, covered in turn by pieces of quarter-inch plywood screwed to the walls through the ozite. The resistance offered by the ozite under pressure eliminated vibrations in this area. The plates must be large so that the entire area does not vibrate as a unit. No internal braces were found necessary.

This woofer is designed to be used in conjunction with additional speakers for the middle and upper frequencies. Tests have indicated that although the response of the woofer continues to rather high frequencies, the "presence" is decidedly enhanced by the use of a low cross-over frequency. The writer uses a cross-over of 300 cps.

An alternative arrangement used earlier by the writer consisted of a two-way speaker in place of the woofer. In order to avoid directional effects, a curved reflector was placed in the upper corner of the room directly above the horn to disperse the middle and upper frequencies into the room. Results were excellent though rather unusual. The apparent source of the sound was, as might be suspected, at the upper corner of the room. The arrangement was finally discarded partly because members of the writer's family objected to the "buzzard's nest" way up in the corner.

It will be noted that the exponential "curve" is obtained by the use of straight boards, each one attached to two shelves. In contrast to the other speaker described, this one is almost exactly exponential in shape. The results are better at extremely low frequencies. This is attributed to the shape and to the fact that the entire structure is larger.

In order to eliminate leaks, rubber weatherproofing strips were used to seal the corners of the horn. These strips are seen in *Fig.* 9. Their use eliminates the problem of precise fitting and also eliminates the transmission of vibrations from certain members of the structure to adjacent panels. The shelves, uprights, and other partitions are assembled with screws—nearly two gross.

Since this was originally an experimental unit, plywood was used to lower the cost. Obviously, solid woods or hardwood veneers can be used to improve the appearance, but their use would, of course, greatly increase the cost.

Fig. 8. Side view of the vertical horn. The damping plate eliminates all objectionable vibration from the horn walls.

The two horn speakers described are merely examples of what can be done. There is no doubt that other constructors can make further improvements by added refinements in design and construction. It is earnestly suggested that anyone who desires to construct a horn speaker should first make a cardboard model. The model should incorporate all the main features of the desired structure. Its use permits the constructor to anticipate difficulties and to discard an inferior design before the speaker is started. The writer constructed several such models before building each unit. All but the last models were discarded for reasons of appearance, acoustic difficulties, or difficulties in construction.

Fig. 9. Vertical horn with side removed. The constructional features of the horn and speaker chamber are clearly shown.

The added time spent with cardboard, shears, and scotch tape was a small price to pay for the effectiveness of the completed speakers. With one exception noted earlier, no changes were required to achieve good acoustic results. It's cheaper to make mistakes on cardboard.

The writer's complete speaker installation consists of the two horn units described in this article, one bass-reflex unit, and high-frequency speakers. These several loudspeakers so diffuse the sound that visitors invariably ask, "Where is the sound coming from?" Like many others, the writer prefers diffused sound to that which comes from a point source.

In conclusion, two warnings should be given the prospective builder of a horn speaker. First, the improvement in low-frequency response is invariably accompanied by an increase in motor rumble. That phonograph motor which used to be "quiet as a mouse" is likely to take on the character of a roaring lion unless it is well made or unless some sort of noise suppression is used which is effective at the lower frequencies. As every audio enthusiast knows, improvement in one element of a reproducing system is likely to make deficiencies in other elements the more obvious. The second warning has to do with the fact that the larger of the two units described in this article is assembled with screws rather than with glue. The reason is—well, do you remember the story about the man who built a boat in his basement and then couldn't get it outside?

A Distributed-Source Horn

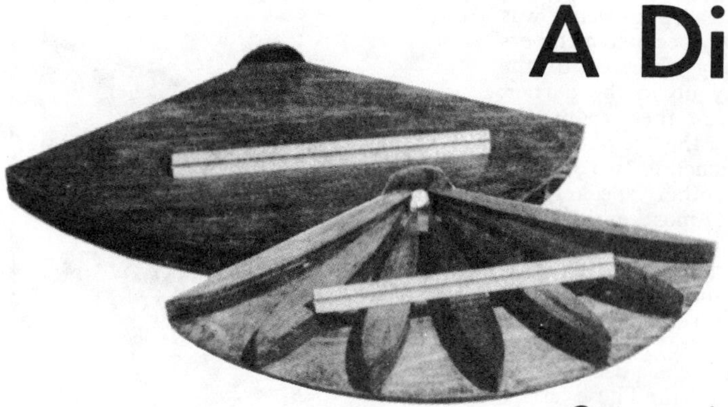

BOB H. SMITH

Constructional data for a high-frequency horn which is particularly well suited for home building by the audio enthusiast.

Fig. 1 (upper). Sound appears to originate from the entire area of the horn rather than from the throat as in a conventional horn. Fig. 2 (lower). With the top removed, the simplicity of construction is apparent.

•

One obvious difference between reproduced music and the original is the difference in spatial distribution of the sound. This difficulty can only be eliminated through the medium of a binaural system, which is not economically feasible at this time. Thus, until now we have been restricted to point-source reproductions. The sound emerging from a distributed source horn (DSH) appears to originate from an area rather than a point and is thus a step closer to the desired spatial distribution than conventional multicellular horns. In addition, the DSH described here provides much broader directivity patterns, is much more easily constructed, and takes up less space than the multicellular type.

The explanation of the apparent distributed source is evident from the field plot shown in *Fig.* 3. In the vertical plane the lines of flow do not diverge until they reach the mouth of the horn, thus to an observer a few feet from the horn they appear to be originating at the mouth of each cell of the horn. Of course, in the horizontal plane the lines of flow diverge at the throat of the horn and so in this plane the apparent source is at this point. Thus, there are seven apparent sources of sound in the DSH described here. The sources tend to blend together and give the impression of the sound originating from the area of the horn rather than at each point.

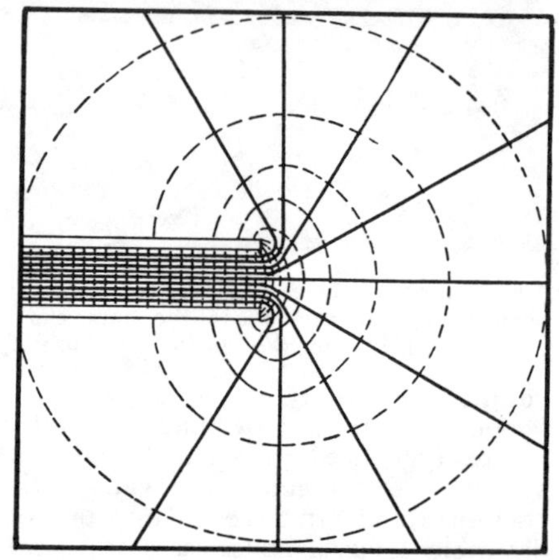

Fig. 3. The apparent source of sound in the vertical plane is just slightly in front of the mouth of each cell, as indicated by the lines of flow (solid lines). In the horizontal plane the source is near the throat.

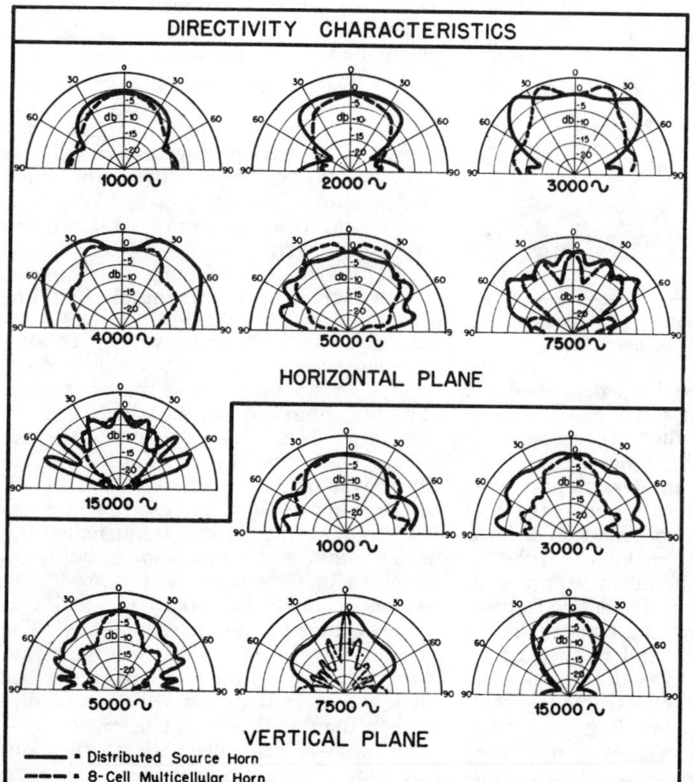

Fig. 4. The directivity patterns of the DSH (solid lines) are considerably broader than those of the conventional 800-cps multicellular horn (dotted lines).

•

The broad directivity pattern in the vertical plane is a result of the fact that the vertical dimension of the mouth is small compared to wavelength through most of the frequency range of interest. In the horizontal plane the directivity pattern is broad because a large portion of a circular wave is generated. The directivity characteristic of the DSH is shown by the solid line in *Fig.* 4. The broken line represents the characteristic of a standard eight-cell multicellular horn.

Figure 5 shows the dimensions of the DSH tested. It has sufficient mouth area to prevent reflection for frequencies greater than 750 cps. The expansion is approximately exponential and provides a cut-off frequency of 375 cps. Thus, the mouth area is the limiting factor and the horn should not be used below 750 cps.

•

The construction of the DSH is much simpler than that of the multicellular horn. Since there is no vertical expansion the islands may be cut from a flat sheet of one-inch plywood with a band saw. The top, bottom, and flange for mounting the driver are cut from ¼-inch plywood and the transition from circular to rectangular cross section near the throat is accomplished by means of plastic wood.

The type of finish is unimportant acoustically; the one shown in *Figs.* 1 and 2 was finished with orange shellac to better illustrate the construction. If the horn were to be mounted on top of the cabinet, or otherwise exposed, it could be made of one of the veneered plywoods. If it is to be mounted within the cabinet no finish would be required.

Plywood is a very satisfactory material for the horn; it provides adequate strength and contains sufficient mechanical resistance to damp any resonances and prevent appreciable motion of the horn walls. Thus, the performance of the horn is strictly a function of the geometry. Since the cut-off frequency, determined by the rate of taper, is well below the lowest frequency for which the horn is intended to be used, the dimensions of the islands are not critical. Slight variations affect performance only near cut-off.

If the horn is to be mounted within the cabinet, the top and bottom pieces need not be cut off along the arc but can be extended to the walls of the cabinet without impairing the performance of the horn. In this case, however, the inside width of the cabinet should be equal to the maximum width of the horn. (See *Fig.* 6.)

Fundamentally the design considerations are the same as those for any exponential horn—i.e., the mouth area must be large enough to prevent reflection at the lowest frequency for which the horn is intended to be used and the rate of taper must be chosen to provide a satisfactory low-frequency cut-off. In addition, there are the directivity considerations which are primarily a function only of the geometry of the mouth of the horn. They are: (1) The smaller the vertical dimension the broader will be the vertical directivity pattern. (2) The larger the arc of the mouth the broader will be the horizontal pattern. One is limited in the first case by the fact that if the vertical dimension is made too small the viscosity losses will be appreciable. Since they are proportional to the square root of the frequency, attenuation may occur at the higher frequencies. In the second case difficulty may be encountered in exciting each of the horn throats equally at the higher frequencies if the arc is made too large. Experimental investigation indicated that neither of these difficulties were appreciable for the horn described in this paper.

The required mouth area is given by the following expression:

$$S_m = \frac{\pi}{36}\left(\frac{c}{f_o}\right)^2 = \frac{1.6 \times 10^7}{f_o^2} \text{ sq. in.} \qquad (1)$$

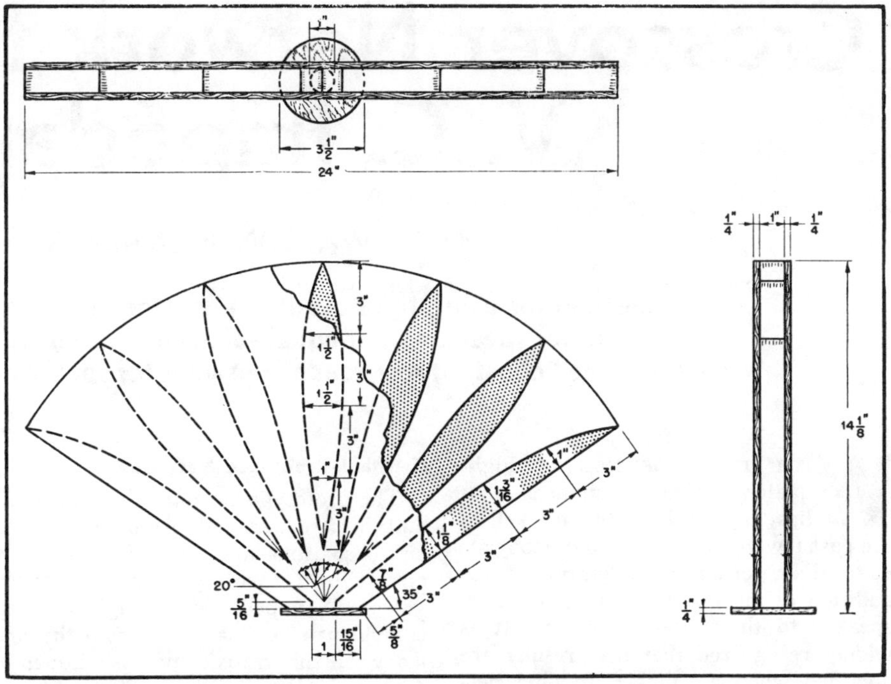

Fig. 5. Construction of the DSH is extremely simple since there is no vertical expansion.

where f_o is the lowest frequency for which the horn is to be used and c is the velocity of sound.

The length of the horn is given by:

$$L = \frac{5}{\theta t}\left(\frac{c}{f_o}\right)^2 = \frac{9.2 \times 10^8}{\theta t f_o^2} \text{ inches} \qquad (2)$$

where θ is the angular arc of the mouth in degrees, and t is the thickness of the mouth in inches.

The cut-off frequency f_c must be less than or at most equal to f_o, and is given by:

$$f_c = \frac{\theta t f_o^2}{20\pi c} \ln \frac{S_m}{S_o} = \frac{\theta t f_o^2}{3.7 \times 10^5} \log_{10} \frac{S_m}{S_o} \qquad (3)$$

where S_o is the area of the throat.

The distance in inches for the cross sectional area of the horn to double is given by:

$$d = \frac{750}{f_c} \text{ inches} \qquad (4)$$

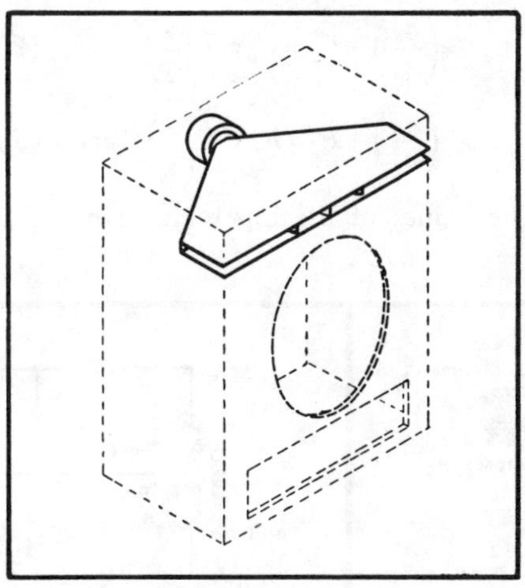

Fig. 6. The top and bottom pieces need not be cut along the arc, but may be cut off at the edges of the cabinet without impairing the performance of the horn.

The design procedure is as follows:

1. Choose f_o, S_o, t, and θ.
2. Calculate S_m from equation (*1*).
3. Calculate L from equation (*2*).
4. Calculate f_c from (*3*). In the event that f_c is greater than f_o choose a smaller value of t.
5. Calculate d from equation 4.
6. Draw a straight exponential horn starting from the cross section S_o and doubling the area each d inches.
7. Choose the number of cells required to fill the chosen arc θ without allowing the mouth of each cell to have a dimension greater than about 5 inches.
8. Draw the center lines of each cell.
9. At equal intervals along the axis of the straight exponential horn measure the cross sectional dimension. Divide this equally between each of the cells, thus determining the dimensions of the islands. (i.e. superimpose this information on the center lines of step 8).
10. Complete the drawing of the horn.

Sometimes vertical expansion is necessary, such as for a 500-cps DSH. Such a horn is intended to be used with a driver having an opening one inch in diameter. If the vertical dimension of the mouth of the horn were held to one inch in order to eliminate vertical expansion, the width of the horn would have been excessive. The greater vertical dimension causes this horn to be a little more directional in the vertical plane than the 750-cps horn. Therefore, the throat tapers gradually from 1 to 1¼ inches in the first five inches from the flange, and thereafter continues at a vertical dimension of 1¼ inches.

Crossover Network for Unequal V. C. Impedances

WATSON F. WALKER

A method for using the output transformer to feed a frequency-dividing network for speakers of unequal voice coil impedances with results identical to the "constant resistance" network for speakers of equal impedances.

With the profusion of high-quality sound equipment that has appeared on the market in the past few years, many audio hobbyists have given serious consideration to the addition of an auxiliary high-frequency speaker to their present system. It is widely recognized that best results are obtained when a suitable dividing network is used between an amplifier and a multiple speaker system in order that each speaker operate over the frequency range to which it is best adapted. The standard designs for such networks usually require, however, that both the high- and low-frequency speakers have the same nominal voice coil impedance. Unfortunately, dividing networks of the "constant resistance" type do not maintain their desirable properties when connected to loads of other than their characteristic impedance. The prototype constant resistance circuit could be used, of course, if the impedance of one of the speakers could be made to equal that of the other. In almost all broad-band impedance matching problems, a transformer or the equivalent is required to change the general impedance level of a network if insertion loss is to be avoided. It is well known, however, that the addition of more audio transformers in a high-quality system may degrade performance in terms of frequency response and distortion. In any event, transformers of high quality degrade the pocket book. It is the purpose of this article to show how the output transformer can be used to perform this job along with its usual function.

Constant-Resistance Network

The prototype constant-resistance circuit for frequency division and its design equations are given in *Fig.* 1. A circuit for use with speakers of different voice-coil impedances along with its equivalent circuit as seen from the primary of the transformer is shown at (A) and (B) in *Fig.* 2. The introduction of the transformer into the circuit gives complete control over the values of the circuit elements in *Fig.* 2 (B) by virtue of the turns ratios of the transformers. Comparing the circuit of (B) with that of the prototype, the two may be made identical if the following relationships are obtained:

$$\left(\frac{n_3}{n_1}\right)^2 L_1 = L_o = \left(\frac{n_3}{n_2}\right)^2 L_2 \qquad (1)$$

$$\left(\frac{n_1}{n_3}\right)^2 C_1 = C_o = \left(\frac{n_2}{n_3}\right)^2 C_2 \qquad (2)$$

$$\left(\frac{n_3}{n_1}\right)^2 R_l = R_o = \left(\frac{n_3}{n_2}\right)^2 R_h \qquad (3)$$

The values of the input impedance R_o, the low and high frequency speaker impedances R_l and R_h respectively, and the crossover frequency f_o are chosen in accordance with the demands of the circuit external to the crossover network proper.

In practice, the transformer used would be an output or line matching transformer and R_o would be the plate-to-plate load impedance required for the output tubes or the line impedance. R_l and R_h are the nominal voice-coil impedances of the two speakers and f_o can be whatever is required by the speaker characteristics.

Where it is desired to adapt this method to use with an existing amplifer, a simplification of the design procedure results because of the following relationship derived from Equation (3):

$$\frac{R_l}{R_h} = \left(\frac{n_1}{n_2}\right)^2 \qquad (4)$$

In other words, where the output transformer provides taps for impedances equal to those of the speakers (e.g. both 16-ohm and 8-ohm taps to accommodate

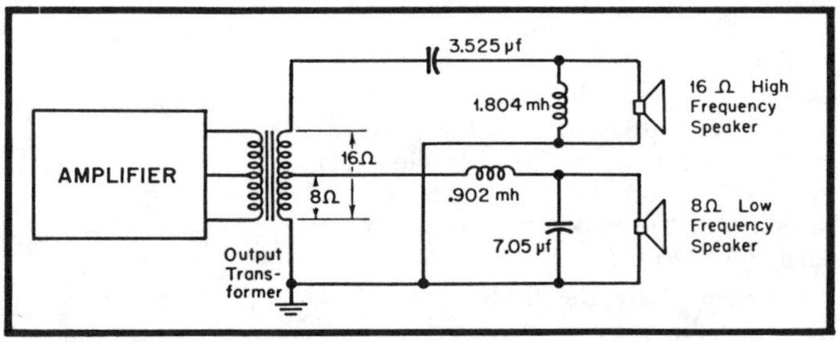

Fig. 3. Typical circuit arrangement for two speakers of unequal impedances.

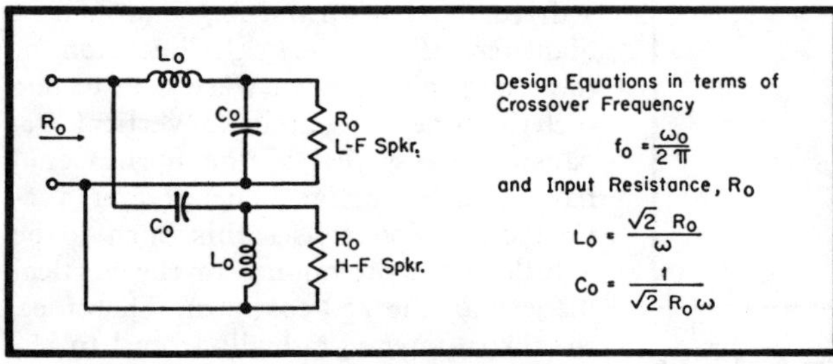

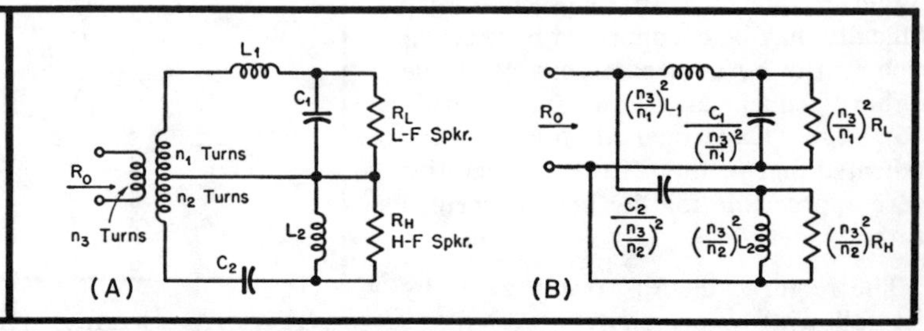

Fig. 1 (left). Prototype constant-resistance crossover network, based on loads of equal impedances. Fig. 2 (right). (A) Adaptation of prototype circuit for unequal impedances by the use of a transformer. (B) Method of proving that (A) provides correct loading.

Constant-Resistance Dividing Networks

BOB HUGH SMITH

A presentation of the design parameters for dividing networks which use two coils of equal inductance with two identical capacitors.

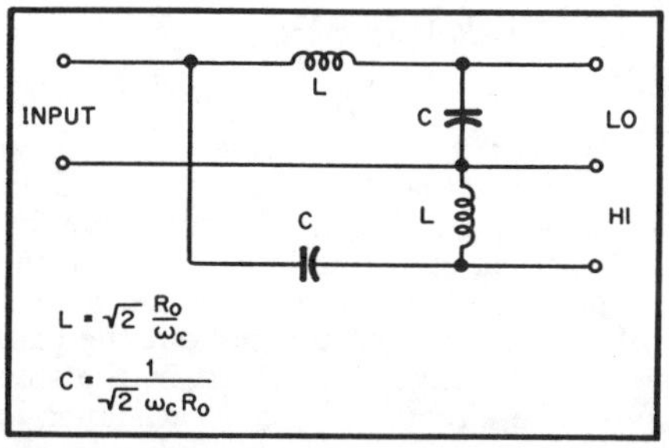

Fig. 1. Schematic of parallel configuration of constant-resistance dividing network.

THERE ARE TWO principal types of dividing networks in extensive use—the constant-resistance network and the *"m"* derived half section network, and the first type has several minor practical advantages over the second. These are: (1) It requires about 15 per cent less inductance in the low-pass section, and hence, for a given wire size the insertion loss is about 7½ per cent less. (2) It requires about 15 per cent less capacitance than the second. (3) When feeding from a zero-impedance source each section of the first type is down 3 db, while for the second type each section is down 4 db, thus the second type results in a one-decibel dip at crossover. (4) The input impedance of the first type is exactly equal to the characteristic impedance of the filter and contains no reactive component, providing the network is terminated in pure resistances. The last two points are of more academic than practical significance.

The design procedure for the constant-resistance network is really quite simple. The values of inductance and capacitance are read from the chart, *Fig.* 2, once the characteristic impedance of the filter and cross-over frequency have been selected. Next, the wire size is chosen from the following considerations: (1) If the wire size is too small, the insertion loss of the low-pass section will be excessive, and (2) if too large a wire size is chosen, the inductances become inconveniently bulky. A satisfactory compromise may be made by choosing a wire size which will result in a coil resistance equal to 10 per cent of the characteristic impedance of the filter. This produces about one decibel loss through the low-pass section and is undetectable on complex signals. The approximate length of wire needed may be obtained from *Fig.* 3, and the wire size may be selected from *Fig.* 4.

Next the coil is wound to approximate size using the length of wire already determined, and then turns are either removed or added as required to provide the exact inductance. If an impedance bridge is not available, the scheme shown

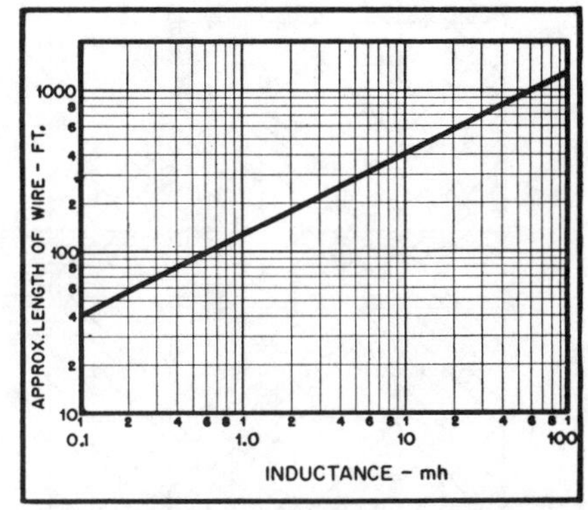

Fig. 3. Chart showing length of wire for given inductance. Example: If the required inductance is 5 mh, approximately 290 feet of wire will be needed for the coil.

in *Fig.* 5 may be used. When the inductance is of the correct value, the voltmeter indicates the null to occur at the crossover frequency. If the null is below the crossover frequency, the inductance is too high and turns must be removed; if above, the inductance is too low and turns must be added. This method has the advantage, over the impedance-bridge method, that it automatically compensates for error in the capacitance. Thus, it is not necessary to have precision capacitors, but they should be of either

Crossovers for Unequal Impedances (Cont'd.)

speakers of these impedances), the proper turns ratios are already available, and the following design equations are applicable:

$$L_1 = \frac{\sqrt{2}\,R_l}{\omega_o} \quad ; \quad C_1 = \frac{1}{\sqrt{2}\,R_l\,\omega_o} \qquad (5)$$

$$L_2 = \frac{\sqrt{2}\,R_h}{\omega_o} \quad ; \quad C_2 = \frac{1}{\sqrt{2}\,R_h\,\omega_o} \qquad (6)$$

As a practical example, suppose it is desired to use an 8-ohm low-frequency speaker with a 16-ohm high-frequency speaker with a crossover frequency of 2000 cps. We have then:

$$f_o = 2\times 10^3; \quad \omega_o = 2\pi f_o = 4\pi\times 10^3$$

$$R_l = 8 \text{ ohms}; \quad R_h = 16 \text{ ohms}.$$

$$L_1 = \frac{\sqrt{2}\,R_l}{\omega_o} = \frac{2\sqrt{2}}{\pi}\times 10^{-3} \text{ henries.}$$

$$= 0.902 \text{ millihenries}$$

$$C_1 = \frac{1}{\sqrt{2}\,R_l\,\omega_o} = \frac{1}{\sqrt{2}\times 8\times 4\pi\times 10^3} \text{ farads}$$

$$= 7.05\ \mu f$$

$$\frac{L_2}{L_1} = \frac{R_h}{R_l}; \quad L_2 = \frac{R_h}{R_l}L_1 = 2\,L_1$$

$$L_2 = 2\times 0.902 = 1.804 \text{ mh.}$$

$$\frac{C_2}{C_1} = \frac{R_l}{R_h}; \quad C_2 = \frac{R_l}{R_h}C_1 = 0.5\,C_1$$

$$C_2 = 0.5\times 7.05 = 3.525\ \mu f.$$

and the circuit would appear as in *Fig.* 3.

It would be more convenient once these calculations were made to select stock sizes of capacitors such as 4 μf and 8 μf in this case, and readjust the values of L_1, L_2 and f_o in accordance with this change, since odd-size capacitors are not generally available. The resulting shift in the crossover frequency (approximately 13 per cent here) would ordinarily cause no difficulty. Design data for air-core inductances of the type required here are available in other standard reference works.

This method of handling the crossover problem yields results identical with those wherein the speaker impedances are equal and costs no more to build than the conventional constant-resistance crossover network.

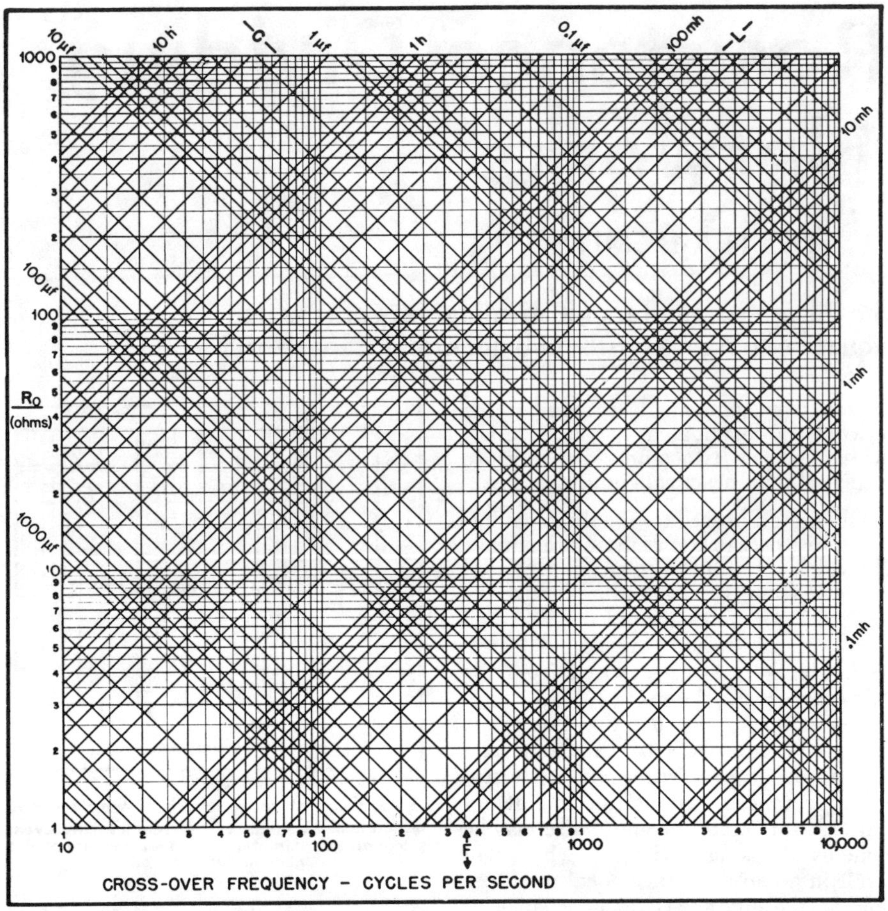

Fig. 2. Chart for determining constants for network. Example: Suppose the characteristic impedance is 11.5 ohms and the desired crossover frequency is 500 cps, then the inductance should be 5.0 mf and the capacitance should be 20 µf.

the oil or paper type and should be non-inductive.

The following example may clarify the procedure of producing a dividing network. Suppose it is desired to design a crossover network for a 16-ohm two-way speaker system. One might select 1,000 cps as the crossover frequency and 16 ohms as the characteristic impedance. *Figure* 2 indicates that the inductance should be 3.5 mh and the capacitance should be 7.0 µf. Referring to *Fig.* 3, it is seen that a coil having an inductance of 3.5 mh will require about 240 feet of wire. *Figure* 4 indicates that the resistance of the coil will be about 1.6 ohms if #18 wire is used. Since it is easier to remove turns from the coil than to add more to it, about 275 feet of wire is measured out and wound into a coil. The process is then repeated, since two such coils are required. Two 7.0 µf capacitors are connected in the circuit of *Fig.* 5, using a 15-ohm resistor for R_0. Suppose it is found that the null occurs at 800 cps; then turns are removed until the null occurs at 1,000 cps. The coil is taped and checked again in the circuit of *Fig.* 5 to be sure that the tape has not pressed the wires closer together and thus increased the inductance. Next the coils and capacitors are mounted and wired, completing the network.

The gain of the low-pass section is:

$$K = 20 \log \frac{\frac{f_c}{f}}{\sqrt{\left(\frac{f}{f_c} - \frac{f_c}{f}\right)^2 + 2}}$$

and of the high-pass section:

$$K = 20 \log \frac{\frac{f}{f_c}}{\sqrt{\left(\frac{f}{f_c} - \frac{f_c}{f}\right)^2 + 2}}.$$

These expressions are plotted in *Fig.* 6. It is evident that the slope of the curve in each stop band is 12 db per octave.

The effect of the coil resistance in the low-pass section is to provide almost uniform attenuation throughout the pass band but to have little effect in the stop band. The effect of the coil resistance in the high-pass filter is negligible in the pass band and also in the stop band down to the frequency for which the coil reactance becomes equal to the coil resistance. Below this frequency the slope becomes asymptotic to 6 db per octave. Thus, for the case in which the coil resistance is 10 per cent of the characteristic impedance of the filter, the high-pass section drops 12 db per octave from crossover down to 7 per cent of the crossover frequency, and from then on is asymptotic to 6 db per octave. Actually, this is hardly significant for a practical network since at 7 per cent of crossover the attenuation is already 46 db.

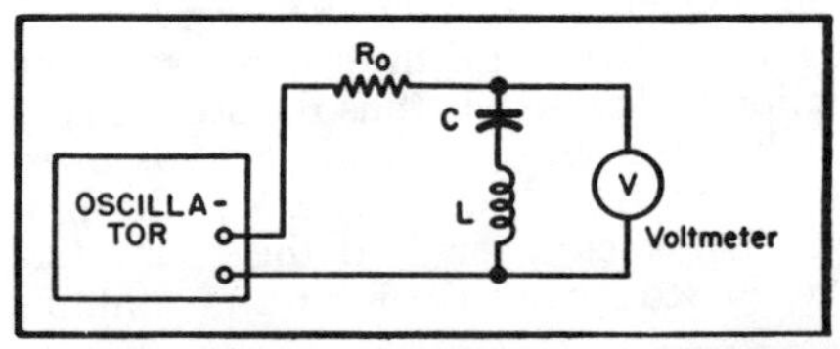

Fig. 5. Equipment set-up used to adjust inductance. Turns are added or removed to obtain a null indication on the voltmeter at the crossover frequency.

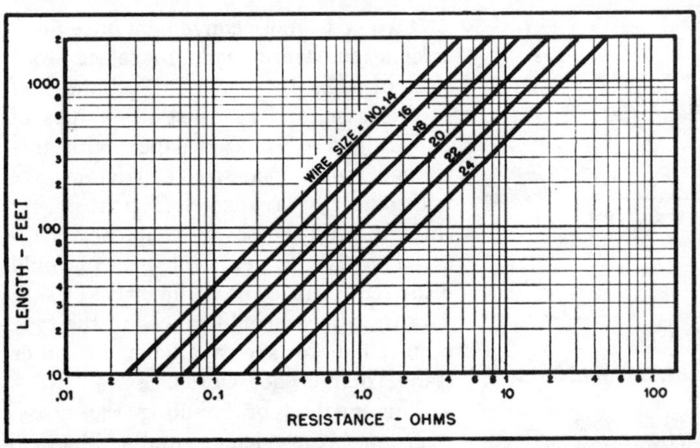

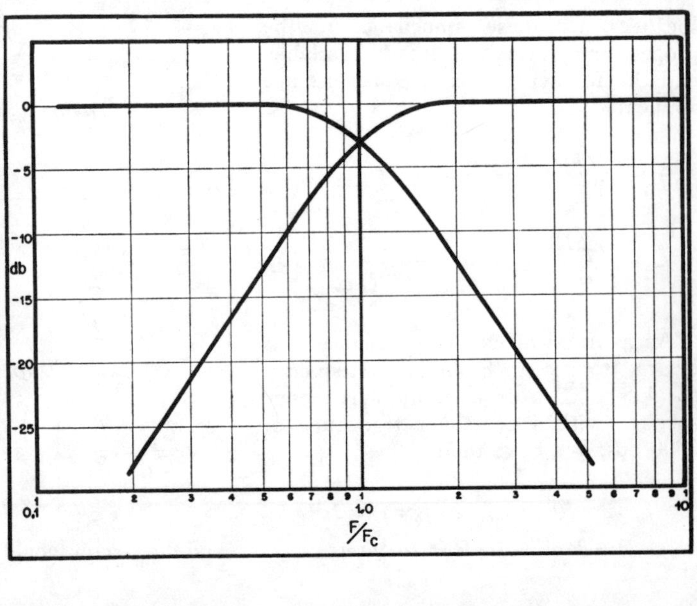

Fig. 4 (above). Chart showing length of wire vs. resistance for various gauges. Example: A coil resistance which is 10% of the characteristic impedance results in coils of practical size. Since the impedance is 11.5 ohms and the approximate length of wire is 290 ft., #16 wire should be used.

Fig. 6 (right). Frequency response of constant resistance network.

Design for Smooth Response

VERN YEICH

A simple and effective method of mounting a small loudspeaker

Fig. 1. Front view of the symmetrical corner mounting for an 8-inch speaker of good quality.

IN THE BROADCASTING FIELD there has always been a striving for a quality of reproduction suggested by the word "presence." This word quite accurately describes the goal of early broadcasters and manufacturers.

One phase of the development of presence has been the trend toward extended frequency response. This idea was the first to get consideration in the early days of radio, and has ever since received by far the greatest amount of attention. And it true that in order for the reproduction to be a facsimile of the original, flat over-all response is necessary. Although the widest possible range is preferred by most people, facsimile reproduction is not requisite to a convincingly real reproduction of music, and hence a feeling of presence. For instance, we are aware that for a given orchestra, the balance of high- and low-frequency tones will vary from one concert hall to another. The difference will be even more pronounced when the same orchestra is heard outdoors. And although we may thoroughly enjoy our "concert under the stars," the farther we are seated from the orchestra shell, the greater is the effect. We might say, then, that the transmission response of this "real" system is not flat. It will likely be found quite smooth, however. This, perhaps, explains why the small plastic radios have been tolerated; their frequency discrimination is not altogether a new experience.

As a matter of fact, this line of thought seems to lead to the conclusion that the remaining factors detrimental to a feeling of presence might better get a little more attention. A review of the literature shows that the following factors are most important. Some cause wave-shape distortions, and most of them cause response irregularities. The main offenders are: inadequate design in the magnet-voice coil system, non-linear suspensions, cabinet resonances, transient distortions caused by the many resonances in the moving system, irregularities caused by phase relations between units of a multiple system, and room transmission response.

From what has been said, it seems reasonable to expect that an entirely satisfactory reproducing system could be built around a single direct-radiator of superior design, whose response is very smooth, and only as extended as the development of the art permits. This assumes that good associated equipment is used, and that the speaker is properly baffled, and properly located in a room of good acoustic design.

With this in mind, work was begun on an experimental system using an eight-inch speaker engineered by one of the country's leading acoustic laboratories, and built by a manufacturer certainly suited for the work. It should be expected that its frequency response is as smooth and extended as any similar speaker available. *Figure* 1 shows the set up. In the corner of a room with its axis looking down the middle of the solid angle formed by the two walls and ceiling, the speaker enjoys the ideal location for acoustic loading, and room coverage. Complying with the manufacturer's recommended volume enclosure, the triangular baffle provides an extremely rigid, resonance-free enclosure with no parallel walls, and no diffracting edges. The speaker was flush mounted to minimize response irregularities caused by speaker-opening cavity resonances. The damping characteristics of the speaker, combined with those of a triode feedback amplifier afforded closest coupling between amplifier input and speaker output.

It seems that the only additional factor which is within the layman's means to help is the transmission response of the listening room, and that is another and important subject. The room used in this case was only fair.

In listening tests the completed speaker system using a Williamson amplifier fed from WWSO's monitoring equipment showed remarkably clean, smooth reproduction. As a result of this experiment, it is felt that smoothness of response is definitely a key factor in the search for realism or a feeling of presence. This experiment, moreover, showed very definite advantages for a symmetrical corner position for better loading and room coverage.

Construction

Building this type of corner enclosure is simplicity itself. It is only necessary to cut a triangular piece of plywood, saw out a hole in the center, and mount the speaker. For the W. E. 755 speaker shown in *Figs.* 1 and 2, the enclosure should have a cubic content of 2 cu. ft., which requires a triangle 39 in. on each side. Table I shows the dimensions for enclosures with various cubical contents.

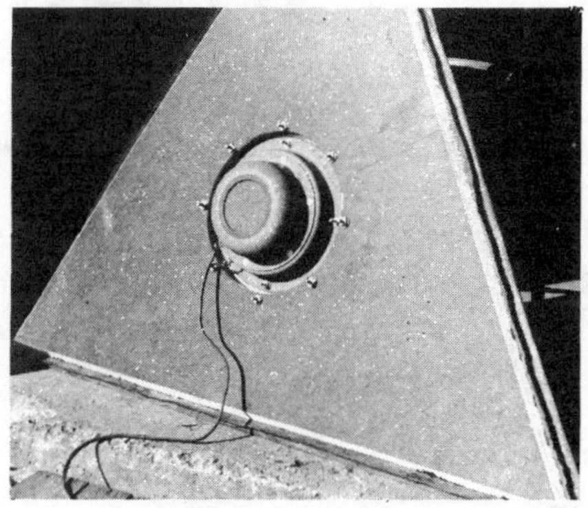

Fig. 2. Rear view of speaker mounting to show use of mounting ring.

To provide for the flush-mount adapter recommended for optimum results, a ring of one-eighth-in. sheet aluminum is turned or sawed out, with the hole equal to the free cone opening in the speaker to be used, and wide enough to extend about one inch larger than the speaker all around. This type of mounting reduces the thickness of the material in front of the speaker to a minimum.

If ⅝-in. plywood is used for the triangular piece, it is desirable to line the inner side with some absorbent material such as Celotex or Ozite. For a small speaker—such as the 755—the entire baffle could be constructed from a single piece of Celotex. For permanent fastening to the corner between the ceiling and the walls, it is suggested that strips be attached to both walls and ceilings so that the baffle can be screwed to them.

Since plywood comes in widths of 48 in., it will be necessary to make the baffle out of two pieces if a volume of more than 4 cu. ft. is required. However, the 755 will give good quality, and there are some other types which would serve quite well in the same installation —notably the Permaflux Royal 8, the Altec 400B, the Jim Lansing 208, or others of similar characteristics. Larger models can be used, if desired, with probable improvement in low-frequency reproduction.

TABLE I

Volume cu. ft.	Side of Triangle inches
2	39
3	44½
4	49
5	53
6	56¼

The Interview Amplifier

C. G. McPROUD

Design and construction of a small amplifier intended for use with a standard tape recorder for one specific use—interviews.

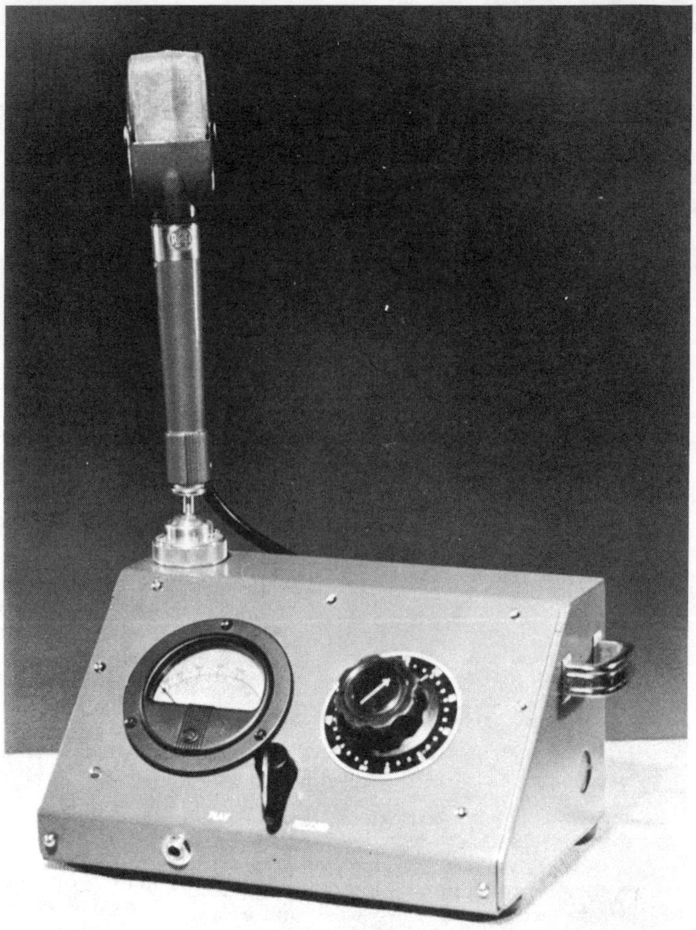

Fig. 1. External view of the amplifier unit with attached microphone on ball-and-socket swivel head.

Some months ago it became necessary to instruct a non-technical writer and radio commentator in the intricacies of operating a standard tape recorder. It was planned to use this recorder to interview agricultural authorities throughout the country on subjects of interest to the farmer for later broadcasting. This individual was at the time the publisher of Æ—one Ladd Haystead, who in his more lucrative moments serves as counselor to the Committee on Agriculture of the American Petroleum Institute, in addition to handling public relations work for a number of other important clients.

Mr. Haystead—an outstanding authority on agriculture—is a complete stranger to electronic devices. His preoccupation with the mechanics of maintaining a satisfactory recording level, operating the necessary controls and push switches, and watching the VU meter, detracted from his role as interviewer to the extent that he was almost unable to "draw out" his speaker, usually a man not familiar with the microphone. Since the equipment was to be used for only one purpose, it seemed desirable to arrange a unit which served essentially as an enlarged microphone stand, yet incorporated the amplifier, the VU meter, a gain control, and as few other controls as possible.

The equipment to be described fulfills these requirements satisfactorily. Including the recorder mechanism, it is still too large for convenience, but it is capable of producing high-quality tapes which are dubbed to disc masters, pressed, and heard over some 325 stations. In addition to providing an easily-controlled amplifier unit, the device is so arranged that it is relatively impossible to make any mistakes in interconnecting the three separate units.

Units of System

The basic equipment consists of a standard tape recorder, without any modifications. This unit—a Magnecord PT6-AH—is in wide use throughout the country, and in case of failure of any kind, it was thought that the nearest broadcast station could serve as a repair center. To date the only trouble encountered is the loosening of the function switch knob—which was duly tightened at a Santa Fe, New Mexico, station. The amplifier consists of three stages, the VU meter, and switching to permit recording or playback. A Magnecord equalizer is used for recording, and the playback equalization is incorporated in the first stage of the amplifier in a manner almost identical to the amplifier normally used with the Magnecorder. Since the recorder is used only at the 7½-in. speed, the equalizer is "permanently" mounted in the amplifier case. The function switch arranges the circuits for the desired operation; for recording, the microphone is connected to the input transformer, the output is fed through the equalizer to the tape head, the VU meter is connected across a part of the output winding, and the speaker is disconnected. For playback, the VU meter is disconnected, the head is connected to the input transformer, the frequency response is modified to provide the necessary bass boost, and the speaker is turned on.

Physically, the amplifier is mounted in a cabinet 9 in. wide, 6 in. deep, and 5 in. high, furnished as a standard unit by The Langevin Co. The RCA ribbon microphone, type KB-2C, is mounted on a ball-and-socket joint originally intended as a swivel head for a camera tripod, but modified somewhat for this purpose. When in use, the microphone is raised to an upright position; for carrying, it is folded down against the case and held secure by the drawstring of a small velvet bag being tied to the case handle. The carrying case for the amplifier contains the power supply,

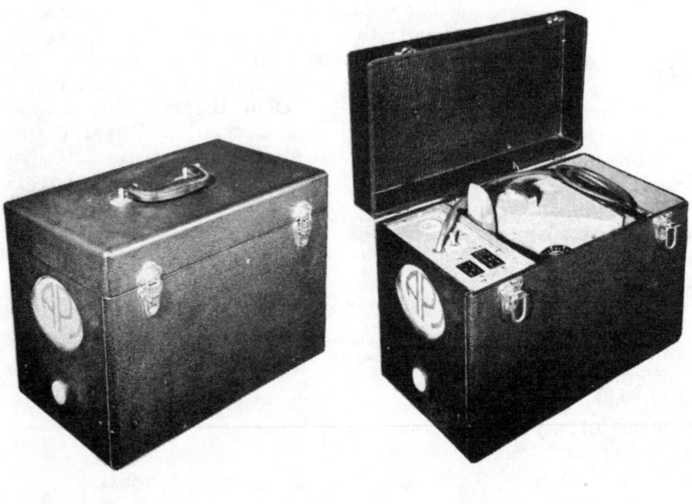

Fig. 2. The carrying case provides space for power supply, amplifier, tape reels, monitor phone, etc.

space for the amplifier unit, and a compartment for several reels of tape, the monitor headphones, and a stop watch. *Figure* 1 shows the amplifier unit in operating position, and *Fig.* 2 shows two views of the carrying case. The speaker is built into the power supply box, and the grille in the end of the case protects the cone from damage. The cabling consists of a short lead from the microphone to the amplifier, a lead from the amplifier to the recorder, and power cables from both recorder and amplifier. Power is furnished from 115-volt a.c. lines.

Electronic Requirements

In the record position, the amplifier was required to have adequate gain to work from the microphone. Since the recording head is specified as having an impedance of 60 ohms, it was most convenient to use the microphone strapped for 50 ohms output, and to use a 30 to 50-ohm winding on the input transformer. The output impedance designed to feed the Magnecord equalizer and the head is 500 ohms and the necessary recording equalization is most readily obtained from a standard equalizer, which can be changed from one speed to another without too much trouble in case such a change becomes necessary. It is not expected that such a change will be made in the field, and neither the extra equalizer nor the 15-in.-per-sec. capstan and idler roller is carried with the equipment.

For playback, the input impedance remains at 50 ohms, a satisfactory match for the 60-ohm head. The gain needed for playback is of the same order of magnitude as that for recording, but the low end must be boosted appreciably. This boost is about 20 db from 1000 cps to 100, following the 6 db/octave slope. The speaker used has a voice-coil impedance of 3.2 ohms, and it is fed from a 2-ohm tap on the output secondary.

Since there is no need for using the amplifier as a public-address system, the function switch has only two positions—record and playback. In either position, all circuit switching is accomplished with one operation. No provision is made to indicate on the amplifier panel as to whether or not the recorder is turning, since it is normally used in the same room with the amplifier—usually within six or eight feet of the amplifier. The standard VU meter is connected across a portion of the output secondary to give the correct indication without the use of any multiplier.

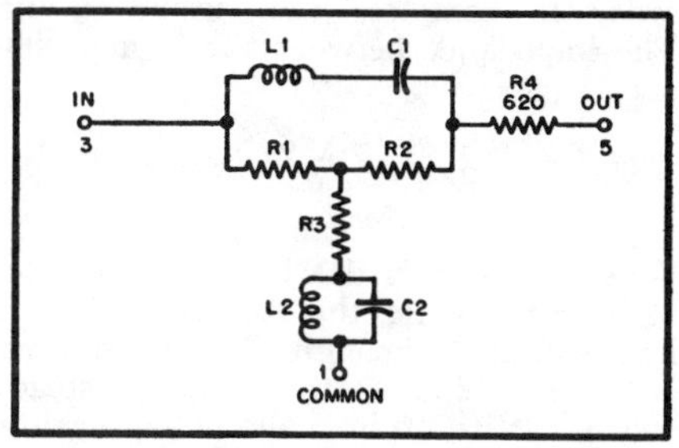

Fig. 4. Schematic of high-frequency equalizer used for recording. Low-frequency equalization, provided in amplifier, is used for playback.

The power supply provides the necessary plate and filament voltages for the amplifier as well as for the bias oscillator in the recorder case. The function switch feeds voltage to the bias oscillator only in the record position. Separate filter circuits in the power supply provide adequate isolation and improve regulation of the amplifier voltage in the two conditions of operation.

Amplifier Design

The amplifier is of straightforward design, being simplified from the basic Magnecord PT6-J amplifier unit. Less power is required than is provided by the original unit, so the output stage is a single 6V6. It is driven by a 6J7 with the gain control in its grid circuit, while the first stage is a low-noise pentode, the 5879. The over-all schematic of amplifier and power supply is shown in *Fig.* 3.

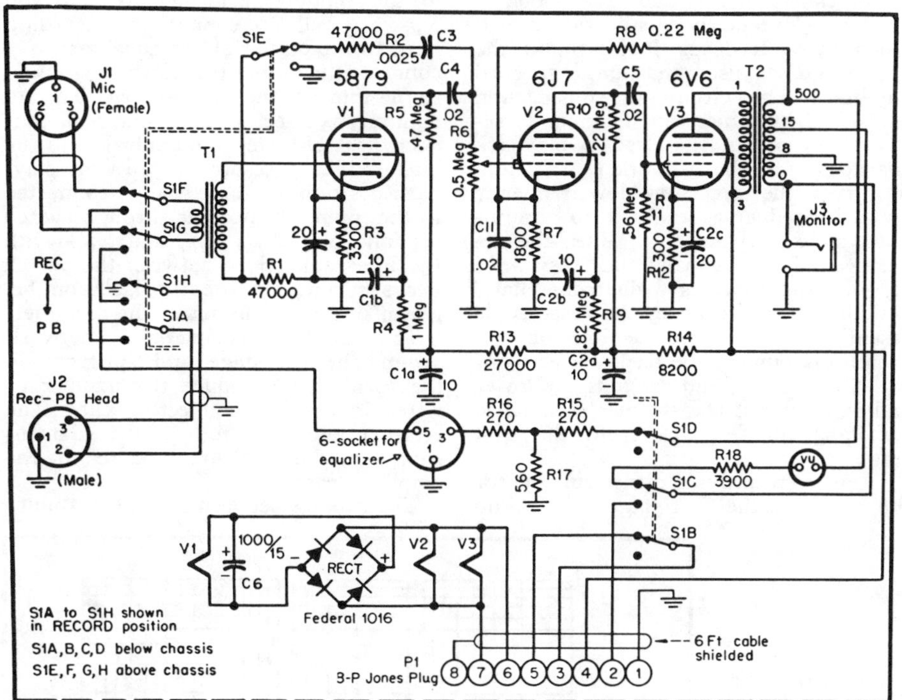

Fig. 3. Over-all schematic of amplifier and power supply.

By measurements on the original amplifier provided by the manufacturers of the recorder, it was observed that when the VU meter indicated "zero" level, a 3.0-volt signal was supplied to the input terminals of the equalizer, with a response curve which was essentially flat to that point. The required equalization—of the high end only—is furnished by the fixed equalizer, having a constant-impedance configuration, as shown in *Fig.* 4. The equalizer consists of a series resonant circuit and a parallel resonant circuit combined with three resistors to fix the amount of equalization, which is of the order of 22 db. The equalizer is followed by the "constant-current" resistor of 620 ohms, and the entire unit is encased in a housing with a 6-pin plug in the base. To isolate the output tube from the equalizer, an 8-db pad was inserted between the transformer and the equalizer socket.

The output transformer used is provided with taps at impedances of 8, 15, and 500 ohms. Between the 8 and 15-ohm taps, the impedance is 2 ohms. To simplify the switching, the 8-ohm tap is grounded, and the VU meter is connected from the "0" tap to the 15-ohm tap, using the conventional 3900-ohm resistor between the transformer and the meter to preserve ballistic action.

Checking the calculations, it is seen that at zero on the meter, 3.0 volts should be applied to the input terminals of the equalizer. Having determined the necessary output connections to simplify the switching, a few calculations are in order to complete the output circuit. The impedance between the 8- and 500-ohm taps is

$$Z_x = (\sqrt{500} - \sqrt{8})^2$$
$$= (22.40 - 2.83)^2 = 19.57^2$$
$$= 382 \text{ ohms.}$$

When the standard VU meter indicates zero on the scale, the voltage applied to its terminals through a 3900-ohm resistor is 1.228 volts (using a steady tone). With this voltage appearing across the 15-ohm winding, the voltage across a 382-ohm winding is determined by the following means.

$$\frac{E_1}{E_2} = \frac{\sqrt{Z_1}}{\sqrt{Z_2}}$$

$$E_1 = \frac{E_2\sqrt{Z_1}}{\sqrt{Z_2}} = \frac{1.228\sqrt{382}}{\sqrt{15}}$$

$$= \frac{24.0}{3.88} = 6.2 \text{ volts.}$$

The ratio between the required 3.0 volts and the indicated 6.2 volts is 2.064, which represents 6.3 db. There is, however, a mismatch between the 500-ohm pad provided between transformer and equalizer, and the voltage appearing across the 500-ohm load and the 382-ohm source is somewhat higher than that calculated, being some 7.4 db higher than the required 3.0 volts. The 8-db pad was assembled from standard preferred-value resistors to approximate the necessary loss and impedance. Final matching of VU meter indication and the signal voltage at the input of the equalizer may be done by small changes in the series VU-meter resistor. In this particular instance, however, the 3900-ohm resistor gave a zero indication with 3.05 volts at the equalizer.

The rated primary impedance is 5000 ohms—normal for the 6V6. With a 7-volt signal being required across the 382-ohm secondary, the signal across the primary is

$$E_p = \frac{7.0\sqrt{5000}}{\sqrt{382}}$$
$$= 25.3 \text{ volts.}$$

In calculating amplifier gains, it is usual to assume a gain of approximately 15 times for a pentode output stage. Thus the signal voltage required at the grid of the 6V6, for normal output, is 25.3/15, or 1.69 volts. To determine the needed gain for the first and second stages, it is first necessary to start with the available input signal. The average microphone has an output of the order of -52 db, according to the specifications, for a sound pressure of 10 bars. Normal speech is somewhere in the vicinity of 0.4 bars, however, which is 28 db below 10 bars. Thus the voltage output from a microphone is about 52 + 28, or -80 db, based on a reference of 1.73 volts across 500 ohms, presumably. Referred to 30 ohms, for the microphone to be used, the reference zero is $1.73\sqrt{30}/\sqrt{500}$, or 0.423 volts. Eighty db below this value is .0000423 volts, or 42 microvolts. The input transformer has a step-up ratio of 30/50,000 ohms, which gives a voltage gain of 40.8, the

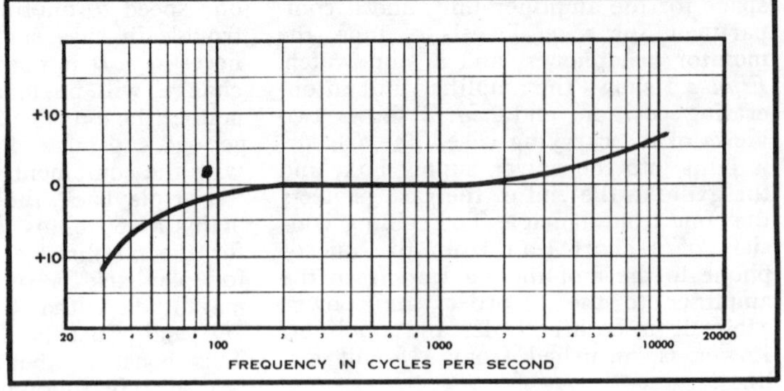

Fig. 5. Frequency response, input to tape output. This involves the amplifier for two operations.

square root of the impedance ratio. Therefore, between the grid of the first tube and the grid of the output stage, it is necessary to have a voltage gain of 1.69/.00172, or approximately 1000. This is easily obtained by using two pentodes, one the low-noise 5879, and the other a 6J7. This will allow for more than enough gain for recording, and for the addition of some inverse feedback. The signal from a playback head is somewhat less than that from a microphone, and an additional 20 db of gain is required for the low-frequency equalization, so both pentodes are used in circuits which give stage gains of approximately 100, as observed from amplifier charts in the Tube Handbook.

The low-frequency equalization is provided by a feedback loop around the first stage, as used in the Magnecord amplifier. The circuit is simple, and lends itself to minor modification, as required, to obtain the desired playback characteristc. The network C_3, R_2, and R_1 gives the necessary low-frequency boost by reducing feedback to a minimum at low frequencies. When the reactance of C_3 equals the resistance $R_1 + R_2$, the curve is 3 db from "flat," and above this frequency the response is essentially flat. One section of the switch disconnects the feedback network R_2 - C_3 from R_1, and grounds the lower end of the input transformer secondary, removing all feedback from the first stage.

In order to provide direct current for the heater of the first stage, a selenium rectifier and a filter capacitor are mounted in the amplifier. In the circuit shown, the voltage at the heater terminals of the 5879 is 6.0 when the a.c. heater voltage is 6.3. The use of d.c. on this heater reduces the hum below audible modulation on the tape.

It is noted that there are quite a number of sections to the function switch. Since the input and the output of the amplifier both have to appear on the switch, some precautions must be taken to avoid unwanted oscillation. The switch is composed of two decks, being shielded by the section of the chassis between them. In the record position, the microphone receptacle is connected to the input transformer by S_{1F} and S_{1G}; S_{1A} and S_{1H} connect the recording head to the output of the equalizer; S_{1D} connects the output transformer winding to the input of the isolation pad. The remaining two sections, S_{1C} and S_{1B}, connect the VU meter and apply B + to the bias oscillator, respectively. In the playback position, the tape head is connected to the input transformer through switch sections S_{1H}, S_{1A}, S_{1F}, and S_{1G}—with the interconnections between the switch decks serving to provide isolation between input and output of the amplifier. S_{1E} connects the equalizing network around the first stage, and S_{1C} turns on the speaker. To reduce the number of wires in the interconnecting cable, the rectifier and filter for the d.c. filament supply to the 5879 are mounted in the amplifier case.

The cabling between the various units

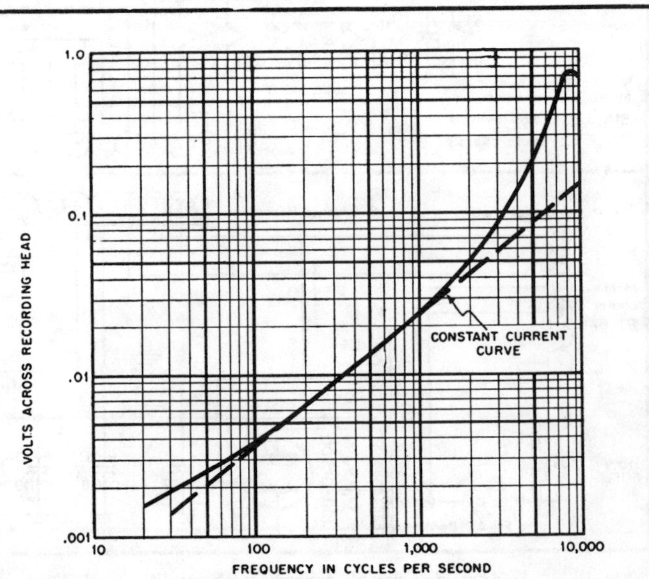

Fig. 6. Voltage across recording head for constant zero indication of VU meter.

is arranged so that there is no possibility of making any incorrect connections. The microphone cable is just long enough to run from the microphone to the jack J_1, and since the microphone remains permanently mounted on the top of the case, there is little reason for disconnecting this lead at any time during normal operation. The power cable from the amplifier to the power supply is attached to the amplifier chassis, and is plugged into the power supply receptacle, using an 8-prong Jones plug. The power supply cable from the recorder terminates in a 6-prong plug which mates with a receptacle on the power supply case. The remaining lead from the recorder terminates in a Cannon plug, which connects into the amplifier chassis.

The power supply cable from the amplifier is a 7-wire shielded cable, with the shield connected only at the plug end. A separate lead is used for the common ground connection. The 6-wire lead from the recorder carries ground, B+, 6.3 volts for the bias oscillator filament, and 115 volts for the recorder motors.

A rubber-covered a.c. line furnishes the main power connection from any convenient 115-volt outlet.

Construction

While any convenient case can be used, the one employed is available as a standard model, and is readily adapted to this application. The front apron is cut down both sides and across the bottom so that a chassis can be slid into position. The front panel is made to overlap the bottom of the case, and the chassis is attached to the front panel permanently. The switch shield and mounting is made by cutting two saw slots and bending a portion of the top surface down, as shown in *Fig. 7*. This requires some nicety of calculation, because the spacing between the chassis and the front panel must be just right to mount the switch, using the spacers and tie bolts of the switch to hold it to the bent-down portion of the chassis, and with the mounting bushing of the switch being just flush with the underside of the front panel. While this may sound difficult, it must be remembered that the spacers between the switch section and the chassis may be cut readily to make the switch fit properly.

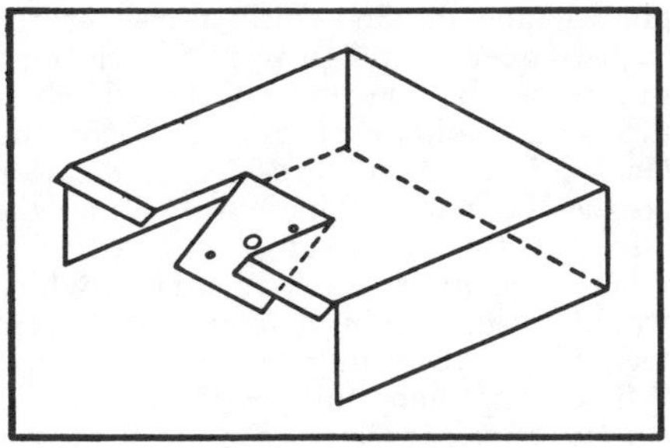

Fig. 7. Sketch of chassis arrangement to provide for shielding between two decks of record-playback switch.

The input transformer is mounted under the chassis on a bracket, making the connections reasonably near the switch for ease in wiring the playback equalizer components. The gain control is above the chassis, and permits a short lead direct to the grid of the 6J7, as shown in *Fig.* 8. The coupling capacitor C_4 is in the small space between the gain control and the front of the chassis.

The general underchassis arrangement may be seen in *Fig.* 9. The 5879 heater supply is under the VU meter, which requires some odd cut outs in the chassis to miss the terminals. In planning for the parts placement, the original drawings were made in three dimensions, with the case angles projected, so that everything clears—but not much. Most of the wiring is made directly from point to point, with resistors and capacitors connected to socket terminals where possible.

The capacitor C_2 is the one visible on the corner of the chassis in *Fig.* 8. R_{16}, R_{17}, and R_{18} are mounted on the three unused terminals of the 6-hole socket for the equalizer. The socket for the 5879 is a Vector, 1½ in. high, and mounts R_3, R_4, and R_5. This socket is riveted to a metal electrolytic capacitor mounting plate, and is flexibly mounted on the chassis by means of four grommets, using the Amphenol kit available for this type of mounting. When a socket is so mounted, flexible leads must be used to make connections to it—the so-called "antenna hank" is ideal for this purpose. R_1 is directly mounted on the input transformer terminal panel.

The microphone is mounted on the ball-and-socket camera tripod top by means of an adapter, since the microphone handle has a 5/8-27 thread and that of the swivel is 1/4-20. The lower part of this type of ball-and-socket unit in removed and discarded. Four holes are drilled in the ring for 3-48 screws, and the unit is then attached to the top of the case, with a coil spring under the ball. This spring is of the type used for spring mounting a record changer, and is about 5/8-in. in diameter. There is enough friction to hold the microphone in any position it is placed, yet it can be laid down against the case for carrying. A slot in the ring of the ball-and socket head permits the microphone to be laid flat against the case in only one position; in all other directions, the microphone can only be lowered to about 30 deg. from horizontal.

The power supply is enclosed in the 3½ × 6 × 8 "Minibox," most parts being mounted on a shelf attached to the case The 4 × 6-in. oval speaker is in this case, and protected by ¼-in. hardware cloth, as shown in *Fig.* 10. To conserve space, the half-shell of the transformer extends outside the case. Two ventilating plugs are installed on the bottom of this case, and one on the top over the rectifier tube. To permit free flow of air, another is installed in the carrying case directly below the speaker grille. The carrying case was made to order, and the power supply is simply dropped into the compartment provided—making a fairly tight fit. Since it was decided that it might be possible to damage the recorded tape if it were too close to the power transformer and chokes, the compartment for the amplifier is at the center of the case, with the tape being spaced away from the power supply by

Fig. 8 (left). Top view of amplifier chassis, to show placement of major components. Fig. 9 (right). Under-chassis view of the amplifier.

Fig. 10. External view of power supply case.

the amplifier. Since this arrangement places the two heavier sections at the center and the end, it is advisable to order the case without a carrying handle. After determining the center of gravity, the handle can then be mounted. The case may not appear symmetrical, but it does carry easier.

The unit as constructed has already been across the country nine times, and has made quite a number of recordings by this time. There is sufficient gain for all microphone work within the requirements of interview service—the gain control normally being used at about 20 on the dial plate, which is a standard attenuator scale used to give this unit a professional appearance. Actually, the calibrations of this scale approximate the attenuation of the volume control, in db. Adequate gain is also provided for playback, and the response is essentially flat from 100 to 7500 cps. *Figure* 5 shows the response from microphone input to tape output—with a rise at the high-frequency end to provide crispness in the speech. The curve of *Fig.* 6 shows the actual voltage across the recording head for constant zero-level indication on the VU meter. This curve shows the effect of the equalizer in its deviation from the 6-db/octave curve which would result from a constant-current feed of the recording head.

The amplifier unit, with its power supply and carrying case, has proven its value for use by non-technical personnel, and therefore justifies this particular design. All flexibility has been eliminated to make the operation as straightforward as possible and to reduce the possibility of error. However, since there are no modifications to the standard Magnecorder, it is obvious that this amplifier could be used for specific applications where its simplicity and compactness was desirable, yet the recorder can be brought back to the studio and used with the standard amplifier whenever necessary.

Parts List for Amplifier and Power Supply

C_1, C_2	10–10–20/450–450–25, electrolytic
C_3	.0025 μf, mica
C_4, C_5	.01 μf, 600 v. paper
C_6	1000 μf, 15 v. electrolytic, with insulating tube and mounting clip
C_7, C_9, C_{10}	40 μf, 450 v. electrolytic
C_8	20 μf, 450 v. electrolytic
C_{11}	.02 μf, 150 v. hearing aid type, paper
J_1	Cannon XL–3–13 receptacle
J_2	Cannon XL–3–14 receptacle
J_3	Single-circuit jack
J_4	Jones S-408-AB receptacle
J_5	Jones S-406-AB receptacle
L_1, L_2	Choke, 8 H. at 40 ma. Thordarson T-20C52
LS	4×6 in. loudspeaker, 3.2-ohm voice coil
M	VU meter, B scale, Simpson Model 45
P_1	Jones P-408-CCT cable plug
R_1, R_2	47,000 ohms, ½ watt
R_3	3300 ohms, ½ watt
R_4	1.0 meg, 1 watt
R_5	0.47 meg, 1 watt
R_6	1-meg volume control, audio taper
R_7	1800 ohms, ½ watt
R_8	0.22 meg, ½ watt
R_9	0.82 meg, 1 watt
R_{10}	0.22 meg, 1 watt
R_{11}	0.56 meg, ½ watt
R_{12}, R_{19}	300 ohms, 5 watt, Ohmite Brown Devil
R_{13}	27,000 ohms, 1 watt
R_{14}	8200 ohms, 2 watt
R_{15}, R_{16}	270 ohms, ½ watt
R_{17}	560 ohms, ½ watt
R_{18}	3900 ohms, 1 watt
RECT	Federal 1016 Selenium Rectifier
S_{1A-H}	8-pole, 2-pos. wafer switch, Centralab 1418
S_2	DPST toggle switch
T_1	30/50,000 input transformer, shielded, Triad HS-5
T_2	Output transformer, secondary impedances 500, 15, 8, 2 ohms, UTC S-14
T_3	Power transformer, 300–0–300 v. at 90 ma; 5 v. at 3 amps; 6.3 v. at 2.5 amps.
Case for amplifier	Langevin Remote Control Cabinet, Type 1-A
Case for power supply	Bud Minibox, CU-2109

Universal Amplifier for Magnetic Tape Recorder

C. G. McPROUD

The description of an amplifier which combines the ability to serve in professional applications as well as to work with a conventional home audio system. In addition, it doubles as a broadcast remote amplifier.

ABOUT fifteen years ago, the writer acquired an urge to have a broadcast remote amplifier — for some reason which still remains obscure—and a number of futile starts were made to fulfill that desire. An extremely small unit which would be quite flexible was contemplated — but was never built in that form.

Following the purchase of a professional tape recorder a few months ago, the need for a record-and-playback amplifier again resurrected the urge for a small flexible unit, and the design to be described was the result. Before launching into a description of the amplifier, its various functions will be discussed.

Primarily, the unit was to serve as the record and playback amplifier to work with a home music system of conventional design. Thus it required two channels, one for recording and one for play-back, and both able to work simultaneously so that the recording could be monitored from the tape. It was therefore necessary to have a high-impedance input for the record amplifier, and an output which could be arranged to feed a high-impedance circuit following the playback preamplifier and equalizer.

It was also desired to be able to mix a microphone with the signal from a radio or phonograph. This function has some drawbacks, but it seems to work satisfactorily in practice.

Fig. 1. Compact, yet with a multitude of capabilities, this unit works with a standard tape recorder to turn out professional work.

For occasions where it might be necessary to record from two microphones at the same time, two low-impedance inputs were considered necessary. Thus the amplifier could be used to mix any two signal sources at mike level.

By suitable connections, the record feature was to be eliminated so that the amplifier could serve as a broadcast remote unit, mixing two microphones, and feeding a balanced line through a 6-db H pad, ungrounded.

As the circuit developed, it was found that the two-channel feature could be used in a small studio, if desired, as a program amplifier and an audition amplifier at the same time, or as a program or audition amplifier and a talk-back amplifier, also at the same time.

And it was also found, as the circuit began to take shape, that it could be used as a reverberation generator, employing the delay of the tape machine—due to the spacing between the record and playback heads—to provide the time delay necessary for reverberation.

One more feature was worked into the circuit—that of being able to plug in three microphones and switching so that any two of the three could be used simultaneously, or, if recording and monitoring at the same time, two microphones could be plugged in, and either one could be selected at will by means of a switch.

A number of other features were incorporated—mainly to increase flexibility. Among them were switching for the VU meter, making it possible to feed the recorder or the line at three levels as well as to measure bias and erase currents, and to permit the meter to be used for other purposes by employing a plug wired in a certain manner. The output key was arranged so

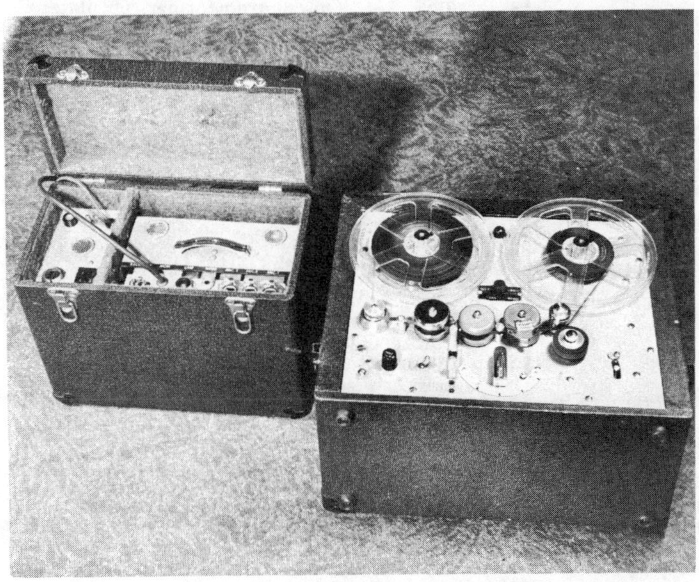

Fig. 2. The amplifier in its carrying case with the portable power supply, and the recorder for which it was designed.

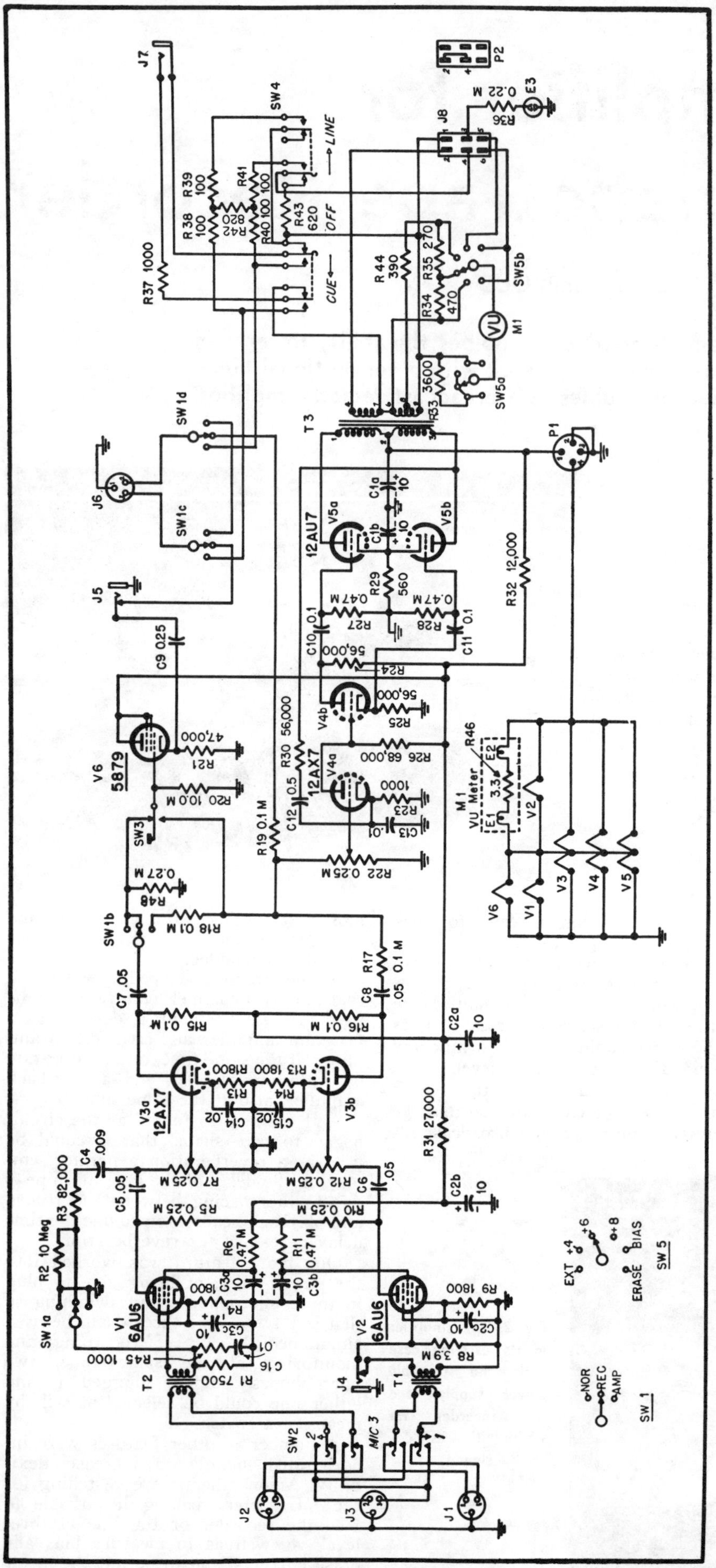

Fig. 3. Complete schematic of the amplifier chassis. Equalization for recording is furnished by a passive equalizer mounted in the recorder case. For use with this amplifier, the playback head must be shunted with a 33-ohm resistor.

PARTS LIST

C_1	10-10/450, electrolytic
C_2, C_3	10-10/450, 10/25, electrolytic
C_4	.009 μf, mica
C_5, C_6, C_7, C_8	.05 μf, paper, 600 v.
C_9	0.25 μf, bathtub, 400 v.
C_{10}, C_{11}	0.1 μf, 600 v.
C_{12}	0.5 μf, upright can, 600 v.
C_{13}, C_{16}	.01 μf, paper, 400 v.
C_{14}, C_{15}	.02 μf, paper, 400 v.
E_1, E_2	Dial light, #291, 0.17 a, 2.9 v.
E_3	NE-51 Neon glow lamp
J_1, J_2, J_3	Cannon XL-3-13 receptacle
J_4, J_5	Switchcraft 12A jack
J_6	Amphenol 91-PC3F receptacle
J_7	Switchcraft 11 jack
J_8	Jones P-306-AB plug
M_1	−20—0—+3 VU meter, G. E.
P_1	Amphenol 91-MC3M, on 6-ft. 3-wire cable
P_2	Jones S-306-FHT, shorted between terminals 2 and 4
R_1	7500, ½ watt
R_2, R_{20}	10.0 meg, ½ watt
R_3	82,000, ½ watt
R_4, R_9, R_{13}, R_{14}	1800, ½ watt
R_5, R_{10}	0.25 meg, IRC Type DCF
R_6, R_{11}	0.47 meg, 1 watt
R_7, R_{12}, R_{22}	0.25 meg pot, Ohmite CA 2541
R_8	3.9 meg, ½ watt
R_{15}, R_{16}	0.1 meg, 1 watt
R_{17}, R_{18}, R_{19}	0.1 meg, ½ watt
R_{21}	47,000, 1 watt
R_{23}, R_{37}, R_{45}	1000, ½ watt
R_{24}, R_{25}	56,000, 1 watt, 5%
R_{26}	68,000, 1 watt
R_{27}, R_{28}	0.47 meg, ½ watt
R_{29}	560, ½ watt
R_{30}	56,000, ½ watt
R_{31}	27,000, 1 watt
R_{32}	12,000, 1 watt
R_{33}	3600, 1 watt, 5%
R_{34}	470, ½ watt
R_{35}	270, ½ watt
R_{36}	0.22 meg, ½ watt
$R_{38}, R_{39}, R_{40}, R_{41}$	100, ½ watt
R_{42}	820, ½ watt
R_{43}	620, ½ watt, 5%
R_{44}	390, ½ watt
R_{46}	3.3 ohms, ½ watt
R_{48}	0.27 meg, ½ watt
$Sw_{1a, b, c, d}$	Mallory 3143J switch
Sw_2, Sw_4	Stromberg Carlson 172-A key
Sw_3	Switchcraft 203 push button
Sw_5	Mallory 3126J switch
T_1, T_2	Input transformer, 30 : 150,000 ohms; Western Electric D-95495
T_3	Output transformer, pp plates to line, 20,000 : 600 ohms; ADC 315B
V_1, V_2	6AU6
V_3, V_4	12AX7
V_5	12AU7
V_6	5879

TRANSFORMER SUBSTITUTIONS*

Make	T_1	T_2
Chicago	BI-2	BO-2
Peerless	K-221-Q	S-220-Q
Stancor	WF-21	WF-36
Triad	HS-5	HS-52
UTC	A-11	HA-114

* While these transformers meet the electrical requirements, they are not necessarily of the same physical dimensions, and may not substitute directly in the same sized chassis. They should, however, perform satisfactorily in the circuit.

as to terminate the amplifier, when used as a remote unit, and connect the monitoring headphones to its output for rehearsal and level setting; or to connect the phones across the line to receive cue from the station on a remote job; or to connect the amplifier to the line through the 6-db pad with the headphones monitoring the output. Another switch was connected so as to permit monitoring of the radio program to be recorded—when using the amplifier with a home system—direct from the input, rather than from the tape.

Circuit Description

The completed amplifier, shown in *Fig.* 1, was built in a case which is exactly like several audio test equipment units used by the writer, and which happened to be available. It is a trifle too small for ease in construction—measuring 8½ in. wide, 8 in. deep, and 5½ in. high. However, after it was completed, its small size was an advantage. The power supply is separate, and the equalizer for the recording head is placed within the recorder case—where there is considerably more room. Two power supplies are used, actually—one is permanently mounted in the home equipment, and the other is installed in a carrying case semi-permanently, along with space for the amplifier itself, as shown in *Fig. 2* with the recorder. For the information of anyone who may be interested enough to inquire, the lettering on the panel was done quite easily with *Tekni-Cals,* a variety of decalcomania designed for radio purposes.

The schematic, *Fig.* 3, shows two preamplifier stages, two gain controls, and the two separate voltage amplifiers, the mixing network and the master volume control, and the output amplifier—a push-pull 12AU7. In addition, a cathode follower, V_6, is used to feed the second channel back to the radio amplifier, or for the audition or talk-back application.

For ease in understanding the operation of the circuit in the three positions of the SELECTOR switch, the block diagrams of *Fig.* 4 should be studied. In normal operation of this amplifier with a home system, a high-impedance circuit —preferably between the control amplifier and the main amplifier—is opened, and the output of the control amplifier and the input of the power amplifier are brought to the recording amplifier by means of a two-conductor plug which is inserted into J_6 in this unit. If the recording amplifier is to be removed from the circuit, the plug is simply shorted by inserting it into a receptacle which has its terminals wired together.

When connected to the recording amplifier, however, it is necessary to short this pair of leads while the recorder is not being used. Thus the SELECTOR switch has one position labeled NOR—for normal—which shorts the radio line, and the power supply may be turned off, putting the entire equipment effectively out of the circuit. At the same time, the amplifier is set up in a two-channel configuration. The main channel, consisting of V_2, V_{3b}, V_4, and V_5,

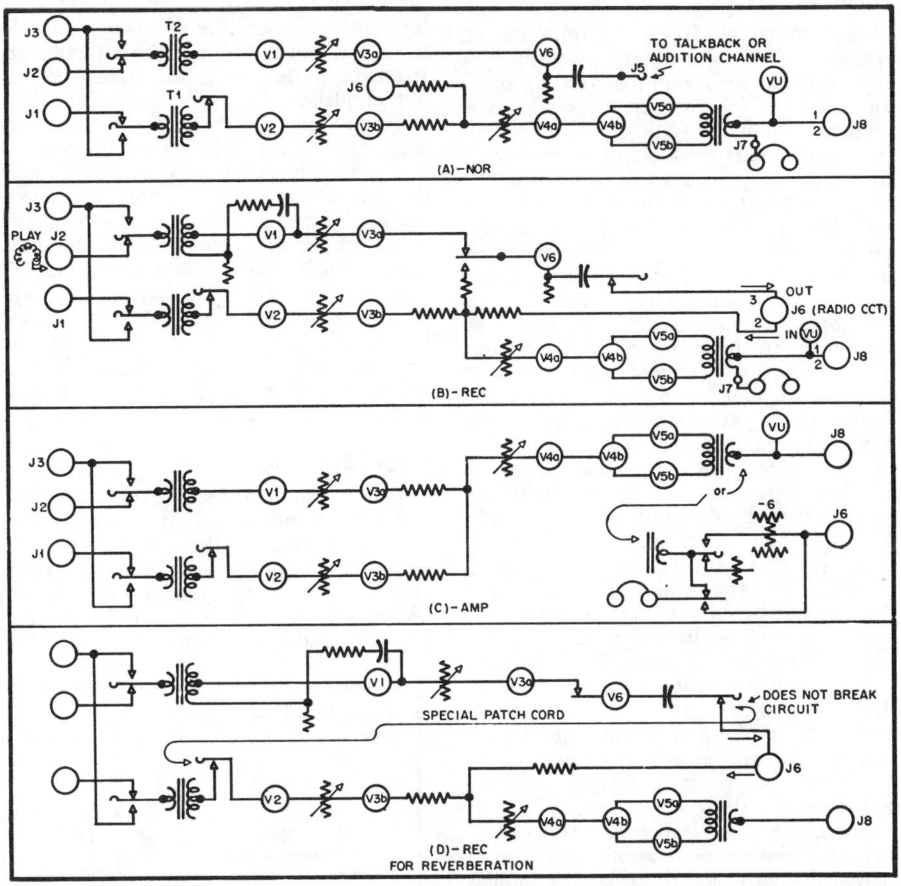

Fig. 4. Block schematic of the amplifier for the three positions of the selector switch, $SW_{1a,\,b,\,c,\,d}$, and when connected for reverberation. When connected to a home music system, J_6 is shorted so the unit is out of the circuit when the selector switch is at NOR.

is connected straight through, and may be used as desired, making the output connections to terminals of J_8.

The second channel consists of V_1, V_{3a}, and V_6, and feeds out from a cathode follower to any suitable high-impedance load. Both channels are "flat," and gain is adequate for low-level microphones. J_2, in the main channel, will accept a high-impedance microphone or some other source which may be operated unbalanced. An unbalanced signal of around zero level may be fed in at J_6 and mixed with the mike if desired.

When using the amplifier for audition and program at the same time, the second channel is simply connected to a monitoring amplifier with a high-impedance input, and any signal introduced into T_2 will be available at the output of V_6, at J_5.

System Philosophy

There are two basic types of equalization which are necessary in any magnetic recording system—that caused by the finite width of the gap, and that due to the magnetization characteristic of the tape itself. The former must compensate for the loss in high frequencies as the wavelength of the recorded signal approaches the width of the gap, and is of the order of 20 to 25 db at the frequency where the wavelength is equal to the gap. The latter requires a boost at the rate of 6 db per octave as the frequency is lowered, with a turnover somewhere in the vicinity of 3000 cps.

In most professional machines, the high-frequency equalization is introduced in the recording circuit, principally because a boost in playback would decrease the signal-to-noise ratio. And in order to avoid overload of the tape, the low-frequency equalization is introduced in the playback amplifier, in much the same manner as with magnetic pickups. Many of the non-professional machines employ the same amplifier circuit for both recording and playback, introducing half of the required equalization in each operation. This simplifies the amplifier design, but runs the risk of overloading the tape on lows; and it reduces the signal to noise ratio by boosting the highs in playback.

Since this machine was to be used for professional applications, the first method of equalization is employed. The high-frequency equalization is provided by a passive network between the output of the amplifier and the recording head (not shown in the diagram, but located inside the recorder case), and the low-frequency boost is obtained by working the 500-ohm head into a 30-ohm transformer primary which is shunted by a 33-ohm resistor. This furnishes most of the low-end boost, the final adjustment being provided by a feedback network around the first stage when the circuit is switched to REC, which is the correct position for playback.

Input Switching

With only two input amplifier stages, it seems obvious that only two separate inputs could be accommodated at once. In order to increase this facility even slightly, input switching was arranged to permit a third circuit to be connected to either of the transformers. A key switch was chosen which is, in effect, two separate DPDT switches, since two circuits are made in the neutral position of the switch, and each throw of the switch affects only one of the circuits. The outside contacts are paralleled and connected to the third microphone jack. The two pairs of arms are connected to the transformer primaries, and the inside contact pairs are connected to the first and second jacks respectively. Thus in the center position, jacks 1 and 2 are connected to the two transformer primaries; when thrown to either side, the third jack is connected to either the first or second transformer, as desired, opening the other input circuit. In use, the associated mixer dial is turned to minimum, the switch thrown, and the dial then turned to the required position.

When the selector switch, Sw_1, is in the REC position, two playback heads may both be plugged in, one into J_2 and one into J_3, and either machine may be used for playback by throwing the switch to the desired position.

Since it was desired to use this equipment for professional applications, it is provided with low-impedance inputs, with 30 ohms being chosen so as to simplify equalization in the playback channel. The input transformers have a relatively high step-up ratio, and it is probable that more modern units would give better performance. However, one of the units chosen was already in the "junk" box, so it was natural that another of the same type would be secured.

In order to provide for a high-impedance input for any purpose for which it might be required, the secondary of the input transformer is connected to the contact of a closed-circuit jack, J_4, the spring being connected to the grid of V_2. The 3.9-meg resistor R_8 provides a grid return if crystal microphones should be used. The first stage employs a 6AU6 to provide sufficient gain, and it is coupled to the gain control in the grid circuit of the second stage. The latter is one half of a 12AX7 (the other half occupies a similar position in the second channel, with separate cathode and plate resistors). The small by-pass capacitor serves a frequency-correcting element. The plate of V_{3b} is coupled to the mixing network—composed of R_{17}, R_{18}, and R_{19}—and the main gain control, R_{22}. The mixing network couples the high-impedance radio input, when used, to the main gain control without appreciable effect on the input stages. When two microphone stages are being used, they are coupled together and to the gain control by the same mixer network.

The output amplifier, V_4 and V_5, will be recognized as being similar to the Williamson amplifier input circuits, and it serves equally well in this application. V_{4a} provides considerable gain, and is directly coupled to V_{4b} as a phase splitter, which drives the two halves of V_5 as a push-pull output stage. Feedback from the plate of V_{5b} stabilizes the output amplifier. It would have been possible to derive the feedback from the secondary of the output transformer, but this would have necessitated grounding the secondary, which is not desirable if the equipment is to feed a balanced line when acting as a remote amplifier. The tape recorder selected, a Presto RC-7, was designed to be fed from a 500/600-ohm transformer, and also some such impedance is desirable to feed a remote line. Actually, the equalization for the recorder is quite simple with this type of feed.

Output Switching

When used to feed a recorder, the 600-ohm winding of the transformer is fed directly to the output jack, J_8, ground being obtained in the recorder chassis. This ground is common to the power supply, so the circuit is properly grounded when so connected.

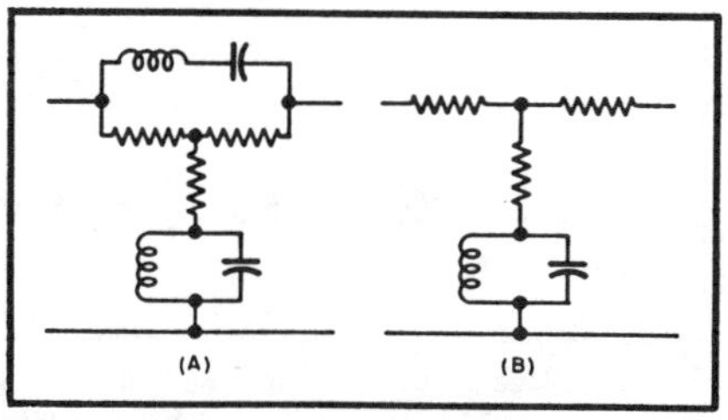

Fig. 5. Equalizer configurations: (A) as originally designed; and (B) as finally used for the 15-in./sec. speed. The 7.5-in/sec. equalizer is like (A).

When used to feed a remote line, a dummy plug with terminals 2 and 4 shorted is placed on J_8 (which is a male Jones plug, actually) so as to connect the "high" side of the output circuit to the output switch, Sw_4. In the center or OFF position of this switch, the amplifier is terminated by R_{43}; in the LINE position, the amplifier is connected to a 6-db pad composed of resistors R_{38} to R_{42}, inclusive, and thence to the line jack, J_6, through two sections of Sw_1; in the CUE position, the monitor jack is connected directly to the line jack so that cue may be heard from the station on the monitor phones. The output switch is the same type of switch used for the input switching.

VU Meter Switching

In order to provide for measuring the output level when feeding a remote line, or for the correct recording level, as well as to measure bias and erase current in the recorder, the VU meter switch, Sw_5, has six positions. Two positions provide for measuring bias and erase currents, and are labeled B and E respectively. Three output levels are provided— +4, +6, and +8—being so arranged that the level fed to the line at the output of the 6-db isolating pad is indicated at zero on the VU meter. The particular recorder used requires a voltage applied to the input of the recording equalizer of approximately 3.5 volts for full modulation (which is presumably 10 to 15 db below tape overload). The sixth position of the switch is marked EX, and connects the meter directly to terminals 1 and 6 of J_8 (with the 3600-ohm series resistor which is always in the circuit). Another plug, with leads connected to these terminals, may be used to check the meter from an external source, or to permit measurements on other circuits with the same meter.

For a given voltage across the meter, the voltage across the entire secondary will be twice that value if the meter is connected across half the winding, and if connected across less than half the winding, the total voltage would be still greater. Thus with a meter indication of 0 VU, which corresponds to a level of +4 due to the meter calibration, a level of +10 would appear across the entire winding when the meter is connected across half of the secondary. The isolating pad has a loss of 6 db, so the level at its output would be +4, as indicated by the switch. Similar calculations and a voltage divider between the center tap of the secondary and a tap of lesser impedance permit adjustment of output to obtain additional levels of +6 and +8 at the line terminals.

For recording, the +4 position provides a level of 2.6 volts to the input of the recording equalizer; the +6 position provides 3.6 volts, and +8 provides 4.5 volts, giving a range of recording levels. A small arrow is placed on the panel at the +6 position to indicate that step is normal for recording. For piano recording, a lower level is desired, and the +4 step gives a cleaner tape. To date, no recording application has been found where the +8 position was found desirable.

The Recording Equalizer

The recorder unit of the Presto RC-7 contains the bias-erase oscillator with suitable resistors to provide indications on the standard VU meter and two leads are brought out for this purpose. In addition, the "constant-current" resistor of 500 ohms is in this unit, as is a trap circuit to keep the bias voltage out of the recording amplifier. One side of the recording circuit is grounded in this unit, and this provides ground to the entire output circuit since the same power supply feeds both sections. Therefore, in order to place the recording equalizer in series with the line, it is only necessary to bring out three leads to connect to the equalizer—input, the "high" line from the amplifier; output, to the "constant-current" resistor followed by the trap and the recording head; and ground.

The equalizer configuration originally planned was that of a constant-impedance network, as shown at (A) in *Fig.* 5. Final adjustments of frequency response required a few changes, and the final configuration for the 15-in./sec. equalizer for a flat frequency response from input to tape output is as shown at (B); the 7.5-in./sec. equalizer follows the original configuration. The

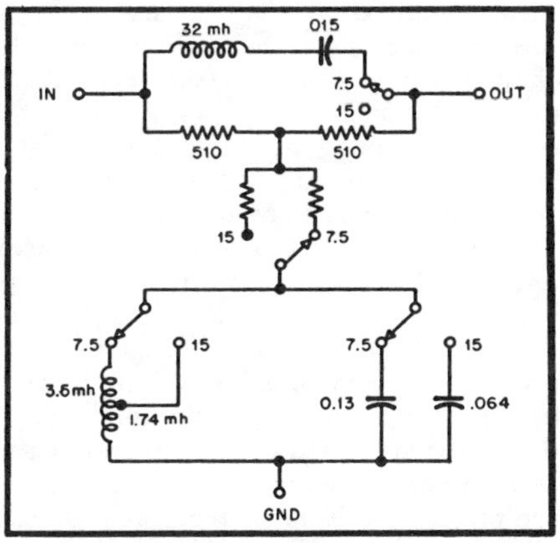

Fig. 6. Schematic of complete equalizer as used with this amplifier and the Presto RC-7 recorder.

complete equalizer is shown in *Fig.* 6, with the necessary switching for the two speeds.

The coils used for these equalizers were made from unmounted toroids, with the windings being adjusted to exact values by removing turns. The toroid coils used were wound in two sections, making it quite simple to obtain the correct values. Complete equalizers of this type are used in the Magnecorder, and are available in separate shield cans with a plug-in base, as described in an earlier article,[1] and could undoubtedly be employed with the Presto recorder if the effort of making the equalizer is considered too troublesome. However, the equalizer as constructed functions satisfactorily and occupies relatively little space.

In most professional-type circuits, the high-frequency equalization is provided in the recording function, while the low-frequency equalization is provided in the playback function. The high-frequency equalization compensates for gap width, and requires a boost of about 18 to 22 db at 7500 and 15,000 cps for the low and high speeds respectively. The curve of the equalizer is adjusted by the relative inductances in the series and shunt elements of the equalizer, and the amount of equalization is adjusted by the resistor network. It will be noted that the series resistors used are each of 510 ohms, and that the shunt resistors are employed to make final adjustments for frequency response. This construction does not follow accepted network practice rigorously, but the performance seems to justify the means taken. Frequency response is within ± 1 db from 30 to 15,000 cps at 15 in./sec., and within ± 2 db from 30 to 8000 cps at 7.5 in./sec.

The entire network, with the switch, is assembled in a Bud Minibox, 2¼ × 2¼ × 4, and is installed inside the recorder case, as shown in *Fig.* 7. The cable from the recorder-oscillator chassis is plugged into the receptacle on the equalizer box. If the recorder is to be used with its regular Presto amplifier, the two terminals of the plug are shorted. The ground lead is carried through the shell of the plug, and serves to ground the equalizer can. A bracket for the extra idler and capstan sleeve is mounted on the side of this can to keep them out of the way—their normal position is on the top panel of the recorder.

In order to permit the adjustment of bias current—so as to allay any suspicions that it might possibly not be at the optimum value—a padder capacitor having a range from 200 to 600 μμf was installed in parallel with a 200-μμf fixed capacitor in place of the 560-μμf capacitor (C_{208} in the Presto diagram). No change in performance has been noted, however.

The Playback Channel

The second channel is almost identical with the first when the selector switch is in the AMP position, except that no provision is made for a high-impedance input. However, in the REC position of the switch, the input stage is connected so as to provide the remainder of the low-frequency equalization necessary for playback from the tape, and to feed the playback signal out to the radio circuit through J_6.

Most of the low-frequency equalization is provided by working the 500-ohm playback head into a 30-ohm resistance termination across which the input transformer primary is bridged, as shown by curve A in *Fig.* 8. The required total equalization is shown in curve B. The small amount of additional equalization is obtained by the feedback network composed of C_4, R_5, and R_1, with some high-frequency correction obtained from C_{16} and R_{45}. This circuit is similar to that used in the Magnecorder amplifier, and previously used by the writer in the Interview Amplifier. However, the amount of equalization is considerably less in this application. The entire equalization set-up for this amplifier was determined in the following manner:

With the selector switch in the AMP position, both input channels were equalized to provide flat response to the top of the main gain control, R_{22}. Then the output amplifier was equalized for flat response to the secondary of the output transformer—this being necessary if the amplifier were to be used as a remote pickup unit. Then the playback channel was adjusted for flat response from a standard tape obtained with the Presto recorder. Finally, the recording equalizer was adjusted to provide flat playback from the tape recorded through this unit. It will be noted that considerable high-frequency compensation was required. It is believed that this is occasioned by the use of extremely small shielded wire which carries high-impedance circuits throughout the amplifier. This will become evident when the construction is described. Again judging from results, it should be stated that the over-all IM distortion of this unit from microphone input jack to tape output—thus including the tape itself—is less than 3 per cent at full recording level, i.e. with a 0-VU indication on the meter with the level switch set at + 6, and with a playback output of 1.0 volts.

The second stage in the second channel is identical with that in the first channel, but the output connects to one section of Sw_1. In the NOR and REC positions of this switch, the output of the second stage is fed to the grid of a cathode follower, V_6. Sw_3 in this grid circuit permits choice of monitoring direct from the mixing network, or from the tape playback. The 5879 was chosen because of its low heater drain, and because it was less microphonic than the 6C4 which has the same drain. The seemingly high value of cathode resistor

Fig. 7. Equalizer in its shield can is mounted inside the recorder case, and supports bracket for idler and capstan sleeve.

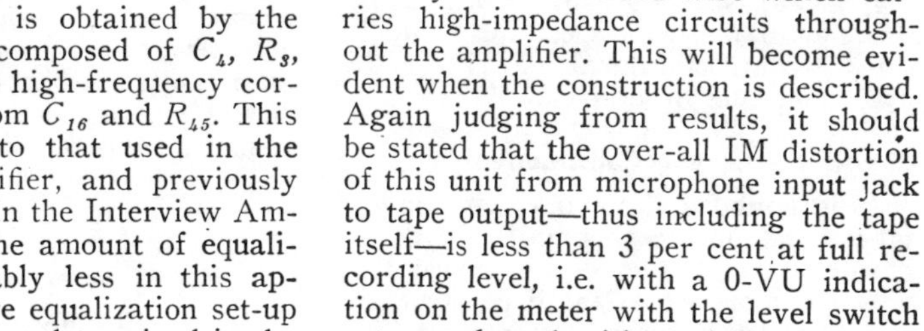

Fig. 8. Curves of playback equalization: (A) that due to feeding the 500-ohm playback head into a 33-ohm resistor; (B) total equalization obtained with addition of feedback network around first stage.

[1] C. G. McProud, "The interview amplifier." See page 110.

for V_6 was determined by IM measurements. In the original construction, a 10,000-ohm resistor was used, but IM distortion in the playback channel was over 4 per cent. As the value of this resistor was increased, the distortion decreased to less than 1 per cent.

The output of the cathode follower stage is fed through a coupling capacitor C_9 to the closed circuit jack J_5 and thence through the REC position of Sw_{1c} to J_6. In the NOR position of Sw_1, the two terminals of J_6 are shorted, and in the AMP position they are connected to the output pad. Thus a single jack serves two purposes.

In connecting this unit to a radio system, usually at high impedance, it is suggested that the output of a control unit be fed into terminal 2 of J_6, and that the input of the power amplifier be connected to terminal 3. Thus this unit intercepts the circuit between the control unit and the main amplifier. This increases the length of shielded leads, but should cause no frequency discrimination if the output of the control amplifier is of sufficiently low impedance, such as would be obtained with a cathode follower. Then in the NOR position of the switch, the recording amplifier is out of the circuit; in the REC position the incoming signal is fed to the mixing network, is recorded, and the playback from the tape is fed back to the line and the main amplifier. Sw_3 permits choice of direct or recorded playback monitoring. There is a loss in level of approximately 10 db, which can be compensated by an additional switch in the main amplifier if desired.

Chassis Layout and Construction

It is doubtful if anyone who undertakes to duplicate this unit will do so physically, and it is recommended that only the more experienced constructors make any very close attempt to follow the mechanical plans closely. Not that there have been any particularly undesirable features in the operation of the amplifier, but in the fact that the extremely small size is conducive to undesirable capacitances between circuits, and if small shielded wire is used, there is the possibility of introducing excessively large shunt capacitances. Therefore, in addition to describing the construction employed in the original model, a few additional suggestions will be offered which would be followed by the author if the amplifier were to be rebuilt.

Since the external case was already available when this unit was planned, and since it was of a desirable shape and size, the chassis layout was made to fit. It will be noted from *Fig.* 1 at the beginning of the article that the external case is of unusual shape—having a vertical apron on which are mounted the monitor-phone jack and the push-button switch which selects direct or recorded monitoring. Another plane of the front panel is at an angle of 45 deg. to the vertical and provides space for the mixing pots. A third plane at 67½ deg. from the vertical mounts the VU meter, key switches Sw_2 and Sw_4 the two rotary switches Sw_1 and Sw_5, and the recording indicator light E_3. All connections to the circuit are made on the rear apron of the chassis, except for the monitoring phones. *Figure* 9 shows the underside of the chassis, and indicates the chassis construction. The top of the chassis is bent up to meet the panel just back of the two keys, and thence follows the contour of the panel to the bottom. To simplify bending the chassis into this shape, the two sides are separate pieces riveted in place after the top is shaped. The panel is attached to the chassis by means of the mounting nuts on the pots and the jack and switch on the front apron. The two keys are attached directly to the front panel by their own mounting screws, and slots in the chassis top clear these parts.

Figure 10 shows one of the unique features of the construction. The sockets for the three low-level tubes and the cathode follower are mounted on a small aluminum channel which is, in turn, mounted to the chassis by eight soft rubber grommets—four on the channel strip and four on the chassis top—with machine screws through the centers of the grommets. The resistor board (shown in *Fig.* 9) is solidly mounted to the channel strip, and connections made from the tube socket terminals directly to the resistor board with bare wire. All connections to the strip or to the sockets mounted on it are made with very flexible wire to avoid transmitting vibration to the tubes. With the additional mass of the resistor board and resistors mounted on it, the tubes are effectively isolated from any chassis vibration. Sockets for V_4 and V_5 are solidly mounted on the chassis top, and a similar resistor board is mounted directly under these sockets for the associated components.

Perpendicularly mounted on the front apron are three electrolytic capacitors—C_1, C_2, and C_3—directly under the three mixing pots. The input transformers are located along the rear of the chassis, as is the output transformer, for most efficient utilization of space, and to keep the leads between tubes and transformers as short as possible. If this unit were to be re-designed, it is felt that a more suitable arrangement of the pots would be on a bracket below the chassis and nearer to the tubes, with short flexible shafts extending from the pots to the respective knobs.

Study of *Fig.* 10 will provide some indication of the wiring plan. Narrow aluminum strips were riveted to the sides of the chassis to hold the wiring in place, and after completion, these strips were bent around the wires and pressed tightly closed over strips of empire cloth tape to prevent abrasion. All shielded wiring is done with very small leads such as that used for phonograph pickups. The small resistor board at the lower center of *Fig.* 10, mounts C_7, C_8 and the three 0.1-meg resistors in the mixing network. C_9 and C_{12} are visible in *Fig.* 11, as are the three trans-

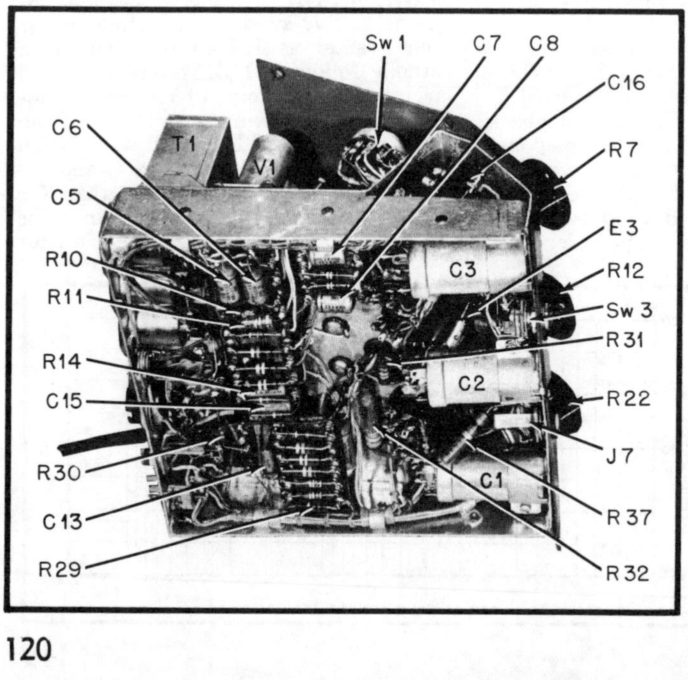

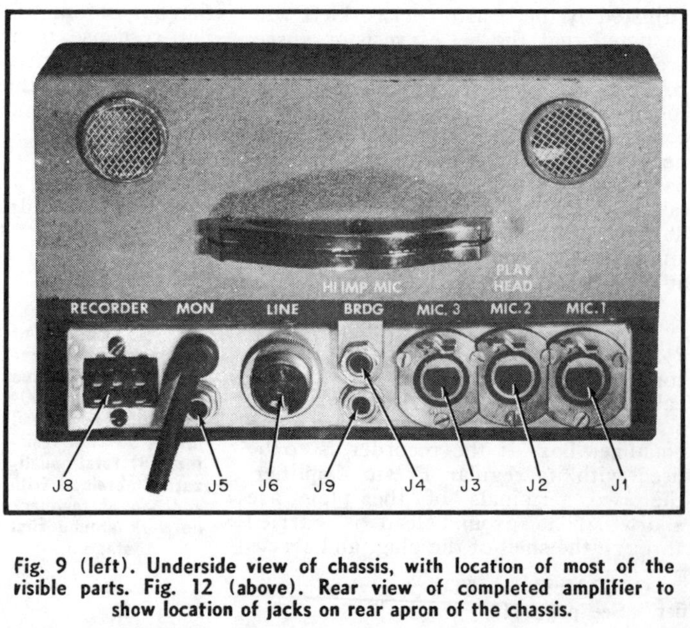

Fig. 9 (left). Underside view of chassis, with location of most of the visible parts. Fig. 12 (above). Rear view of completed amplifier to show location of jacks on rear apron of the chassis.

formers. The locations of most of the other parts are shown on *Fig.* 9.

The rear of the case with the chassis in place is shown in *Fig.* 12. The two ventilating openings are considered desirable, and the handle is provided to facilitate removing the amplifier from its portable case, shown in *Fig.* 1. Note that the three Cannon receptacles, J_1, J_2, and J_3, have been trimmed to permit more compact mounting. The rear of the chassis is held firmly in the case by the bracket extending down over the jack, J_4.

Patching Facilities

Mention was made of the use of this unit for adding reverberation to a recorded program. To do this requires the use of a special patch cord, shown schematically in *Fig.* 13. This cord is inserted into jacks J_5 and J_4, with the plug which goes into J_5 being filed flat on the side which contacts the spring of the jack. This must be done carefully, so that contact is still made with the jack spring, yet without breaking the circuit through the normal contact. For ideal results, it is probable that this cord should have attenuation, without frequency discrimination. It was found, however, that a more realistic effect was obtained with the patch cord circuit shown. Using 1/2-watt resistors and an Erie Ceramicon for the capacitor, it is possible to construct this network small enough so that it will be completely enclosed in a large metal-shell plug. The tip of the common PL-55 surplus plug may be filed to the required flatness without coming apart.

The jack marked BRDG on *Fig.* 12 was originally intended to connect to the top of R_{12}, for bridging high-impedance circuits, but with this connection it was found that the tube noise resulting from the open transformer primary was objectionable. The spring of this jack is now connected to the arm of Sw_{1d}, and for dubbing from one machine to another, a patch cord—without any built-in network—is connected between J_5 and this new jack, later numbered J_9. A shorting plug for J_6 could be used with equal success, if desired. When dubbing, Sw_1 should be in the REC position.

Power Supplies

Two power supplies have been built to use with this unit—one remaining in the fixed installation, while the other is mounted in the carrying case for field use. Both have similar characteristics, and are similar in design, although differing in detail.

The fixed unit is shown in *Fig.* 14, with a schematic as shown in *Fig.* 15. In the original unit, the transformers used were obtained at surplus, but the accompanying parts list employs standard jobber items.

This power supply uses a tube rectifier for the plate supply, which must furnish 35 ma at 300 volts for the amplifier, and about 40 ma at the same voltage for the oscillator in the recorder case. It must also furnish 6.3-volt a.c. for the heater of the oscillator. The amplifier heaters all operate from 12-volt d.c. obtained from filament windings on the transformer and selenium rectifiers. A study of *Fig.* 1 will show a rather unique arrangement of the heaters. This was used so as to minimize the change in voltage on any of the tubes in case of failure of one of them. The VU meter is illuminated by two pilot lamps connected in series—a 3.3-ohm resistor serving as the connecting lead. This is easily done with the General Electric meter, and with the lamps specified, the current through this portion of the heater circuit is equivalent to that of a tube. It will be noted that this arrangement of series-parallel heaters simplifies wiring, in addition to its more obvious advantage in maintaining reasonable voltage on any heater in case of a failure.

Two outputs are provided for this power supply—the Amphenol 91PC-4F provides power to the amplifier while the Jones S-406AB furnishes a.c. at both 6.3 and 115 volts to the recorder, as well as 300-volt d.c. and ground.

The portable power supply is somewhat smaller, and employs selenium rectifiers for both heater and plate voltages. It is constructed in a Bud Minibox, 3-1/2×6×8. The power receptacles, a.c. switch, fuse, and input cord are all mounted on one of the ends of this box, and all parts are attached to this same section of the box. The schematic for the portable power supply is shown in *Fig.* 16. Since space is important, and since selenium rectifiers are used for the B-supply, it is suggested that the specific parts named be employed for this unit. Internal construction is straightforward, and can be laid out by the builder to suit his own requirements.

For field use, a carrying case was built by a local luggage shop, being constructed of 3/8-in. plywood and covered with a material which matches the recorder case. The case is a fairly tight fit for both amplifier and power supply, and is lined with thick felt to prevent jarring within the case. The partition between the power supply and amplifier compartments extends to about two inches from the bottom so as to provide ventilation through two vent-holes in the side of the power supply. Circulating air currents can thus pass through the unit and out through the vent hole in the top panel.

Fig. 11. Sockets for low-level tubes are mounted on a channel strip which is flexibly mounted on the chassis to reduce microphonics.

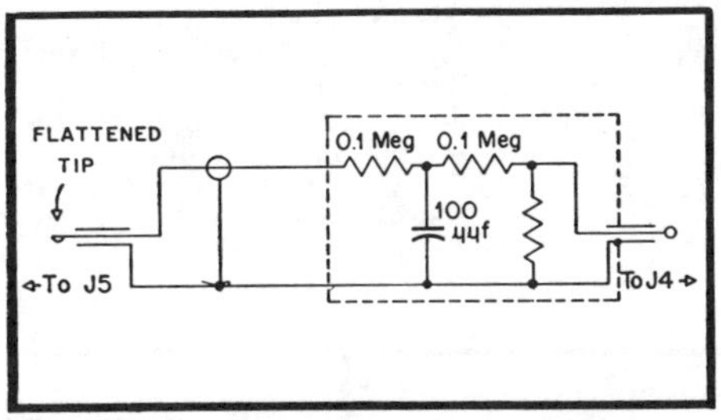

Fig. 13. Schematic of patch cord used for making artificial reverberation machine out of the amplifier.

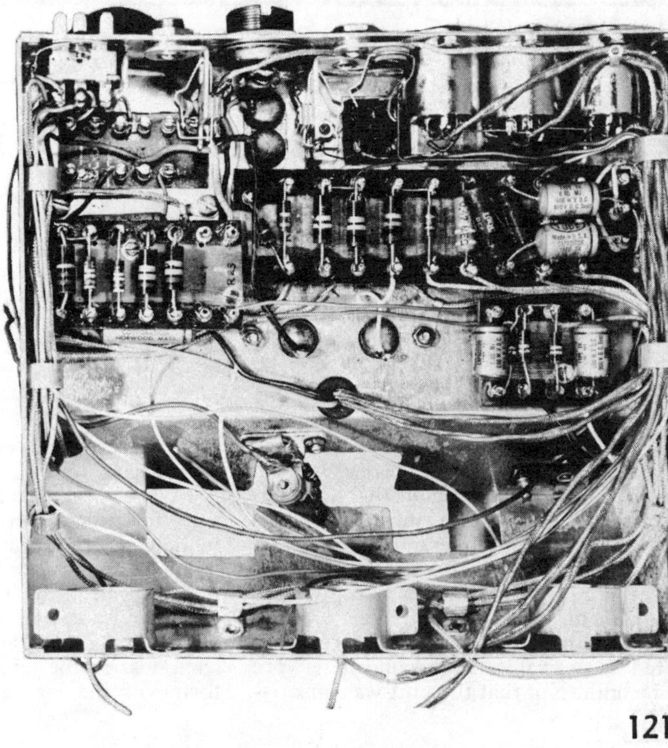

Fig. 10. Bottom view of chassis during construction. Note wiring clips along sides, which aid while building as well as after completion.

Fig. 14. External view of power supply used in the fixed installation.

Conclusion

The design and construction of a unit such as this is a somewhat complicated procedure, and was not done in a week. However, this equipment has been in use for almost a year and no trouble has developed. Those experienced in miniaturizing equipment such as is done for military uses will undoubtedly be amazed at all the waste space in the amplifier; on the other hand, anyone who has not attempted extra-small construction may be confused by the lack of space. Unless your requirements indicate a particularly compact construction, it is suggested that an electrically similar unit would be much easier to build in a case two inches larger in each dimension. Now that it is complete, however, the writer has no hesitation in recommending the unit to anyone with similar desires in equipment.

PARTS LIST
Fixed power supply—Fig. 15

C_1	2 μf, 600 v, oil filled
C_2, C_3, C_4	40 μf, 450 v, electrolytic
C_5	2000 μf, 15 v, electrolytic
J_1	Amphenol 91PC-4F receptacle
J_2	Jones S-406-AB receptacle
L_1	6 H, 100 ma filter choke
L_2	20 H, 40 ma filter choke
R_1	1000 ohms, 10-watt, Ohmite Brown Devil
SR_1, SR_2	Federal 1017 selenium rectifiers, 600 ma, 26 volts
T_1	350–0–350 at 90 ma; 5 v at 2 a; 6.3 v at 3 a.
V_1	6X5

PARTS LIST
Portable power supply—Fig. 16

C_{21}, C_{22}, C_{23}	40 μf, 450 v. electrolytic
C_{24}	2000 μf, 15 v. electrolytic
F_{21}	3-amp fuse, Type 3AG
J_{21}	Amphenol, 91PC-4F receptacle
J_{22}	Jones, S-406-AB
L_{21}	8 H, 80-ma. choke, UTC R-18
L_{22}	8 H, 40-ma. choke, UTC R-14
R_{21}	22 ohms, 1 watt
SR_{21}, SR_{22}	Federal 1017 selenium rectifier, 600 ma, 26 volts
$SR_{23, 24, 25, 26}$	Federal 1002 selenium rectifiers, 75 ma, 130 volts
Sw_{21}	DPDT toggle switch
T_{21}	Merit P-3051 (260-0-260 at 70 ma; 5 v at 3 a; 6.3 v at 3 a.)

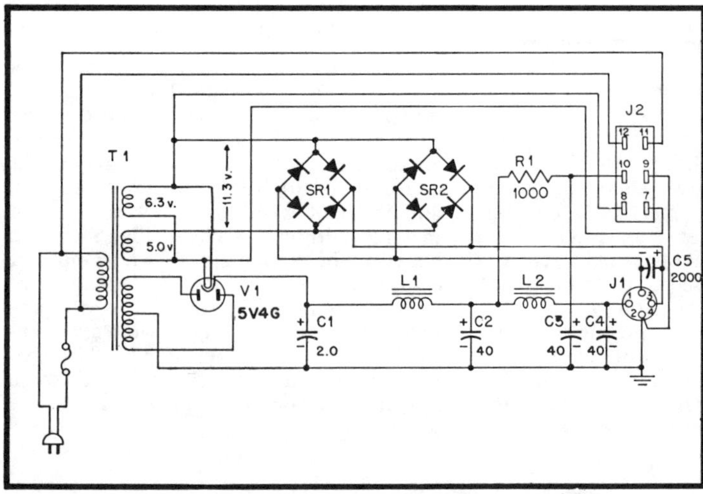

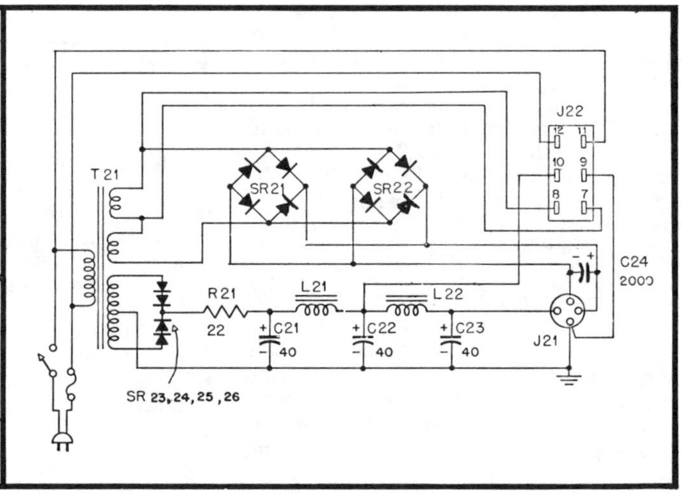

Fig. 15 (left). Schematic of power supply shown in Fig. 14. Fig. 16 (right). Schematic of portable power supply used for field use.

Improvements on the Universal Amplifier

Description of the second model of the unit described in the previous article—with certain minor refinements and slightly improved performance.

ONE OF THE DIFFICULTIES in building only one of a piece of equipment is that there is no opportunity to improve on layout, circuit, or mechanical design in the same fashion that a manufacturer would before introducing the instrument or equipment to the market. The Universal Amplifier described in the preceding article was no exception to this difficulty, as was mentioned. Certain changes and improvements were indicated as desirable if a second unit were to be built. Not that the unit was unsatisfactory as it was—but although it performed well and turned out good recorded tapes, there were a few slight deficiencies.

In the first place, the transformers used for inputs were relatively old, and while their frequency response was adequate for most applications, there was a 4-db peak at 10,500 cps which caused a higher hiss level than was considered desirable on playback. Also, there was some loss in the circuit for direct monitoring of the incoming signal, as compared to the NOR position of Sw_1, necessitating an increase in the gain setting of the amplifier which followed. Furthermore, it was felt that the difficulty of changing the volume controls—in case they became noisy—was too great.

Accordingly, the entire amplifier was rebuilt. The same outside case was used, as well as the same panel, VU meter, and switching arrangements. The input transformers were changed to Triad HS-5 and the output transformer was changed to Triad HS-52. The Triad

HS-5 transformers were designed to work from a 30/50-ohm microphone into a grid; frequency response is extended, and the amplifier is now free of the hiss level which accompanied the first model. This was not considered too troublesome at the time of the first articles on this amplifier, because it was felt that very few who might wish to duplicate the unit would go to the expense or trouble of obtaining the Western Electric transformers specified for the input circuits.

All three of the Triad transformers are mounted in the same size of case, and the total chassis space required was somewhat less than with the three previously employed. This permitted the use of a 4-section electrolytic capacitor for plate-supply decoupling and for bypass across the cathode resistor of the output stage, since space enough for this capacitor (a Cornell-Dubilier UP-22245C) was available along the rear of the chassis. The feedback capacitor C_{12} is mounted between the output transformer and the filter capacitor.

With all of the transformers and these two capacitors mounted along the rear of the chassis, space was available for the mounting of the tubes in a row, thus making it possible to locate all six of them on a single rubber-mounted aluminum channel, basically similar to that used for the first four tubes in the original construction. The resistors and capacitors related to the tubes are mounted on a resistor board which is spaced about ¼ in. below the socket terminals. This resistor board was made up from a punched phenolic strip and a number of terminals supplied by NAALD. With these terminals and the punched strips, and using the small staking tool designed for applying them, it is possible to make up any kind of terminal strip desired. For this particular application, a total of 31 resistors and capacitors are mounted on a single strip just below the tube sockets, thus making for short leads from the socket terminals to the associated components.

The volume controls are mounted on two brackets—one holding R_{12}, and the other holding R_7 and the inclusive control, R_{22}. Short flexible shafts are used between the knobs and the controls—with panel bearings and shaft extensions used at the panel to provide a good bearing for the knob shafts. The flexible shafts can be seen in *Fig.* 3.

The principal change in circuitry is in the cathode-follower section, V_6 in the original circuit, and V_{6a} and V_{6b} in the revised arrangement. This provides some additional gain so that when the unit is used with a home radio system, the output at terminal 3 of J_6 is at the same level as the signal fed into terminal 2—the loop circuit between the FM tuner and the remainder of the system. One section of the 12AU7 is employed as an amplifier, with about 10 db of gain; the other section is a cathode follower, and feeds the signal out at a low impedance. J_{10} has been added, with the voltage divider R_{49} and R_{50}, so that an ordinary patch cord may be inserted between J_{10} and J_4 to permit the use of the amplifier as an

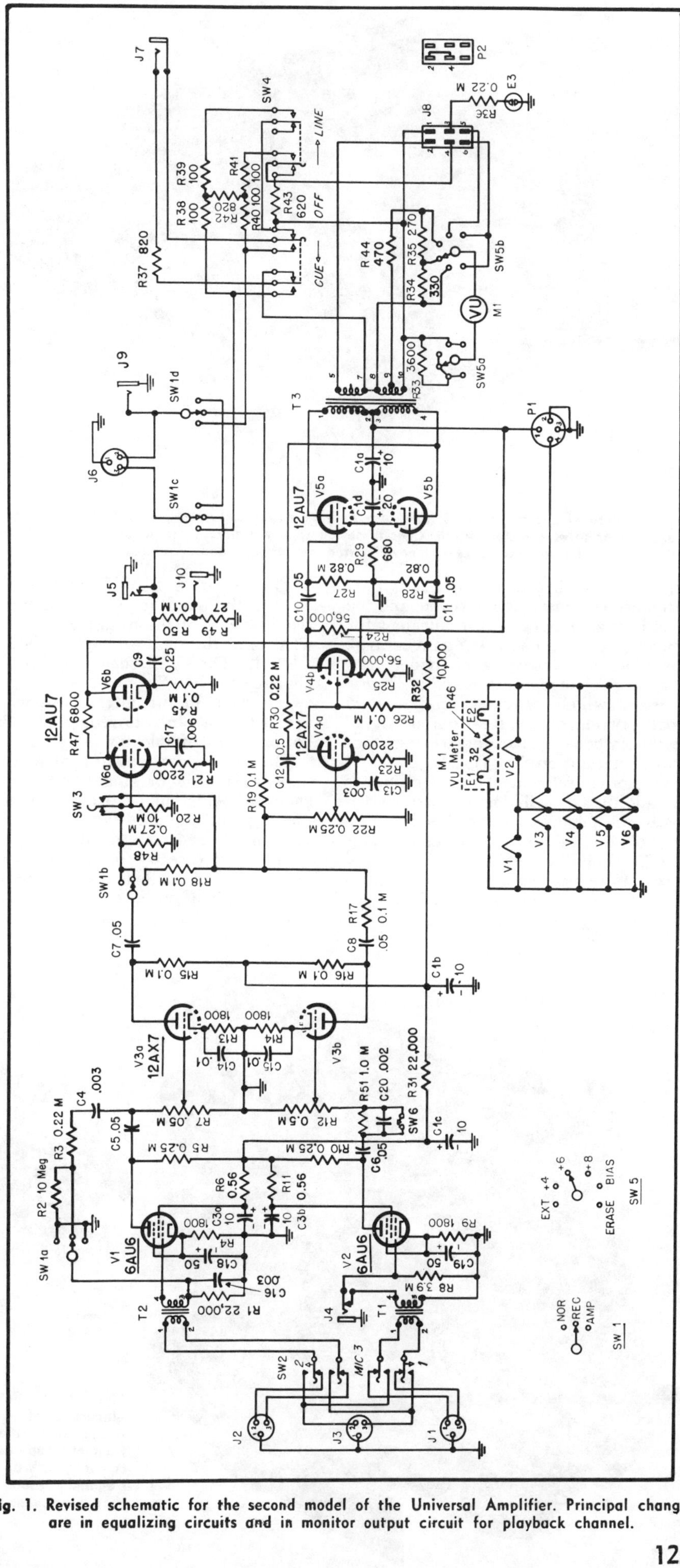

Fig. 1. Revised schematic for the second model of the Universal Amplifier. Principal changes are in equalizing circuits and in monitor output circuit for playback channel.

Fig. 2. View of chassis removed from case to show new transformers, 4-section filter capacitor, and mounting of selector switch. Chassis design is essentially the same as in the first model, but all six tubes are now located on rubber-mounted channel strip.

artificial reverberation generator. Performance is somewhat improved, and the level available at J_{10} is just sufficient for convenient operation of the gain control, R_{12}. Note that the voltage divider consists of a 0.1-meg resistor and a 27-*ohm* resistor, which provides only a very small portion of the total output signal at the cathode of V_{6b} for feeding back into the circuit on the grid of V_2.

One other refinement in the circuit is the addition of a dialogue equalizer between the plate of V_3 and the gain control of the microphone channel. This equalizer consists of R_{51} and C_{20}, and serves to reduce the gain approximately 10 db at 100 cps, with a gradual slope up to normal gain at 500 cps.

The use of a 12AU7 instead of the 5879 in the monitor circuit changes the requirements of the heater circuit somewhat, and the new wiring is shown in *Fig.* 1. The VU meter lights were changed to No. 47's, and the resistor between them in the meter case was increased to 32 ohms to reduce the illumination to a suitable intensity.

The basic construction plan is the same as in the original model—minor changes being made to accommodate the components used in the improved version. The 4-section filter capacitor is mounted directly on the chassis in holes punched, drilled, and filed by hand —rather than on a standard phenolic mounting wafer—in order to save space. The 2-section capacitor C_{3a} and C_{3b} is mounted on the right side apron of the chassis just behind the front panel as shown in *Fig.* 3. The two cathode bypass capacitors C_{18} and C_{19} are mounted on the resistor board, and are C-D BBR 50-6 capacitors, 50 µf at 6 volts, and quite small.

The components of the dialogue equalizer, R_{51} and C_{20}, are mounted directly on Sw_6, and the components of the tape playback equalizer are mounted on a small resistor board just above the dialogue equalizer switch. By using two Mallory extension bushings, Sw_1 has been lowered to a position where the terminals are about flush with the chassis deck. A small resistor board, located between R_7 and R_{22} and mounted on the bracket which holds the two pots, provides for R_{17}, R_{18}, and R_{19}, the mixer network, as well as for R_{49}. R_{50} connects directly from one of the terminals of the capacitor C_9 to the ungrounded end of R_{49}. R_{30} is connected directly from terminal 4 of the output transformer to a terminal of C_{12}.

A small shield made of tin was required to eliminate all traces of oscillation due to the proximity of J_4 and J_9. Before this shield was installed, the amplifier would go into oscillation when R_{12} was turned up past 50 per cent rotation. The shield eliminated the trouble completely.

The new model has all of the operating conveniences of the original, together with improved frequency response and the added advantage of the dialogue equalizer on the microphone channel. If necessary, a similar equalizer could be wired into the second channel, but the writer's requirements have not made it necessary yet. Perhaps that will come in the third version—although to date there have been no indications that this second model will be rebuilt.

Fig. 3 (left). Bottom view of second model of the amplifier to show placement of volume controls and short flexible shafts from knobs on panel to the controls. This permits considerably shorter leads from the tube-mounting channel and the associated circuits to the controls, with appreciably less need for equalization to maintain high-frequency response. Fig. 4 (above). View of right end of the chassis to show placement of tube channel on which all six tubes are mounted.

BOOKS *from Old Colony Sound Lab*

BUTTERWORTHS

B-1 LOUDSPEAKER AND HEADPHONE HANDBOOK *edited by John Borwick.* This comprehensive technical volume fully covers the theoretical and practical aspects of loudspeaker and headphone design, operation and performance. Distinguished contributing authors include: Glyn Adams, Peter J. Baxandall, John Borwick, Martin Colloms, Laurie Fincham, R.D. Ford, Mark R. Gander, Stanley Kelly, Peter Mapp, C.A. Poldy, Desmond Thackeray, Floyd E. Toole and J.M. Woodgate. An invaluable reference textbook for the dedicated speaker builder. 573 pp., hardbound. Each **$97.50**

GRAMOPHONE PUBLICATIONS

G-1 LIVING WITH HI-FI [3RD ED., 1983] *by John Borwick.* Audio Editor of England's *Gramophone Magazine,* John Borwick does a basic introduction to "hi-fi" in a comprehensive and helpful survey of what good sound is and how to realize it in a home installation. The book reviews all audio system components, including digital cassette recorders. It is exceptionally helpful about correct installation procedures. Excellent introduction for friends new to audio. 120pp. softbound **$7.95**

HOWARD W. SAMS

S-4 UNDERSTANDING IC OPERATIONAL AMPLIFIERS [3RD ED.] *by Roger Melon* and *Harry Garland.* Basic course in semiconductor electronics. Covering integrated op amp circuitry, design factors, bias current, offset voltage, frequency compensation and slew rate. 212 pp., softbound. Each **$12.95**

S-9 REGULATED POWER SUPPLIES [3RD ED.] *by Irving M. Gottlieb.* Static and dynamic characteristics, regulation techniques, IC regulation. 424pp., softbound. Each **$19.95**

S-10 IC OP AMP COOKBOOK [3RD ED.] *by Walter G. Jung.* Jung's popular classic in a revised and expanded 1986 edition. One hundred pages are added to cover new devices, applications and manufacturer's data sheets. With over two hundred practical circuits with component values, this cookbook is one of Old Colony's best selling technical volumes. It belongs on every electronics bookshelf. 580 pp., softbound. Each **$21.95**

S-11 HOW TO BUILD SPEAKER ENCLOSURES *by Alexis Badmaieff and Don Davis.* The "whys" and "hows" of speaker enclosures. Drawings and instructions cover infinite baffle, bass reflex, and horn types plus combinations. 144pp., softbound. Each **$6.95**

S-13 AUDIO IC OP AMP APPLICATIONS [3RD Ed.] *by Walter G. Jung.* This new, updated version of a classic reference source is probably the best book available on the subject. Mr. Jung is not only a recognized authority on op amps, but has an enthusiast's interest in audio quality and actively pursues new techniques for better sound reproduction. Like his previous books in this series, this latest edition is in "cookbook" style, and includes actual circuits with component values which may be incorporated in working projects. The book also details the most common pin-outs, manufacturer listings and a wealth of other resource material. A must for any audiophile who constructs or modifies equipment. 250pp., softbound. Each $17.95

S-14 ACTIVE FILTER COOKBOOK *by Don Lancaster.* A practical, user-centered volume with everything you need to build your own active filters. Explains the various types and how to select the best for the circuit. 240pp., softbound. Each **$15.95**

S-17 IC TIMER COOKBOOK [2ND ED.] *by Walter G. Jung.* A full introduction to the IC timer, the types, and general usage pointers. Many surprisingly useful audio related uses. An excellent practical and theoretical volume with lots of reference data. 430 pp., softbound. Each **$17.95**

S-18 TTL COOKBOOK *by Don Lancaster.* Everything the beginner will need to know about transistor logic elements and usage. It has become a reference guide for engineers as well. Softbound. Each **$14.95**

S-19 SOUND SYSTEM ENGINEERING [2ND ED.] *by Don and Carolyn Davis.* A thorough introduction to sound systems for halls, studios, outdoor locations and much else; now expanded! This edition went into publication 11/86 it is a hard bound consisting of 665pp., 7½x9¾ replacing the previous softbound issue. Each **$39.95**

S-22 MODERN DICTIONARY OF ELECTRONICS [6TH ED.] by *Rudolph Graf.* This book should be in every library. It has more than 20,000 terms unique to electronics and other closely related fields. From angstrom to zoom lens, you'll find it in this updated dictionary. 1984, 1152pp., hardbound. Each **$39.95**

S-25 ELECTRONIC PROTOTYPE CONSTRUCTION by *Stephen D. Kasten.* Here's a great book for either the beginner looking to try electronic prototyping for the first time or the expert looking for a handy reference guide. Areas covered include wire wrap and related techniques such as solder pad and perfboard assembly. This book will help you through all the pitfalls of PC board design, and will help you put the project together in an attractive but functional package. 1983, 399pp., softbound. Each **$17.95**

S-26 CMOS COOKBOOK (2ND ED) by *Don Lancaster.* CMOS is low cost and widely available, and it uses an absolute minimum of power. It's also fun to work with and very easy to use. This book offers practical circuits and does not dwell on math or heavy theory. Eight chapters cover just about every aspect of CMOS usage. Projects include high-performance op amps, TV typewriter, digital instruments, music synthesizers and video games. 1977, 414pp., softbound. Each **$18.95**

S-27 DESIGN OF OP AMP CIRCUITS by *Howard Berlin, W3HB.* Op amps are a versatile and inexpensive integrated circuit. They can be used for linear amplifiers, differentiators, integrators, voltage and current converters, comparators, rectifiers, oscillators and more. The text includes 37 different uses and applications for op amps. Beginners will find this book helpful. 1977, 221pp., softbound. Each **$12.95**

S-28 HANDBOOK FOR SOUND ENGINEERS: The New Audio Cyclopedia *edited by Glen Ballou.* Thirty-one sections covering just about every aspect of audio engineering from room acoustics, loudspeakers and amplifiers, to soundfield measurements and image projection. A complete audio reference library in itself, the most comprehensive and authoritative work on audio available. 1987, 1250pp., hardbound. Each **$79.95**

S-29 PRINCIPLES OF DIGITAL AUDIO [2ND ED.] *by Kenneth C. Pohlmann.* A state-of-the-art reference text on digital audio technology. Comprehensive material on fundamentals, recording and reproduction is presented in this completely up-to-date handbook written both for engineers and audiophiles. 1989, 350 pp., softbound. Each **$29.95**

S-30 TUBE SUBSTITUTION HANDBOOK [21ST ED.] *by Howard W. Sams Engineering Staff.* This guide contains more than 6,000 receiving tube and 4,000 picture tube substitutes. Includes American to Foreign and Foreign to American equivalents. Tube basing diagrams provide a reference for pinouts. 128pp., softbound. Each **$5.95**

S-31 HOW TO READ AND INTERPRET SCHEMATIC DIAGRAMS *by J. Richard Johnson.* This book teaches the recognition of schematic symbols, their use and function in diagrams, and the interpretation of diagrams for design, maintenance, and repair of electronic equipment. Especially helpful to the novice kit builder. 210pp., softbound. Each **$12.95**

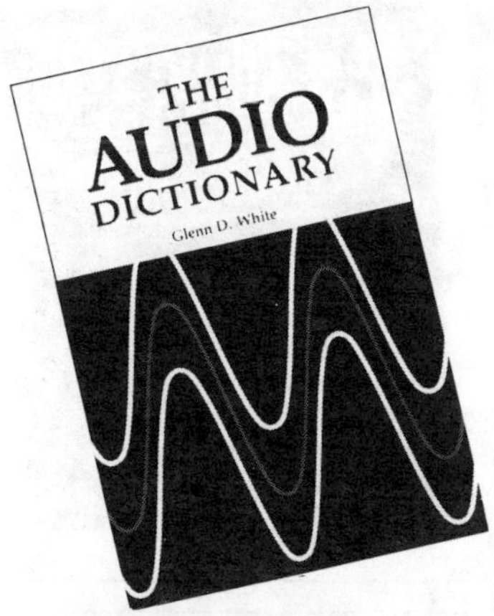

McGRAW-HILL

MH-1 HANDBOOK FOR ELECTRONICS ENGINEERING TECHNICIANS [2ND ED.] *by M. Kaufman and A.H. Seidman.* A comprehensive compendium of electronic facts. Component selection, circuit analysis, power supplies, IC uses and characteristics, op amps, transistors, batteries and tubes. A one-volume encyclopedia on how components work and how to choose the best of them for your application. Includes seven new chapters on topics such as logic analysis, fiber optics, active filters, digital test gear and microprocessors. 752pp., hardbound. Each **$48.95**

MH-2 HOW TO MAKE PRINTED CIRCUIT BOARDS *by Joel Goldberg.* All the topics on making your own circuit boards are covered both in theory and in practical advice: design and layout, artwork preparation, photo layout, silk screening, and etching. Each **$8.90**

MH-3 LEARNING ELECTRICITY AND ELECTRONICS THROUGH EXPERIMENTS *by J. A. "Sam" Wilson and Milton Kaufman.* Using four perfboards and typical "junk box" parts, this book demonstrates theoretical principles in 54 circuits you can build as you learn. Each **$12.95**

MH-4 AUDIO ENGINEERING HANDBOOK *edited by K. Blair Benson.* A complete desktop resource for engineering professionals and students. Topics covered include Digital Audio, Broadcast Transmission Technology, Microphones and Amplifiers, Analog and Digital Disk/Tape, Film Recording and Reproduction, Studio Production and Noise Reduction. Also provides a listing of up-to-date standards and recommended international practices. Among the Handbook's 33 authors are audio industry experts such as: Dr. R. Cabot, R. Dolby, T. Holman, Dr. T. Doi, H.G. de Haan, Dr. J. Bloom and L. Tyler. More than 700 illustrations. 1988, 1056pp., hardbound. Each **$85.00**

UNIV. of WASHINGTON PRESS

U-1 THE AUDIO DICTIONARY *by Glenn White.* This lively and entertaining mini-encyclopedia is strongly recommended for every audiophile's library. Accurate and up-to-date, Mr. White's book covers the terminology and basic concepts relative to sound recording, sound reinforcement, and musical acoustics. Additional material includes: "How to Subdue a Hi-Fi Salesman," historical anecdotes and a discussion on digital audio. 1987, 302 pp., softbound. Each **$14.95**

WILEY BOOKS

W-1 HIGH PERFORMANCE LOUDSPEAKERS [3RD ED.] *by Martin Colloms.* A thorough revision of the 2ND ED.(1980) updates and expands this authoritative volume. The author surveys new developments in the four-year interim and adds some highly original proposals for low frequency alignments, listening tests and much else. Colloms' book is indispensable for the serious speaker builder. 313pp., hardbound. Each **$34.95**

The PENGUINS

P-1 THE PENGUIN GUIDE to STEREO RECORDS & CASSETTES [3RD ED.] *by Edward Greenfield, Robert Layton, and Ivan March.* Drawing on in-depth technical and musical knowledge, this newly revised guide to recorded classical music covers more than 4,000 LPs and cassettes, providing details of title, performers, record number, label and price range. 1,387 pp., softbound. *Please add $1.00 per book for shipping, in addition to postage.* Each **$12.95**

P-1A THE PENGUIN GUIDE TO COMPACT DISCS, CASSETTES AND LPs *by Edward Greenfield, Robert Layton and Ivan March.* An invaluable companion to their best selling Stereo Records and Cassettes Guide, this most recent offering is extensive in its coverage of compact discs. Comparative reviews are included that assess the best versions of material available in all three formats. 1,217 pp., softbound. *Please add $1.00 per book for shipping, in addition to postage.* Each **$14.95**

P-2 DICTIONARY OF ELECTRONICS [2ND ED.]*by E. C. Young.* This remarkably compact reference covers electronics from A-Battery to Z-parameters with succinct, concise definitions and illustrations. A quick reference completely revised and updated with lots of added charts and reference data. 671pp., softbound. Each **$8.95**

P-3 A NEW DICTIONARY OF MUSIC [3RD ED., 1973; first pub. 1958] *by Arthur Jacobs.* Alphabetically arranged entries covering composers, individual musical works, orchestras, performers, conductors, musical instruments, and technical terms. 458pp., softbound. Each **$8.95**

P-8 INTRODUCING MUSIC *by Otto Karolyi.* Even if you've listened to music for years, you may be surprised at how little you know about its fundamentals. A beautifully basic and spare introduction to the grammar and vocabulary of music—enough to understand the language without speaking it. It will deepen almost any non-musician music lover's pleasure in listening. 174pp., softbound. Each **$5.95**

HOUSE OFFERINGS

BKAA1 AUDIO AMATEUR LOUDSPEAKER PROJECTS. A collection of 25 loudspeaker construction projects. These are the complete articles as they appeared in *Audio Amateur* (1970-79). Since 1980, *Speaker Builder* magazine has been the forum for these authors and designers. Although some of the projects are showing their age, many embody ideas that can be developed using our newer knowledge and insights. The electrostatic devices are still unsurpassed by anything done since. The collection is a rich source of both theory and practice as well as design. 135pp., softbound, 8½x11. Each **$20.00**

BKAA2 THE LOUDSPEAKER DESIGN COOKBOOK (3RD ED.) *by Vance Dickason.* Everything you need to build the loudspeaker system you have always wanted but could not afford. Easy ways to pick the exact box size, the ideal drivers, the most pleasing finish and the correct way to feed your music to your new superb loudspeaker system. Proven designs, where to buy low cost parts and how to test the results for yourself. 82pp., softbound, 8½x11. Each **$19.95**

BKAA3 AUDIO ANTHOLOGY (Vol. 1) A fascinating look at sound reproduction as presented in the pages of Audio Engineering magazine. Features 38 articles from May 1947 to December 1949. Compiled by editor, designer and author, C.G. McProud, this series will be of interest to all audio enthusiasts. 124pp., softbound, 8½x11. Each **$16.95**

BKAA3 AUDIO ANTHOLOGY (Vol. 2). More on amplifiers, loudspeakers and enclosures, and phonograph equipment. 45 articles from January 1950 to July 1952. Each **$16.95**

BKX-1 PERCEPTION OF REPRODUCED SOUND *edited by Soren Bech and O. Juhl Pedersen.* A collection of 18 papers from the symposium held at Gammel Avernaes, Denmark, in September 1987. This gathering of professionals reviewed the subjective evaluation of reproduced sound and assessed its importance for the electroacoustic industry. Participant authors include: T. Poulsen, R. Plomp, H. Staffeldt, J. Blauert, F.E. Toole, A. Gabrielsson, S. Bech, N. Kousgaard, M. Williams, V. Hansen and G. Thiele. All offer highly informative presentations. 165pp., hardbound. Each **$35.00**

TAB BOOKS

T-6 ELECTROSTATIC LOUDSPEAKER DESIGN AND CONSTRUCTION *by Ronald Wagner.* An exceptionally well written and illustrated builder's book. 256pp., softbound. Each **$15.95**

T-7 BASIC ELECTRONICS COURSE [2nd Ed.] *by Norman H. Crowhurst.* This highly recommended textbook defines literally every aspect of electrical behavior and state-of-the-art electronics. Included are fundamental building blocks of theory and practice such as Kirchhoff's Laws, properties of resistance, electron flow, magnetic fields and power calculations. 430pp., softbound. Each **$15.95**

T-8 ENHANCED SOUND—22 ELECTRONICS PROJECTS FOR THE AUDIOPHILE *by Richard Kaufman.* Twenty-two cost efficient projects are offered which can increase the capabilities of your audio system, and improve its performance. Discussions on theory, design applications and construction techniques are also included. A sampling of the projects...surround sound decoder, passive image enhancer, auxiliary input switch, tone controls, amplifier bridging circuit, modern infrasonic filter, Linkwitz-Riley crossover, improved dipole, helical antennas, and more. Over 100 illustrations. 1988, 170 pp., softbound. Each **$10.95**

T-9 THE ILLUSTRATED DICTIONARY OF ELECTRONICS [4th Ed.] *by Rufus P. Turner and Stan Gibilisco.* This outstanding offering from TAB Books contains over 27,000 terms and definitions. Clearly written and detailed with over 450 illustrations, it is an excellent reference for hobbyist and researcher alike. 1988, 648 pp., softbound, 7½ x 9. Each **$24.95**

DOVER BOOKS

D-1 REPRODUCTION OF SOUND *by Edgar Villchur.* An elegantly simple 92-page softbound primer on the subject by the man who invented the bookshelf speaker. Loan or give it to friends who want to build a system as good as yours. Each **$6.00**

D-3 MUSIC, PHYSICS AND ENGINEERING [2nd Ed.] *by Harry F. Olson.* A thorough introduction to the physical characteristics of sound and the relationship of sound to musical instruments by the former head of staff at RCA's lab for acoustical and electromechanical research at Princeton, NJ. A classic by one of the giants in the audio field. Good, easy to read chapters on acoustics, mikes and recording, recording and playback systems, as well as an electronic music chapter. 1967, 460pp., softbound. Each **$14.25**

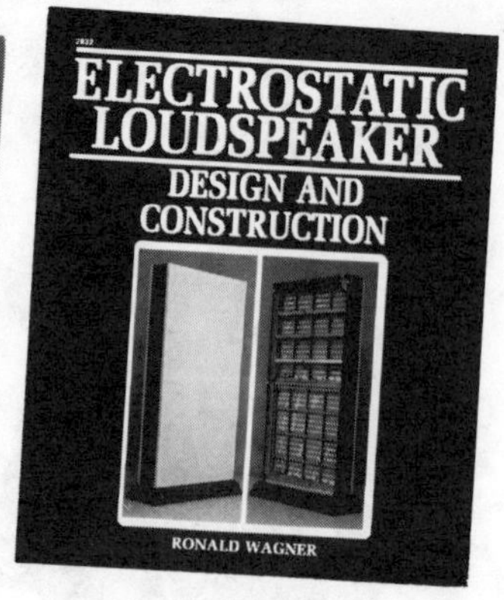

PRENTICE-HALL

PH-1 HANDBOOK OF SIMPLIFIED SOLID-STATE CIRCUIT DESIGN [2nd Ed.] *by John Lenk.* Design rules for basic circuitry including audio amps, integrated circuits, wave forming and shaping plus power supplies. Assumes basic theoretical base. Discusses purposes and types of testing. Each **$39.95**

NATIONAL SEMICONDUCTOR

N-5 VOLTAGE REGULATOR HANDBOOK. A manual which simplifies selection of three-terminal regulators depending on your needs, and which offers information on heatsink selection and design, along with transformer and filter specifications. 312pp., softbound. Each **$7.00**

N-6 LINEAR DATA BOOK 1. First book in a comprehensive 3-volume set of specifications and applications. Topics covered, with example circuits, include voltage regulators, op amps and buffers, voltage comparators and instrumentation amplifiers. 1988 Edition. 1,250pp., softbound. Each **$16.95**

N-7 LINEAR DATABOOK 2. This second volume includes information on active filters, analog switches and multiplexers, both analog-to-digital and digital-to-analog converters, temperature sensors and voltage references. 1988 Edition. 920pp., softbound. Each **$10.95**

N-8 LINEAR DATABOOK 3. The third book in this series focuses on audio, video and radio circuits. Also included is information on an array of special functions. 1988 Edition. 750pp., softbound. Each **$10.95**

N-678 LINEAR DATABOOKS. As above, complete 3-volume set. Each **$35.00**

To order from **Old Colony Sound Lab**, please write each book's number below with quantity of each and price. Total the amounts and remit by check, money order, MasterCard, or Visa. *For charge card orders under $10 add $2 service charge.* Please add $1.75 for postage on the first book; 50¢ per additional book. Orders shipped outside the US please add an additional $4.50 for postage. All remittances must be in US funds. *Please use clear block capitals.* **NOTE:** *Prices are subject to change without notice.* **Books are not returnable for exchange or credit.**

Qty.	Book No.	Price

Postage $ ______

☐ Check/MO ☐ MasterCard/Visa **TOTAL** $ ______

OCSL has added an answering machine for the convenience of our customers. For after 4:00 p.m. EST weekdays, and weekends please call (603) 924-6371 to place your MasterCard or VISA order. Please remember this is an ordering line only.

OLD COLONY SOUND LAB

BOOKS

NAME ______ MAGIC NUMBER ______

STREET AND NO. ______

CITY ______ STATE ______ ZIP ______

MASTERCARD OR VISA NUMBER ______ EXPIRATION DATE ______

OLD COLONY SOUND LAB
PO Box 243, Peterborough, New Hampshire 03458 USA
Telephone Charge Card Orders: (603) 924-6371 or 924-6526
FAX: (603) 924-9467
Weekends and after 4:00 p.m. MC/VISA ordering line: (603) 924-6371